GLOBAL ATMOSPHERIC–BIOSPHERIC CHEMISTRY

ENVIRONMENTAL SCIENCE RESEARCH

Recent Volumes in this Series

GLOBAL ATMOSPHERIC–BIOSPHERIC CHEMISTRY

Edited by

Ronald G. Prinn

Massachusetts Institute of Technology
Cambridge, Massachusetts

Springer Science+Business Media, LLC

Library of Congress Cataloging in Publication Data

Prinn, Ronald G.
Global atmospheric–biospheric chemistry / edited by Ronald G. Prinn.
p. cm. — (Environmental science research; v. 48)
"Proceedings of the First Scientific Conference of the International Global Atmospheric Chemistry (IGAC) Project, held April 18–22, 1993, in Eilat, Israel. "—T.p. verso.
Includes bibliographical references and index.
ISBN 978-0-306-44884-3 ISBN 978-1-4615-2524-0 (eBook)
DOI 10.1007/978-1-4615-2524-0
1. Atmospheric chemistry—Congresses. 2. Biosphere—Congresses. I. International Global Atmospheric Chemistry (IGAC) Project. Scientific Conference (1st: 1994: Eilat, Israel) II. Title. III. Series.
QC879.6.P73 1994 94-22508
551.1'11—dc20 CIP

A Core Project of the International Global Change IGBP Geosphere-Biosphere Programme

Proceedings of the First Scientific Conference of the International Global Atmospheric Chemistry (IGAC) Project, held April 18–22, 1993, in Eilat, Israel

ISBN 978-0-306-44884-3

PREFACE

This volume contains the invited papers and a transcript of the final panel discussion in the First Scientific Conference of the International Global Atmospheric Chemistry (IGAC) Project, held in Eilat, Israel from April 18–22, 1993. The conference was hosted by the Israeli Institute for Biological Research (IIBR) and was the 37th in the prestigious OHOLO Conference series in Israel.

The conference was devoted to the subject of "Global Atmospheric-Biospheric Chemistry" and was a landmark event in this area. It provided the first comprehensive report of progress under IGAC toward improving our understanding of the chemical and biological processes that determine the changing composition of the earth's atmosphere. This work is an essential component of the comprehensive International Geosphere-Biosphere Program (IGBP) devoted to measuring and understanding global changes in the past and present, and predicting the future evolution of our planet.

I want to devote this brief foreword to thanking several people who worked especially hard to make the conference a success and who helped to produce this volume as a record of the event. Paul Crutzen, Amram Golombek, Pamela Matson and Henning Rodhe did sterling service on the conference organizing committee. Special thanks go to Amram Golombek and Dr. Cohen, the Director of IIBR, who hosted the event in Israel. Anne Slinn did an excellent job in producing the Abstract book and helping with administrative matters. Alex Pszenny helped capably to critically review the Abstracts. Linda Kubrick worked tirelessly on travel arrangements and made sure that nobody got lost.

Several organizations provided financial or other support for the First IGAC Scientific Conference, specifically the U.S. National Science Foundation, the IGAC Core Project Office, the U.S. National Oceanic and Atmospheric Administration, the International Geosphere-Biosphere Program, the Israel Institute for Biological Research (37th OHOLO Conference), the European IGAC Project Office, the U.S. Environmental Protection Agency, and the Israel Ministry of Tourism. I thank these organizations for their help and support, which enabled the conference to proceed so well.

Finally, and most important, two people have played essential roles in producing this book: Alex Pszenny critically reviewed the papers enabling many substantive improvements, and very special thanks go to Anne Slinn who carefully checked each paper and produced the final page proofs.

Ronald G. Prinn
Cambridge, MA, 1994

CONTENTS

GLOBAL ATMOSPHERIC-BIOSPHERIC CHEMISTRY

Ronald G. Prinn

Massachusetts Institute of Technology
Cambridge, MA 02139 USA

ABSTRACT

International concerns about the possibility of detrimental future climate change rest to a significant degree on the documented fact that the atmospheric levels of a number of long-lived greenhouse gases are steadily rising. The underlying causes for the currently observed changes in atmospheric composition involve both natural physical, chemical and biological processes and vital human activities like energy and food production. There is also clear evidence for major changes in atmospheric composition over the last 160,000 years contained in ice cores. The ice core data show us that atmospheric changes can be driven by natural processes alone so the current human processes need to be understood in terms of their perturbations to the natural biogeochemical cycles. Human activity is certainly a major contributor to the positive atmospheric trends of CO_2 (0.4% per year), CH_4 (1.2 down to 0.3% per year), and N_2O (0.25% per year) in recent decades. Humans are in addition the sole contributor to the observed positive CF_2Cl_2 and $CFCl_3$ trends (4.0% and 3.8% per year respectively in 1978–1988, 2.6% and 3.2% per year respectively in 1989–1991) and probably also for the observed negative trends in stratospheric O_3 (0.2% to 3% per year depending on latitude). Tropospheric O_3 appears to have increased at least in northern hemisphere midlatitudes over the last century and there are reasons to expect such an increase based on combustion-derived emissions. Similarly, SO_2 levels, while controlled in some industrially developed countries, are increasing in several industrially developing regions due particularly to increased coal combustion. All of these trends are important to radiative forcing of climate change: positive trends for CO_2, CH_4, N_2O, $CFCl_3$, CF_2Cl_2, and tropospheric ozone produce a positive (warming) force while negative trends for stratospheric ozone and positive trends for SO_2 (and thus for the derived reflective sulfate aerosols) produce a negative (cooling) force. Biological processes in terrestrial ecosystems (cultivated or natural) play major roles in the budgets of CO_2, CH_4 and N_2O while marine biology plays a significant role in controlling remote marine atmospheric SO_2 levels and the flux of CO_2 between atmosphere and ocean. The International Global Atmospheric Chemistry (IGAC) Project of the International Geosphere-Biosphere Program (IGBP) is underway and devoted to understanding the

Global Atmospheric-Biospheric Chemistry
Edited by R.G. Prinn, Plenum Press, New York, 1994

complex coupled atmospheric and biospheric chemistry leading to currently observed changes in atmospheric composition and to improving our capability to predict future changes in atmospheric gaseous and particulate composition.

INTRODUCTION

The gaseous and particulate composition of the atmosphere is determined by a complex system involving both internal processes and exchanges with the hydrosphere, pedosphere, and biosphere. The metabolic activities of soil microbes and higher plants on land and phytoplankton in the oceans lead to emission and/or uptake of many chemically and radiatively important species. Fossil fuel and biomass combustion, land clearing, farming and agriculture are key human activities affecting surface fluxes. Most of the internal chemical processing occurs in either the troposphere or the stratosphere where ozone and nitrogen oxide and hydroxyl free radicals are key reactive species. In addition,

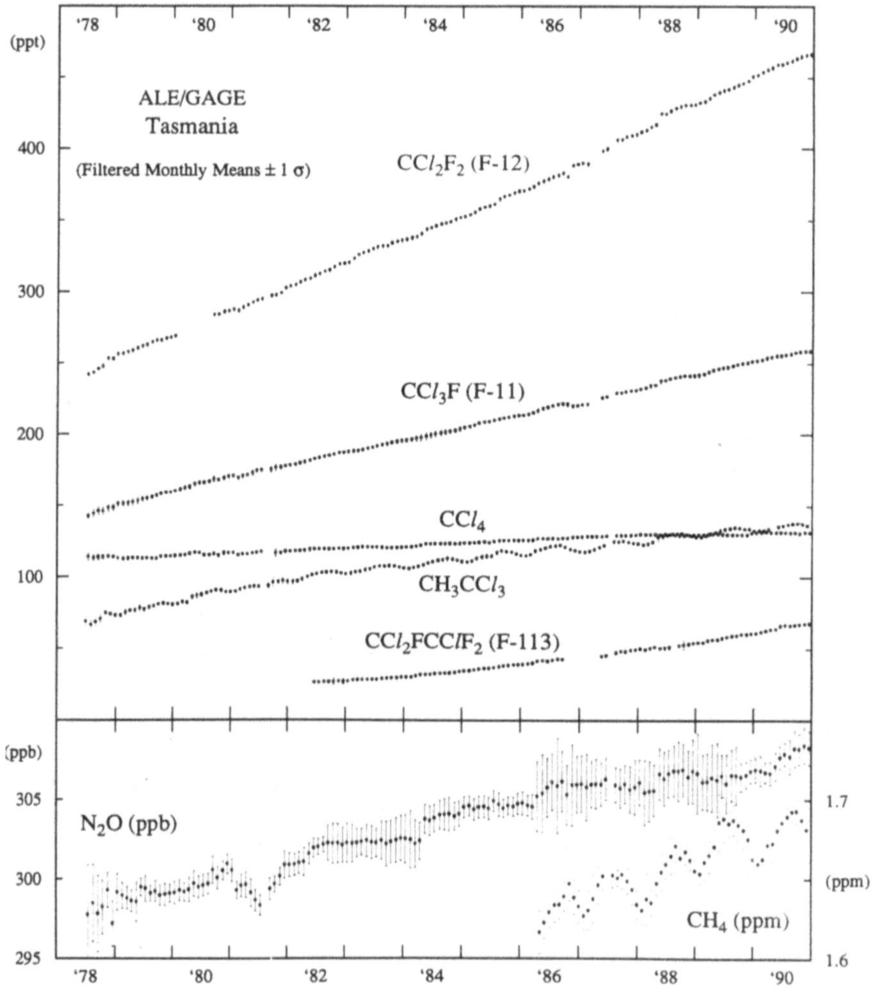

Figure 1. Concentration of CO_2 (parts per million by volume) measured at Mauna Loa Observatory, Hawaii (C.D. Keeling, private communication).

reactions on and within aerosols and cloud droplets are important transformation and loss mechanisms for many gaseous constituents.

Figures 1 and 2 show measurements of the concentrations of most of the climatically important gases at selected observing sites. The trends are all positive and current concerns about the possibility of detrimental future climate change rest to an important degree on this clear evidence of rises in several long-lived greenhouse gases.

Methane, an important greenhouse gas, has major natural biological and anthropogenic sources and is destroyed largely by reaction with the hydroxyl radical in the troposphere. Both its sources and sink are strongly influenced by human activity. Nitrous oxide has a similar range of sources. Through its decomposition it plays a controlling function in the ozone layer and it is also a significant greenhouse gas. The chlorofluorocarbons are purely anthropogenic and play a similar controlling role in the ozone layer. However, the potency of both nitrous oxide and the chlorofluorocarbons as greenhouse gases is offset partially by the ozone they destroy. Ozone, a key chemical and protective ultraviolet shield, has a

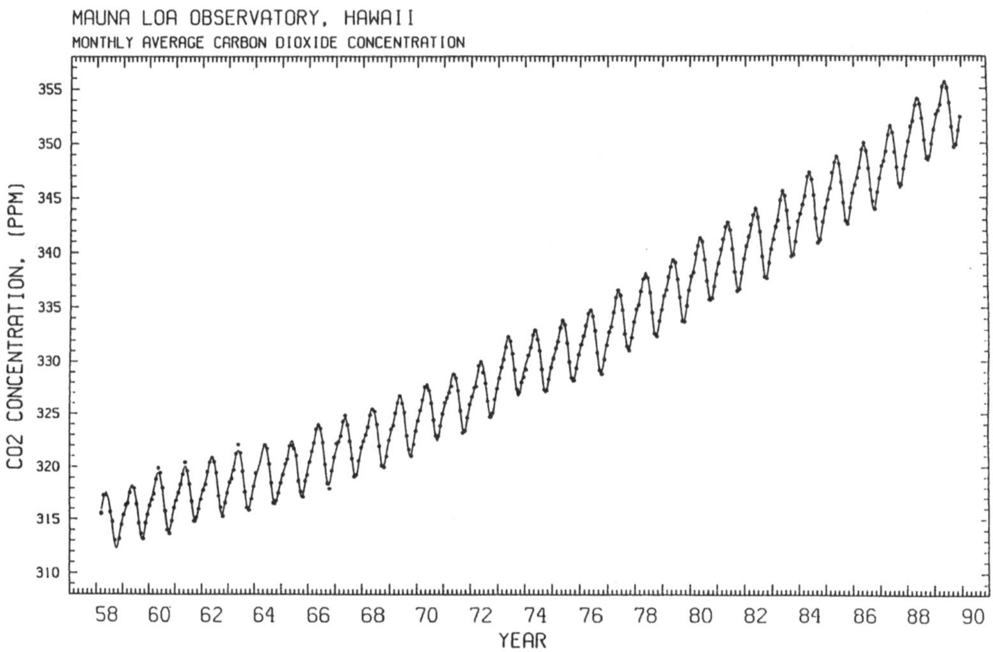

Figure 2. Monthly mean concentrations of CF_2Cl_2, $CFCl_3$, CH_3CCl_3, CCl_4, $CF_2ClCFCl_2$, CH_4, and N_2O measured in real time at the Cape Grim, Tasmania station of the GAGE network (R. Prinn, R. Weiss, P. Fraser, F. Alyea, D. Cunnold, P. Simmonds, private communication).

complex chemistry influenced by many other trace species and is also an important greenhouse gas. Nitric oxide, nitrogen dioxide, hydrocarbons heavier than methane, and carbon monoxide, while not very important directly as greenhouse gases, play a key role through their influence on the ozone, hydroxyl radical, and methane concentrations. Gaseous sulfur compounds, both natural and anthropogenic, are efficiently oxidized to particulate sulfates, which can be very important contributors to the albedo, offsetting somewhat the influence of the greenhouse gases.

A major challenge before us is therefore to quantify the processes controlling atmospheric composition and the links between atmospheric chemistry and climate. The existence of these links is demonstrated by the remarkable record of trace gas composition in the ice cores and the temperature record that goes with it as shown in Figure 3 (Lorius *et al.*, 1990). The remarkable correlation seen on the glacial/interglacial time scale, and sometimes at shorter scales, between the concentrations of CO_2 and CH_4 and the (high latitude) temperature is remarkable. But scientists still lack an acceptable theory to explain why CO_2 and CH_4 should track temperature like this. On one hand, perhaps the greenhouse gases rose and that caused the temperature changes. But the direct radiative forcing due to these trace gas changes is only about a tenth of what is needed to explain the temperature changes. So for this explanation to work, there must exist a number of other positive feedbacks and also, since it is high latitude temperature that is recorded in the ice, perhaps low latitude temperature did not vary as much (Lorius *et al.*, 1990). On the other hand, equally plausible is the hypothesis that climate changed for other reasons thus changing the source and/or sinks of the trace gases causing their levels to change. In this case the trace gases, because they rise with rising temperature, are a positive feedback but not a driver for climate change. Our present lack of understanding about why these climate-composition correlations occurred in the past is an indication of our present ability to predict the future and therefore the urgent need for research to understand the driving processes.

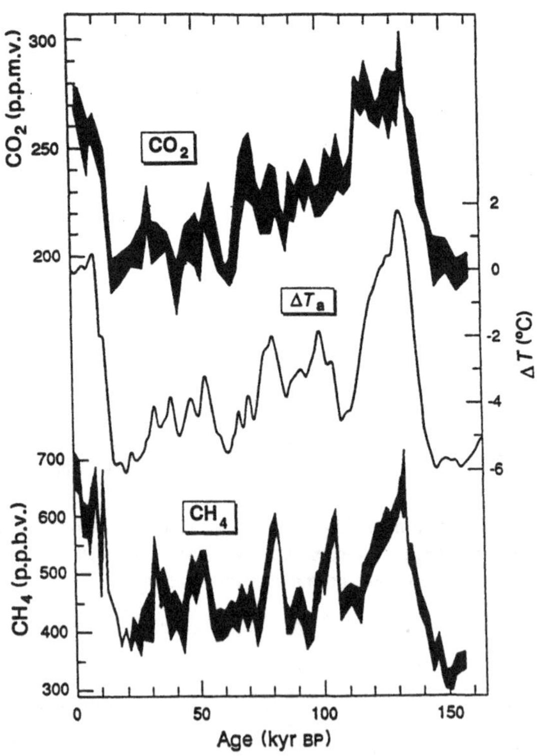

Figure 3. The concentrations of CO_2 (parts per million by number) and CH_4 (parts per billion by number) in air trapped in the Vostok, Antarctica ice core are shown along with the temperature change from present values (°C) deduced from the isotopic composition of the ice. The ice at the bottom of this 2 km deep core is 160,000 years old. Envelopes for gas concentrations denote uncertainty and temperature record has been smoothed (Lorius *et al.*, 1990).

In this paper I will provide an overview of chemical, biological, and anthropogenic processes controlling atmospheric composition with an emphasis on those species important in determining the earth s radiation balance and thus its climate. Despite recent important advances it is very clear that there are many major questions to be answered if we want to understand the basic atmospheric and biospheric processes well enough to predict atmospheric composition. If we cannot do this task then we certainly cannot do the task of predicting climate since atmospheric composition is a fundamental input for the latter predictions. The International Global Atmospheric Chemistry (IGAC) Project was created in 1988 to try to answer these questions about the atmosphere and biosphere and to provide the needed predictive capability for atmospheric composition as its contribution to the International Geosphere-Biosphere Program (IGBP) and the World Climate Research Program (WCRP). This paper will also provide a brief overview of IGAC and some examples of its early accomplishments.

ATMOSPHERIC CHEMISTRY OF CLIMATICALLY IMPORTANT SPECIES

Most radiatively important gases emitted into the atmosphere undergo a series of (usually oxidizing) reactions that can occur in either the gas phase or in water droplets. Because the hydroxyl free radical (OH) is responsible for much of or most of the gas phase oxidation in the troposphere on local, regional and global scales its concentration is one important measure of the capacity of the troposphere to oxidize trace gases injected into it. In Figure 4 is a summary (with much simplification to help clarify) of the chemistry involved in production and removal of OH as we presently understand it. The chemistry is remarkable because it involves compounds that play roles in determining the radiative forcing of climate, atmospheric ozone concentrations, air pollution levels, and acid rain (HNO_3, H_2SO_4) fluxes. The greenhouse gases shown here include H_2O, CO_2, CH_4 and O_3. Conservatively scattering aerosols, which can cool the earth by reflecting sunlight back to space, are represented here by sulfuric acid (H_2SO_4) particles. Primary pollutants emitted mainly as a result of human activity (including fossil fuel combustion, biomass burning, and land use) include hydrocarbons (RH), carbon monoxide (CO) and nitrogen oxides (NO, NO_2). Reactive free radicals or atoms are in two categories: the very reactive species ($O(^1D)$, OH) and the less reactive ones (HO_2, O, NO, NO_2).

Several things need to be emphasized about the hydroxyl radical. Reactions with methane and higher hydrocarbons, carbon monoxide, NO_2 and SO_2 are collectively the primary OH sinks (Crutzen and Zimmerman, 1991; Donahue and Prinn, 1990). The primary OH source involves water vapor, which reacts with the very reactive singlet oxygen atom, ($O(^1D)$). In turn, singlet oxygen comes from destruction of ozone by solar ultraviolet radiation. Typically, OH oxidizes other compounds either by donating oxygen or removing hydrogen leaving an H atom or organic free radical (R) which then attaches rapidly to molecular oxygen (O_2) to form hydroperoxy radicals (HO_2) or organoperoxy radicals (RO_2), which are relatively unreactive. If levels of NO and O_3, which can react with HO_2 to reform OH, are kept low then levels of OH are also kept low. Adding the nitrogen oxides (NO and NO_2), which have many human-related sources, significantly changes this picture. Specifically, nitric oxide reacts with the HO_2 to form nitrogen dioxide (NO_2) and recycle OH. Ultraviolet sunlight then decomposes NO_2 to form NO_3 and to reform NO so the nitrogen oxides are not consumed in this reaction. For most polluted environments, it is this catalytic cycle that pumps the majority of the OH into the system. If the concentration of NO_2 gets too high its reaction with OH to form HNO_3 ultimately limits the OH concentration (Ehhalt *et al.*, 1991). So if we want to predict levels of OH, we cannot just consider the reduced compounds that we put into the air like methane and carbon monoxide. We also have to estimate how much of the nitrogen oxides we put in.

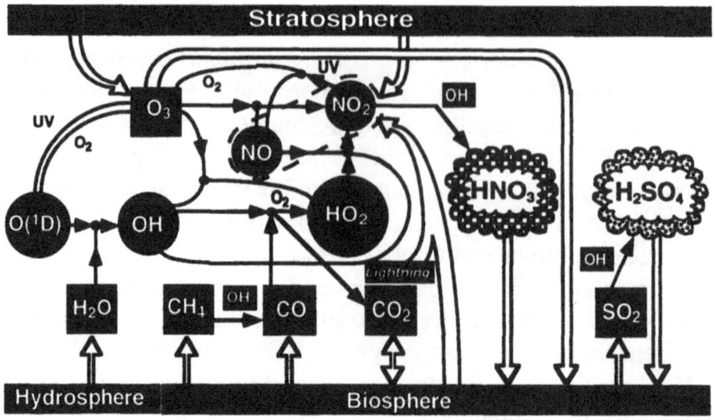

Figure 4. Simplified schematic outlining some of the important reactions involved in oxidation in air rich in nitrogen oxide free radicals (NO, NO_2). In remote marine air where NO and NO_2 are scarce, the major recycling of HO_2 to OH occurs by reaction of HO_2 with O_3 or by formation and photodissociation of H_2O_2. A more complete schematic would also include reactions involved in oxidation of volatile organics (RH) by OH to form CO, organic acids, etc. and in oxidation of $(CH_3)_2S$ and SO_2 by OH to form sulfuric acid aerosols (Donahue and Prinn, 1990; Charlson *et al*., 1987; Crutzen and Zimmerman, 1991).

Despite its importance, tropospheric OH is a difficult species to measure. It has a mixing ratio of about 10^{-14} in daylight and a lifetime of only about a second. Transport occurs on much longer time scales, so OH levels are determined locally and quickly and vary rapidly with changing sunlight, ozone, and nitrogen oxide levels. Low pressure laser-excited fluorescence (Hard *et al.*, 1992) and long-path ultraviolet laser spectroscopy (Ehhalt *et al.*, 1991) are promising techniques for measuring it directly. But even if OH is measured locally, it is so variable with time and over the globe that some integrating mechanism is needed. For determining the global removal rate of CH_4 by OH there is a convenient industrial chemical, methyl chloroform (CH_3CCl_3), whose atmospheric oxidation process is very similar to methane. The methyl chloroform industry provides very accurate estimates of the emissions of this compound over the globe. Also, measurements of methyl chloroform have been made at five globally distributed stations over the last 15 years in the Global Atmospheric Gases Experiment (GAGE). To interpret these measurements (some of which are shown in Figure 2) the estimates of the emissions are used together with a global chemical transport model to determine what weighted global average concentration of the OH free radical will best explain the observations. To do this accurately a long-term record of measurements is needed so the natural circulation-induced variability in trace species concentrations can be understood. Table 1 (Prinn *et al.*, 1992) summarizes the results obtained for OH. The deduced average OH concentrations are heavily weighted toward the lower atmosphere in the tropics where methyl chloroform (and methane) are predominantly destroyed. In addition, because the data record is sufficiently long, one can determine whether there is a trend in OH. The trend, remarkably, is positive: $1 \pm 0.8\%$ per year (Prinn *et al.*, 1992). Table 1 also summarizes the results of taking methane measurements (Blake and Rowland, 1988) and the derived average OH concentrations to deduce the lifetime and global emissions of methane. Including the positive OH trend, these emissions are increasing at about 1.3% per year. If the positive OH trend is ignored, a smaller but still positive rate of emissions increase is deduced.

A positive trend in OH means that the lifetime of methane is decreasing with time. We expect that increasing methane levels in the atmosphere, keeping everything else constant, should drive OH levels down. If that is currently not occurring, then what is happening at the present time to drive (tropical) OH levels upwards? It is known that land disturbance can lead to increased nitrogen oxide emissions from tropical soils (Keller *et al.*, 1991). Increased urbanization in the tropics also increases nitrogen oxide emissions from combustion. These emissions drive up tropospheric ozone (and thus the primary source of OH) and drive up the rate of recycling of HO_2 to OH (the secondary or catalytic source of OH). Increasing temperature by itself can do it. If the tropical oceans are increasing in temperature, we expect increased water vapor in the tropical lower troposphere. Because water vapor is a part of the primary source of OH, that also increases OH. Rising temperature also increases the rate of reaction of methyl chloroform and methane with OH thus lowering the lifetimes of both gases. And finally, decreasing stratospheric ozone can increase tropospheric OH. The singlet oxygen atom that reacts with water vapor to form OH is produced by wavelengths less than 310 nanometers. When stratospheric ozone is decreased, 310 nanometer fluxes are increased, and therefore OH production increases (Madronich and Granier, 1992).

Table 1. Methyl chloroform and methane lifetime (Prinn *et al.*, 1992).

Methyl chloroform lifetime	5.7 + 0.7 / –0.6 years
Weighted global average OH concentration	$(8.1 \pm 0.9) \times 10^5$ radical/cm^3
Methane lifetime†	11.1 + 1.4 / –1.1 years
Global methane emissions	$(470 \pm 50) \times 10^{12}$ gm/year
Rate of increase of methane emissions	
– including OH trend	1.3% per year
– excluding OH trend	0.4% per year

†Includes correction for ocean sink for CH_3CCl_3

If stratospheric ozone changes influence tropospheric OH levels and the methane lifetime, then we also need to include chemicals affecting the stratosphere in our assessment of methane, and other species like it, which are removed by OH. Observed decreases in total ozone are of the order of 0.4 to 0.8% per year in the northern midlatitudes, less than 0.2% per year in the tropics, and largest trends are seen in the southern polar hemisphere associated with the Antarctic ozone hole (Stolarski *et al.*, 1991). It is now known with considerable certainty that these current decreases in stratospheric ozone are linked very strongly to the rise of atmospheric chlorofluorocarbons (Rowland, 1990). The average atmospheric trends of $CFCl_3$ and CF_2Cl_2 were respectively 4.0% and 3.8% per year in 1978–1988 but they decreased to 2.6% and 3.2% per year in 1989–1991 (Cunnold *et al.*, 1994). Emissions controls on chlorofluorocarbons begun with the Montreal Protocol are clearly taking effect. This means, however, that the sequence of emissions of chlorofluorocarbons needs to be included in the predictions of the removal rates of many tropospheric gases.

Figure 5 is a flow diagram (again highly simplified) summarizing how "source" gases like the chlorofluorocarbons and nitrous oxide are transported up to the stratosphere where they produce chlorine-, nitrogen-, and hydrogen-carrying free radicals that destroy ozone. As is well understood, nitrogen oxide free radicals are not only responsible for destroying a significant fraction of the ozone in the stratosphere but also produce ozone in the troposphere through the chemistry outlined in Figure 4. When the ratio of nitrogen oxide free radicals to ozone is low, as it is in the stratosphere, then the tendency is to destroy ozone, while if this ratio is high, as it is in much of the troposphere, then the tendency is to produce ozone (Crutzen, 1978).

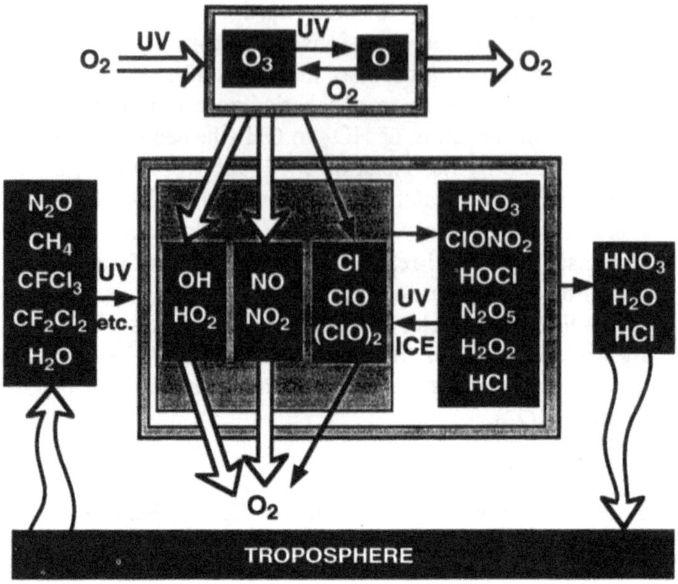

Figure 5. Simplified schematic showing how various "source" chemicals (N_2O, $CFCl_3$, etc.) are emitted at the surface, transported through the troposphere into the stratosphere, and decomposed there by solar ultraviolet radiation (UV) to form destructive free radicals (NO, NO_2, Cl, ClO, etc.). These free radicals can be converted into non-destructive "reservoir" chemicals (HNO_3, $ClONO_2$, etc.), which can be reconverted back to free radical forms by UV or by reactions on polar stratospheric ice particles. Long-lived reservoirs (HNO_3, H_2O, HCl) serve as "sinks" to return elements to the surface to complete the cycle. The various destructive free radicals catalyze the reaction between O_3 and O or O_3 and O_3 to return O_3 to its parent O_2 from which ozone was originally formed by UV dissociation.

The source gases for the destructive free radicals in the stratosphere are themselves greenhouse gases, as is stratospheric ozone itself. This is another coupling of which we need to take account. While increases in the concentrations of the source gases will increase radiative forcing of warming, these increases will at the same time lead to decreases in stratospheric ozone, which lowers the radiative forcing. Ozone, unlike the other gases, absorbs both solar ultraviolet and terrestrial infrared radiation and this must also be accounted for in calculations of radiative forcing. Calculations for the period 1979–1990 show that at higher latitudes the decreased radiative forcing caused by ozone depletion is comparable to the increased radiative forcing by all the other gases (Ramaswamy *et al.*, 1992). We clearly need to take full account of the effect of each source gas on ozone, as well as its direct effect on radiative forcing, in order to determine its net radiative effect.

Suspended particles (water droplets, ice crystals, aerosols) play significant roles in the chemistry and radiative balance of both the troposphere and stratosphere. Aerosols cool the earth by reflecting sunlight back to space or warm the air locally by absorbing radiation. The 1991 eruption of Mt. Pinatubo in the Philippines injected about 20 Teragram (1 Teragram [Tg] = 10^{12} gm) of sulfur dioxide into the stratosphere. The sulfur dioxide cloud spread out and reached hemispheric extent after about 20 or 30 days. Over this same time scale the sulfur dioxide was oxidized and combined with water vapor to form sulfuric acid particles. These particles efficiently reflected sunlight back to space. Calculations using climate models and the observed densities of the sulfuric acid aerosols show that the earth should cool for a few years by up to 6 tenths of a degree Kelvin (Hansen *et al.*, 1992).

Indeed, such a global cooling has been observed since 1991. So there is little doubt that these aerosols formed by oxidation of volcanic SO_2 are an important but intermittent influence on the climate.

Volcanoes are not the only source of sulfuric acid aerosols. When coal is burned without complete desulfurization (as it is in many developing countries) SO_2 is emitted into the lower atmosphere. In the lower atmosphere, the oxidation of the sulfur dioxide again produces sulfuric acid aerosols. These aerosols can reflect sunlight back to space as they do in the stratosphere but they can also provide condensation nuclei for clouds. If condensation nuclei are added to a fixed volume of cloud water, potentially more droplets can form out of the water, which makes the cloud more reflective.

The radiative forcing (cooling) by these effects appears to be substantial. The radiative forcing due to aerosols produced from present-day anthropogenic sulfur dioxide sources (e.g., coal burning) adds to that from background aerosol produced from natural (e.g., oxidation of dimethyl sulfide derived from phytoplanktonsulfur sources (Figure 4). The calculated cooling by anthropogenic aerosols in the northern hemisphere averages about 1 watt per square meter, which, if correct, significantly offsets calculated warming by rising greenhouse gases (Charlson *et al.*, 1991). There is still debate over the magnitude of the aerosol effect but there is little doubt that it is a cooling influence that has to be taken into account. Oxidation of SO_2 is achieved by the OH free radical in the gas phase and also by reactions in water droplets. The faster the oxidation rate of SO_2 the smaller the region over which the sulfuric acid particles are produced, and thus, in general, the smaller the potential effect on incoming sunlight. Also, other combustion-related pollutants (e.g., soot) can be incorporated into and darken the H_2SO_4 aerosols thus lowering significantly their reflectivity. We must therefore specify emissions of all gases affecting OH (and also emissions of aerosol-darkening substances) if we are to define the radiative effects of SO_2 emissions. Biomass burning, as discussed in the next section, is another copious source of tropospheric aerosols whose role in radiative forcing also needs careful assessment.

Particulates also play a very important role in polar ozone destruction. Research over the past several years indicates that winter-time temperatures in the meteorologically isolated air mass lying between 10 and 30 km altitude over the Antarctic become so low (less than 190K) that ice and nitric acid trihydrate crystals can and do condense there. Decomposition of some of the benign "reservoir" species, shown in Figure 5, occurs on and within these crystals producing ClO in the early Antarctic spring (Anderson *et al.*, 1989). The ClO levels become very high because HNO_3 remains in the crystal phase and is therefore not available to contribute to the conversion of ClO back to one of its key reservoirs ($ClONO_2$). The scarcity of O (formed from O_3 photodissociation) in the Antarctic spring means that different catalytic cycles operate there, compared to those in extrapolar regions (specifically the Antarctic cycles catalyze the reaction $O_3 + O_3 \rightarrow 3O_2$; Molina *et al.*, 1987). There is now growing evidence that similar but less severe ozone destruction occurs in the Arctic spring and also that sulfuric acid aerosols may play a role in extrapolar ozone depletion.

BIOSPHERIC CHEMISTRY OF CLIMATICALLY IMPORTANT SPECIES

Biological processes play a major role in determining fluxes of climatically important species into and out of the atmosphere. This role is exemplified by biospheric controls on CH_4, N_2O, CO_2, and marine sulfuric acid aerosols.

Methane is a very important greenhouse gas for three reasons: its large trend, its strong absorption and emission of infrared radiation, and the location of its absorption and emission bands at wavelengths where CO_2 and water vapor (H_2O) do not absorb (so-called window regions). Figure 2 shows that there is a positive trend in methane. Based on these measurements, and those taken by others beginning before 1983 (Steele *et al.*, 1992; Blake and Rowland, 1988), the methane concentration was increasing in the late 1970s at around 1.2% per year, and now in the early 1990s it is around 0.3% per year. There has therefore been a remarkable deceleration in the rate of increase of methane, but notice that the time average trend is still about twice that for CO_2. Figure 2 also shows a very obvious CH_4 annual cycle. In the southern hemisphere the cycle is caused by annual variations in transport and in the OH sink for methane. Variations in both the OH sink and in surface sources contribute to the cycle in the northern hemisphere.

Table 2. Estimated sources and sinks of methane (Tg CH_4 per year) (IGBP, 1992).

Sources		
Natural		
Wetlands	115	(100 – 200)
Termites	20	(10 – 50)
Ocean	10	(5 – 20)
Freshwater	5	(1 – 25)
CH_4 Hydrate	5	(0 – 5)
Anthropogenic		
Coal Mining, Natural and Petroleum Industry	100	(70 – 120)
Rice Paddies	60	(20 – 150)
Enteric Fermentation	80	(65 – 100)
Animal Wastes	25	(20 – 30)
Domestic Sewage Treatment	25	?
Landfills	30	(20 – 70)
Biomass Burning	40	(20 – 80)
Sinks		
Atmospheric (tropospheric + stratospheric) Removal	470	(420 – 520)
Removal by Soils	30	(15 – 45)
Atmospheric Increase	32	(28 – 37)

Table 2 (IGBP, 1992) gives a recent summary of current ideas on the global budget of methane. Note in the natural category that wetlands are a large source of methane. It is the metabolic activity of small microbes, methanogens, living in the oxygen-poor bottom environments in wetlands that produces the methane (Cicerone and Oremland, 1988). Methanogens in similar oxygen-poor environments in the intestines of termites and in the ocean are less important as natural sources. Among the anthropogenic emissions, leaks of fossil methane from the coal mining and natural gas industries are probably the leading ones, followed by methane from methanogens in rice paddies, enteric fermentation involving methanogens in cattle (the latter manifesting itself in methane emission by the belching and flatulation of the cattle), and methane produced by methanogens in garbage dumps and domestic sewage. Thus, there are a large number of sources related to simple microbes and rather basic human activities. The sources of methane are also geographically very patchy and linked to population. This is quite evident from estimates of methane emissions by animals (Lerner *et al.*, 1988). The numbers of most of the (largely domesticated) animals are quite strongly linked to the global population and so global animal emissions are rising and should continue to rise. Similar conclusions about geographical patchiness and links to human population growth are reached about methane emission from rice paddies.

Humans are not the only drivers of change in methane emissions. Emissions can change, for example, because of a climate change affecting the areal extent, inundation cycle, and temperature of wetlands. Figure 6 (Whalen and Reeburgh, 1992) shows the results of some studies of methane emissions in Alaska over the years 1987 through 1990. Note that the flux changes by several orders of magnitude over the year. Also shown in Figure 6 is the soil temperature as a function of time, which is obviously correlated with the emission rates. But temperature alone does not explain the details of the changes in emissions evident here. Other important factors are the degree and history of inundation and aeration of the soil.

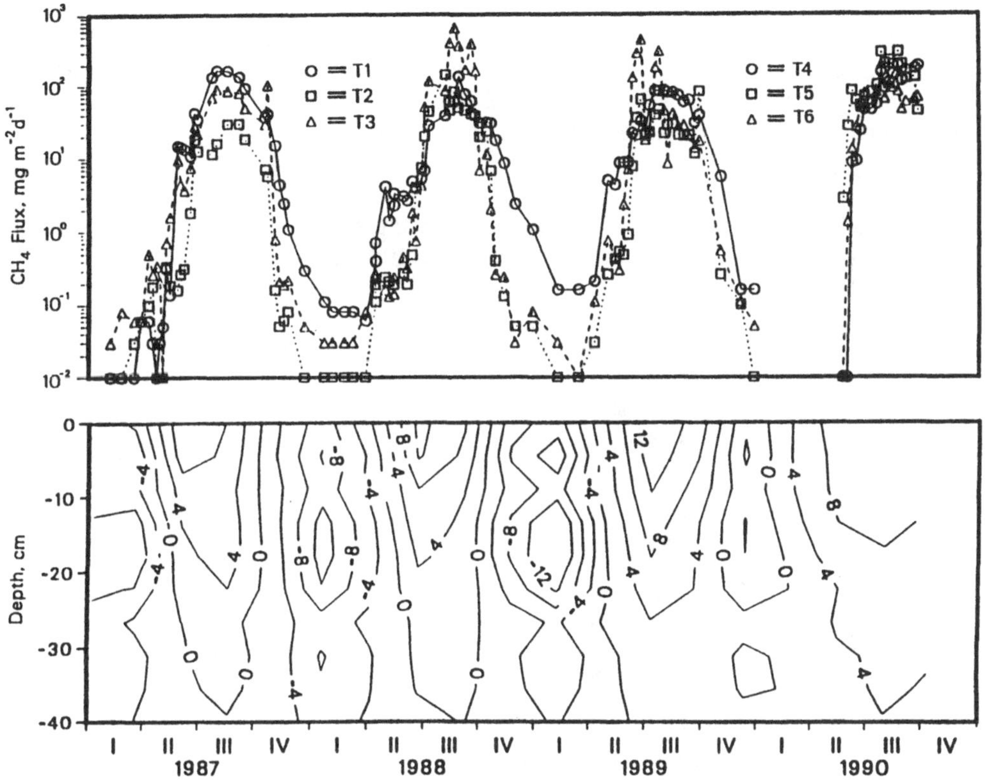

Figure 6. Net methane flux to the atmosphere (top panel, mg/m^2/day) and underlying soil temperatures (bottom panel, °C) for selected Alaskan tussock (T) sites (Whalen and Reeburgh, 1992).

As mentioned above, nitrous oxide, through production of nitrogen oxide free radical catalysts, is at present the largest contributor to ozone destruction in the stratosphere. It is also a long-lived greenhouse gas (but ranking behind carbon dioxide, methane, and the chlorofluorocarbons; Ramanathan *et al.*, 1985). It is naturally occurring in the atmosphere and is presently increasing at about 0.2% to 0.3% per year as evident from Figure 2 (Prinn *et al.*, 1990). The best fit to the observed latitudinal gradient requires between 52% and 68% of the emissions to be tropical, that is they emanate from latitudes between 30°N and 30°S (ibid.).

Table 3 (IPCC, 1992) shows a summary of emission sources compiled in this way. These emissions are compatible with the latitudinal gradients (Prinn *et al.*, 1990) if one allows for the uncertainties in Table 3, which are very large. Note the list of sources is as long as that for methane. It is a complex mix of natural and anthropogenic processes, some of which humans have a direct influence on, and others they do not. Some of the sources, for example nitrogen bacteria in soils (Firestone and Davidson, 1989), are known to be very dependent on climate, as was seen for some of the methane natural sources. But even more so than for methane, the uncertainty in the magnitude of the individual sources for nitrous oxide is too large to make a meaningful ranking of their importance. The major sink for nitrous oxide appears to be photodissociation in the stratosphere (Crutzen, 1978) but nitrogen bacteria in some soils also consume this gas (as well as emit it) so there is uncertainty also in the total sink. To obtain the trend in the gas by subtracting the sinks from the sources in Table 3 leads to a trend so uncertain that we could not even deduce whether the trend is positive (as observed) or negative. Under these circumstances we need to entertain the possibility that the trend in nitrous oxide could change substantially due to source or sink processes not presently understood or recognized. We particularly need to understand better the relevant nitrogen microbial processes in both terrestrial and aquatic environments together with their aerial extent and temporal variations.

Table 3. Estimated magnitude and latitudinal distribution of sources, and magnitudes of sinks of nitrous oxide (IPCC, 1992; Prinn *et al.*, 1990).

Sources (units Tg N per year)	
Natural	
Oceans	1.4 – 2.6
Tropical Soils	3.8 – 4.8
Wet Forests	2.2 – 3.7
Dry Savannas	0.5 – 2.0
Temperate Soils	≈ 0.6
Forests	0.05 – 2.0
Grasslands	?
Anthropogenic	
Cultivated Soils	0.03 – 3.0
Biomass Burning	0.2 – 1.0
Stationary Combustion	0.1 – 0.3
Mobile Sources	0.2 – 0.6
Adipic Acid Production	0.4 – 0.6
Nitric Acid Production	0.1 – 0.3
Source Distribution (units percentages)	
90°N – 30°N	22 – 34
30°N – 0°	32 – 39
0° – 30°S	20 – 29
30°S – 90°S	11 – 15
Sinks (units Tg N per year)	
Removal by Soils	?
Photolysis in the Stratosphere	7 – 13
Atmospheric Increase	3 – 4.5

Carbon dioxide is essentially chemically inert in the lower atmosphere and is cycled between atmosphere, upper ocean, green vegetation, and soils on decadal time scales, and between the upper ocean and deep ocean on a time scale of a hundred years. It is currently increasing at a rate of about 0.34% per year and possesses a significant annual cycle driven by uptake of CO_2 by photosynthesizing vegetation in spring and summer, and release of

CO_2 by decaying vegetation in fall and winter (Figure 1). Classical ideas had the oceans playing the major role as a sink for carbon dioxide. A typical global carbon budget had the current fossil fuel combustion source (5300 Tg C/yr) opposed by an oceanic sink (2300 Tg C/yr) to yield the observed atmospheric increase (3000 Tg C/yr). The land biosphere was generally considered to be a small contributor to the budget, although estimates of the deforestation source ranged from 400 to 2600 Tg C/yr.

However, comparison of the observed CO_2 latitudinal gradient with a global model containing only an ocean sink (Tans *et al.*, 1990) showed poor agreement. A new, significant northern hemispheric sink is needed to fit the data, but the available ocean uptake estimates do not allow the northern hemisphere oceans to provide this sink. Tans *et al.* (1990) concluded therefore that a significant northern hemispheric continental sink (about 2700 Tg C/yr), augmented by a smaller oceanic sink (about 1000 Tg C/yr), was needed to balance the atmospheric accumulation (3000 Tg C/yr), the fossil fuel input (5300 Tg C/yr), and the deforestation input (1400 Tg C/yr). Thus the ocean sink and the *net* biospheric sink (2700 – 1400 = 1300 Tg C/yr) are comparable. If correct, this implies that northern hemispheric continental forest regrowth and soil carbon uptake will need to be understood and appropriately modeled. Efforts in this direction have already begun (Melillo *et al.*, 1993). Of particular interest is the possibility of CO_2 fertilization. Perhaps related to this is the increasing amplitude of the CO_2 annual cycle in recent years (Figure 1), however there is still uncertainty since the available $^{13}C/^{12}C$ data do not indicate a biospheric sink of the size needed (Keeling *et al.*, 1989). On the other hand, recent studies of CO_2 uptake in a Massachusetts forest (Wofsy *et al.*, 1993) are not incompatible with a globally integrated temperate forest sink of 2000 Tg C/yr. The ocean sink for CO_2 is dominated by sinking of dense cold saline water in polar seas. This thermohaline circulation is dependent on rates of polar oceanic cooling and sea ice formation, and on high latitude rainfall, which lowers the density and inhibits sinking, so changes in the CO_2 ocean sink with climate change need to be included also.

Over the oceans the predominant source of cloud condensation nuclei (CCN) is either land-derived CCN and CCN precursors like SO_2 advected off the land, or the CCN precursor dimethyl sulfide (DMS), which is produced in the surface ocean itself by specific phytoplankton. For a fixed volume of cloud water, current cloud microphysical theory implies that the more CCN available, the greater the number of droplets, and the greater the cloud albedo. If this is correct, changes in DMS emissions ought to induce changes of the same sign in remote marine cloud albedo. Thus, climate changes that influence the populations of DMS-emitting plankton could lead to a feedback that could accelerate or decelerate the climate change (Charlson *et al.*, 1987). The DMS is produced from dimethylsulfonium propionate (DMSP) in the tissues of a specific group of phytoplankton (prymnesiophytes) so the dependence of this group on environmental variables needs careful assessment as does the proposed atmospheric effects of the emitted DMS.

ANTHROPOGENIC PROCESSES

Two anthropogenic activities, combustion and land use, play major roles as sources of climatically important species (or their precursors). Combustion of fossil, biomass, and detrital fuels are well recognized sources of CO_2, SO_2, NO, NO_2, and hydrocarbons, whose significance to climate has already been noted. Assessing the current and predicting the future emissions of these gases from such sources is a major challenge with the inventory of current emissions already underway. Land use and land use change are dominating factors in the CH_4 budget (through rice and cattle production and land fills) and the N_2O

budget. The difficulties in determining even current regional emission for these latter two gases were discussed earlier.

The influence of the short-lived pollutants from combustion are not just seen in the industrialized northern hemispheric midlatitudes. Tropospheric column ozone maps constructed by subtracting the stratospheric from the total atmospheric ozone columns, as measured from satellites, gave the first indication of very high ozone levels in the September to November period in the tropical central Atlantic (Fishman *et al.*, 1991). The only serious candidate for the phenomenon is air advected off Africa and South America and containing copious ozone and/or its precursors (NO_x, CO, etc.). The precursors are produced by biomass burning associated with land clearing and nomadic agriculture. A study of the distribution of fires over the globe, recently complied from Space Shuttle data, shows clearly the high frequency of fires in the tropics (Andreae, 1993). Also, the fires are quite seasonal, occurring in the dry season following the wet growth season. It is now clear that biomass burning, in addition to being very important to the land system itself, actually drives much of the tropical atmospheric chemistry. From the chemistry illustrated in Figure 4, the mixture of gases from fires are the very ingredients needed to pump up ozone and OH levels. Current research on biomass burning is tending to support a picture of seasonal burning in many tropical areas pumping chemicals into the air, which, through O_3 production, can increase the oxidizing capacity of the atmosphere.

INTERNATIONAL GLOBAL ATMOSPHERIC CHEMISTRY PROJECT

As noted in the introduction, the IGAC Project was created in 1988 with the goals to understand the atmospheric and biospheric processes controlling atmospheric composition and to provide the needed predictive capability for atmospheric composition as its contribution to the IGBP. Research in IGAC covers all of the areas discussed in this paper.

The Activities of IGAC, which are summarized in Table 4, are collected in seven groups called Regional, Global, or Fundamental Foci (IGAC, 1992). The regional Foci are necessary because of the patchiness of both natural and anthropogenic sources (and sometimes sinks) over the globe. This patchiness makes it impossible to study just one part of the globe and expect to understand the atmospheric-biospheric chemistry of all the other parts. The Global Focus includes global modeling, global inventories, global networks for determining trends, and global surveys for determining chemical distributions (both aerosols and gases). The Fundamental Focus includes basic laboratory studies and calibrations that underpin all the measurements. The subsequent chapters in this book report on the status of several IGAC Activities already underway. The work is done by individuals and groups of scientists funded largely by national programs but cooperating internationally. So IGAC is an international infrastructure for guiding, assessing, facilitating, and prioritizing the work.

Because of the complexity of the surface chemical sources and the needed global models, some of the work in IGAC is done cooperatively with other projects. Figure 7 is an attempt to illustrate how the various parts of IGAC contribute to the broader IGBP and WCRP goal of predicting global changes. In IGBP, IGAC is already interacting with the Joint Global Ocean Flux Study (JGOFS), and the Global Change and Terrestrial Ecosystems Project (GCTE), and will as time proceeds work with several other projects. IGAC interacts with the WCRP for global transport modeling and the Global Atmosphere Watch (GAW) for global network measurements. It cooperates in field programs with these projects and also looks to them to provide other data and models that are needed. For example, to predict atmospheric composition we need ecosystem models because changes in ecosystems will change the trace gas emissions from them. It is the IGAC (exclusive or shared) task to understand what the magnitudes of the emissions are for a given state of an ecosystem, but we look to GCTE and other projects for the predictive ecosystem models. Without strong cooperation we are not going to get the answers we need.

Table 4. Current (1994) IGAC Steering Committee, Foci and Activities, and Convenors of the Coordinating Committees for each Activity.

Steering Committee	R. Prinn (USA, Chair), P. Crutzen (Germany, Vice Chair), P. Matson (USA, Vice Chair), H. Akimoto (Japan), G. Brasseur (USA), R. Conrad (Germany), R.J. Delmas (France), D. Ehhalt (Germany), I. Galbally (Australia), B. Huebert (USA), V. Isodorov (Russia), N. Owens (UK), P. Matrai (USA), H.-U. Neue (Philippines), S. Penkett (UK), H. Rodhe (Sweden), E. Sanhueza (Venezuela), M.-X. Wang (China). *Liaison Members*: J. Miller (WMO), S. Slanina (IUPAC).
FOCUS	***Activity*** **(ACRONYM)†: Convenor(s)**
Marine	*North Atlantic Regional Experiment* (NARE) ***A***: F. Fehsenfeld, USA & S. Penkett, UK *Marine Aerosol and Gas Exchange* (MAGE) ***A***: B. Huebert, USA *East Asian/North Pacific Regional Experiment* (APARE) ***A***: H. Akimoto, Japan
Tropical	*Biosphere-Atmosphere Trace Gas Exchange in the Tropics: Influence of Land Use Change* (BATGE) ***A***: M. Keller, USA & R. Delmas, France *Deposition of Biogeochemically Important Trace Species* (DEBITS) ***A***: G. Ayers, Australia *Biomass Burning Experiment: Impact on the Atmosphere and Biosphere* (BIBEX) ***A***: M.O. Andreae, Germany *Rice Cultivation and Trace Gas Exchange* (RICE) ***A***: R. Sass, USA & H.U. Neue, Philippines
Polar	*Polar Atmospheric and Snow Chemistry* (PASC) ***A***: L. Barrie, Canada & R.J. Delmas, France
Boreal	*High Latitude Ecosystems as Sources and Sinks of Trace Gases* (HESS) ***A***: W. Reeburgh, USA
Midlatitude	*Midlatitude Ecosystems as Sources and Sinks for Atmospheric Oxidants* (METRO) ***P***: W. Chameides, USA *Trace Gas Exchange: Midlatitude Terrestrial Ecosystems and Atmosphere* (TRAGEX) ***A***: K. Smith, UK & G.P. Robertson, USA
Global	*Global Tropospheric Ozone Network* (GLONET) ***A***: J. Miller, Switzerland & V. Mohnen, USA *Global Atmospheric Chemistry Survey* (GLOCHEM) ***A***: D. Ehhalt & O. Schrems, Germany & E. Atlas, USA *Global Tropospheric Carbon Dioxide Network* (GLOCARB) ***A***: C. Keeling, USA *Multiphase Atmospheric Chemistry* (MAC) ***A***: T. Bates, USA & J. Gras, Australia *Global Emissions Inventory Activity* (GEIA) ***A***: T. Graedel, USA *Global Integration and Modeling* (GIM) ***P***: G. Brasseur, USA
Funda-mental	*Intercalibrations-Intercomparisons* (ICIC) *Methane and Halocarbons Intercalibration Experiment* (MEHALICE) ***A***: P. Fraser & P. Steele, Australia *Non-Methane Hydrocarbon Intercomparison Experiment* (NOMHICE) ***A***: J. Calvert & F. Fehsenfeld, USA *Carbon Dioxide Intercalibration Experiment* (CARBICE) ***A***: C. Keeling & P. Tans, USA *Atmospheric Chemistry Education in Global Change* (ACE) ***P***: J. Winchester, USA

† ***A*** = Activity already begun, ***P*** = Activity in planning phase

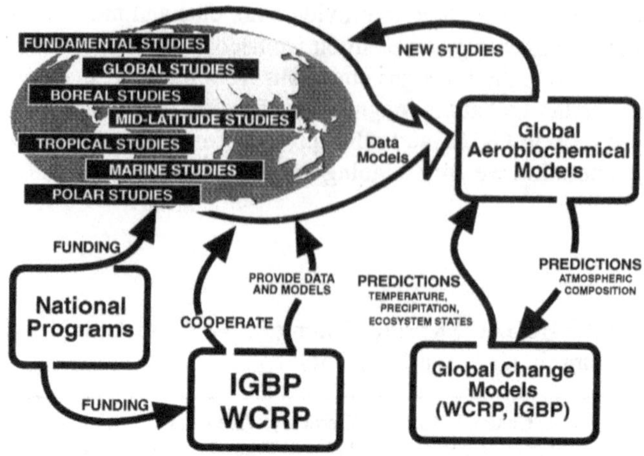

Figure 7. Schematic showing the organizational plan of IGAC. The five regional (boreal, midlatitude, tropical, marine, polar), the fundamental, and the global foci provide laboratory and field data and submodels to contribute to and test the integrating aerobiochemical models. The aerobiochemical models provide predictions of atmospheric composition for use in global (climate and biospheric) change models. The latter global change models in turn predict future environmental variables (temperature, precipitation, ecosystem state, etc.) for use (ideally interactively) in the aerobiochemical models. National programs are the basis for funding, with IGAC playing an international role for prioritizing, cooperating, and integrating. Cooperation with other IGBP projects and with WMO programs is an essential element of IGAC.

In Figure 7, I call the end-product of the IGAC effort "aerobiochemical" models. These are models which, given climate and ecosystem information, will predict the gaseous and aerosol composition of the atmosphere going into the future. These are the models that IGAC can provide to the physical climate modelers. I made the point earlier, however, that the cooperation does not finish there. The predictions of changes in temperature and precipitation made in the climate models are going to require running the aerobiochemical and physical climate models interactively, because the surface fluxes of trace gases like methane, carbon dioxide, and nitrous oxide will change as climate changes. It is unfortunate that it is that complicated, but the complexity needs to be included if we are to have any chance of getting accurate predictions of future atmospheric composition.

I end with a cautionary note. When understanding is incomplete, as it presently is, we need also to be ready for surprises. Atmospheric chemists were not smart enough to predict the existence of the Antarctic ozone hole. Prior to the discovery of this phenomenon the researchers were not including reactions on stratospheric particles in their models. It is these reactions, together with the special nature of polar meteorology, which cause the ozone hole. We need to be cognizant of the possibility of similar critical gaps in our current knowledge of the troposphere, and the probability of surprises and discoveries, which may alter present perspectives and approaches.

ACKNOWLEDGEMENTS

The author's atmospheric chemistry research is supported by NSF Grant 92-16340, and NASA Grants NAGW-474 and NAGW-732. The IGAC Core Project Office is supported by NSF Grant 91-01087.

REFERENCES

Anderson, J.G., W.H. Brune, and M.H. Proffitt, 1989, Ozone destruction by chlorine radicals within the Antarctic vortex: The spatial and temporal evolution of ClO-O_3 anticorrelation based on in situ ER-2 data, *J. Geophys. Res.*, 94:11,465.

Andreae, M., 1993, Global distribution of fires seen from space, *EOS Trans. Amer. Geophys. Un.*, 74:129-135.

Blake, D., and S. Rowland, 1988, Continuing worldwide increase in tropospheric methane, 1978 to 1987, *Science*, 239:1129-1131. Private communication, 1991 for 1987-1990 data.

Charlson, R., J. Langner, H. Rodhe, C. Leovy, and S. Warren, 1991, Perturbation of the northern hemisphere radiative balance by backscattering from anthropogenic sulfate aerosols, *Tellus*, 43:152-163.

Charlson, R., J. Lovelock, M. Andreae, and S. Warren, 1987, Oceanic phytoplankton, atmospheric sulfur, cloud albedo, and climate, *Nature*, 326:655-661.

Cicerone, R., and R. Oremland, 1988, Biogeochemical aspects of atmospheric methane, *Global Biogeochem. Cycles*, 2:299-328.

Crutzen, P., 1978, The role of NO and NO_2 in the chemistry of the troposphere and stratosphere, *Ann. Rev. Earth Planet. Sci.*, 7:443-472.

Crutzen, P., and P. Zimmerman, 1991, The changing photochemistry of the troposphere, *Tellus*, 43:136-151.

Cunnold, D., P. Fraser, R. Weiss, R. Prinn, P. Simmonds, B. Miller, F. Alyea, and A. Crawford, 1994, Global trends and annual releases of CCl_3F and CCl_2F_2 estimated from ALE/GAGE and other measurements from July 1978 to June 1991, *J. Geophys. Res.*, 99:1107-1126.

Donahue, N., and R. Prinn, 1990, Nonmethane hydrocarbon chemistry in the remote marine boundary layer, *J. Geophys. Res.*, 95:18,387-18,411.

Ehhalt, D., H. Dorn, and D. Poppe, 1991, The chemistry of the hydroxyl radical in the troposphere, *Proceedings Roy. Soc. Edinburgh*, 97:17-34.

Firestone, M., and E. Davidson, 1989, Microbiological basis of NO and N_2O production and consumption in soils, *in:* "Exchange of Trace Gases Between Terrestrial Ecosystems and the Atmosphere," M. Andreae and D. Schimel (eds.), J. Wiley & Sons, New York, pgs. 7-21.

Fishman, J., K. Fakhruzzaman, B. Cros, and D. Nganga, 1991, Identification of widespread pollution in the southern hemisphere deduced from satellite analyses, *Science*, 252:1693-1695.

Hansen, J., A. Lacis, R. Ruedy and M. Sato, 1992, Potential climate impact of Mt. Pinatubo eruption, *Geophys. Res. Lett.*, 19:215-218.

Hard, T., A. Mehrabzadeh, C. Chan, and R. O'Brien, 1992, GAGE measurements of tropospheric OH and inferences, *J. Geophys. Res.*, 97:9795-9818.

Intergovernmental Panel on Climate Change, 1992, Climate Change 1992, J. Houghton, B. Callander, and S. Varney (eds.), Cambridge U. Press.

International Global Atmospheric Chemistry (IGAC) Project, 1992, Vols: I. Overview, II. NARE, III. MAGE, IV APARE, V. BATGE, VI. DEBITS, VII. BIBEX, VIII. RICE, IX. PAC, X. PASE, XI. HESS, XII. TRAGEX, XIII. GLONET, XIV. GLOCHEM, XV. GLOCARB, XVI. MAC, XVII. GEIA, XVIII. MEHALICE, XIX. CARBICE, IGAC Core Project Office, MIT, Cambridge, MA 02139, U.S.

Keeling, C.D., S.C. Piper, and M. Heimann, 1989, A three dimensional model of atmospheric CO_2 transport based on observed winds: 4. Mean annual gradients and interannual variations, *in:* "Aspects of Climate Variability in the Pacific and the Western Americas," D.H. Peterson (ed.), Geophysical Monograph 55, American Geophysical Union, Washington, D.C., pgs. 305-363.

Keller, M., D. Jacob, S. Wofsy and R. Harriss, 1991, Effects of tropical deforestation on global and regional atmospheric chemistry, *Climate Change*, 19:139-158.

Lerner, J., E. Matthews, and I. Fung, 1988, Methane emission from animals, *Global Biogeochem. Cycles*, 2:139-156.

Lorius, C., J. Jouzel, D. Raynaud, J. Hansen, and H. LeTrent, 1990, The ice-core record: climate sensitivity and future greenhouse warming, *Nature*, 347:139-145.

Madronich, S. and C. Granier, 1992, Impact of recent total ozone changes on tropospheric ozone photodissociation, hydroxyl radicals, and methane trends, *Geophys. Res. Lett.*, 19:465-468.

Melillo, J., A. McGuire, D. Kicklighter, B. Moore, C. Vorosmarty, and A. Schloss, 1993, Global climate change and terrestrial net primary production, *Nature*, 363:234-240.

Molina, M., T. Tso, L. Molina and F. Wang, 1987, Antarctic stratospheric chemistry of chlorine nitrate, hydrogen chloride and ice: release of active chlorine, *Science*, 238:1253-1257.

Prinn, R., D. Cunnold, P. Simmonds, F. Alyea, R. Boldi, A. Crawford, P. Fraser, D. Gutzler, D. Hartley, R. Rosen, and R. Rasmussen, 1992, Global average concentration and trend for hydroxyl radicals deduced from ALE/GAGE trichloroethane (methyl chloroform) data for 1978-1990, *J. Geophys. Res.*, 97:2445-2461.

Prinn, R., D. Cunnold, R. Rasmussen, P. Simmonds, F. Alyea, A. Crawford, P. Fraser, and R. Rosen, 1990, Atmospheric emissions and trends of nitrous oxide deduced from 10 years of ALE/GAGE data, *J. Geophys. Res.*, 95:18,369-18,385.

Ramanathan, V., R. Cicerone, H. Singh, and J. Kiehl, 1985, Trace gas trends and their potential role in climate change, *J. Geophys. Res.*, 90:5547-5566.

Ramaswamy, V., M. Schwarzkopf, and K. Shine, 1992, Radiative forcing of climate from halocarbon-induced global stratospheric ozone loss, *Nature*, 355:810-812.

Rowland, S., 1990, Stratospheric ozone depletion by chlorofluorocarbons, *Ambio*, 19:281-292.

Steele, P., E. Dlugokencky, P. Lang, P. Tans, R. Martin, and K. Masarie, 1992, Slowing down of the global accumulation of atmospheric methane during the 1980s, *Nature*, 358:313-316.

Stolarksi, R., P. Bloomfield, R. McPeters, and J. Herman, 1991, Total ozone trends deduced from NIMBUS-7 TOMS data, *Geophys. Res. Lett.*, 18:1015-1018.

Tans, P., I. Fung, and T. Takahashi, 1990, Observational constraints on the global atmospheric CO_2 budget, *Science*, 238:1253-1257.

Whalen, S., and W. Reeburgh, 1992, Interannual variations in tundra methane emission: a four year time series at fixed sites, *Global Biogeochem. Cycles*, 6:139-160.

Wofsy, S., M. Goulden, J. Munger, S.-M. Fan, P. Bakwin, B. Daube, S. Bassow, and F. Bazzaz, 1993, Net exchange of CO_2 in a mid-latitude forest, *Science*, 260:1314-1317.

ATMOSPHERIC CHEMISTRY AND COMPOSITION OF AIR OVER THE NORTH ATLANTIC OCEAN

Stuart A. Penkett[1], Frederick C. Fehsenfeld[2] and Joseph M. Prospero[3]

[1]School of Environmental Sciences
University of East Anglia
Norwich, UK
[2]Aeronomy Laboratory
NOAA
Boulder, Colorado, USA
[3]Rosenstiel School of Marine and Atmospheric Science
University of Miami
Miami, Florida, USA

ABSTRACT

The North Atlantic Regional Experiment (NARE) Activity of IGAC seeks to study chemical processing and transport over the North Atlantic Ocean of pollutants emitted from the surrounding continents. It has various components with research activities based on measurements of aerosol composition, and of ozone and its precursors.

The results obtained on the chemical composition of aerosols over the past several years clearly show that the entire North Atlantic atmosphere is heavily impacted by anthropogenic sources. For example, mean aerosol NO_3^- concentrations are 5 to 17 times higher than those measured in the remote South Pacific; non-seasalt-SO_4^{2-} values are 2 to 7 times higher. High concentrations of aerosols (and some gases) can be clearly related to the transport of pollutants from continental sources. The measured concentrations of sulfate aerosols equal or exceed those used in recent calculations that suggest that anthropogenic aerosols could have a significant cooling effect on climate in the northern hemisphere (Langner and Rodhe, 1991; Charlson *et al.*, 1992). Nonetheless, at some locations at various times of year, natural sources do play an important role in the cycles of a number of important aerosols, gases and precipitation constituents.

Ozone is produced photochemically from anthropogenic pollution in North America and is transported to the North Atlantic. To evaluate the significance of this transport, ozone and carbon monoxide (CO) levels were measured during 1991 and 1992 at sites on the Atlantic coast of Canada. A strong positive summertime correlation between carbon

Global Atmospheric-Biospheric Chemistry
Edited by R.G. Prinn, Plenum Press, New York, 1994

monoxide and ozone was attributed to photochemical production of ozone from anthropogenic emissions. In contrast, during the winter when ozone is titrated by NO and other anthropogenic emissions, carbon monoxide and ozone were anti-correlated. The measured relations between the concentrations of ozone and carbon monoxide provide the basis for an estimate of ozone exported from North America: approximately 100 billion moles of ozone per summer. This amount is greater than that reaching the lower troposphere in this region from the stratosphere, the primary natural source of ozone. This conclusion supports the contention that ozone derived from anthropogenic pollution has a hemisphere-wide effect at northern temperature latitudes.

Observations of ozone at two locations in the free troposphere at an altitude of ~3 km over the North Atlantic (Izania [~25°N] and off the west coast of Scotland [50°N]) show a marked seasonal oscillation, with a maximum of about 55 ppbv in late spring (May–June) and a minimum of about 35 ppbv in winter (December–January). The same seasonal pattern is exhibited by the hydrocarbon chemical precursors of ozone. These results suggest that the increase in ozone between January and May is in part due to increased production from photochemical processes within the troposphere. A further feature of interest, which comes from close examination of the Atlantic free tropospheric hydrocarbon data, is that the distribution of hydrocarbons, particularly with respect to the ratios of the "normal" to the "iso" isomers of pentane, hexane, heptane and octane, can be interpreted in terms of an efficient NO_3 chemistry preferentially removing the "iso" isomer during winter months.

INTRODUCTION

The North Atlantic Regional Experiment (NARE) was established by the International Global Atmospheric Chemistry (IGAC) Project to study the chemical processes that are occurring in the remote marine environment of the North Atlantic. The rationale behind NARE is that the industrial regions on the continental rim of the North Atlantic are major global sources of gases that influence the oxidizing capacity and the radiative balance of the earth's atmosphere. In this region, the gases move from the continental coastal areas into the marine environment where they are processed to form both gaseous and aerosol products (see Figure 1). This combination of a well defined source region enclosing a vast, relatively source-free area, provides an excellent regime for the study of the mechanisms involved in the chemical processing of many compounds, and the factors affecting the persistence of these compounds and their oxidative products in the marine atmosphere.

The NARE experiment takes place against a background of an increasing trend in tropospheric ozone, evidence for the presence of products of industrial emissions such as sulfate and nitrate aerosols at remote Atlantic sites, the presence of ozone precursors in the Atlantic free troposphere, and the co-presence of ozone and carbon monoxide in air being transported from North America into the Atlantic.

The goals of NARE are:

- To assess the long range transport of photochemically active compounds and/or their products and determine the impact of this transport on hemispheric air quality.
- To ascertain the effect of these compounds on the oxidative properties and radiation balance of the atmosphere.
- To estimate the amounts of these compounds that are deposited in this marine environment and to determine the impact of this deposition on surface sea-water chemistry and marine biological processes.

A proposed additional goal of NARE is to study the impact of in-situ emissions from subsonic aircraft on the oxidizing capacity of the atmosphere over the North Atlantic.

In achieving these goals, NARE can only provide part of the answer on a global scale, and close links will be established between other regional experiments investigating the atmospheric chemistry of mid to high latitudes of the northern hemisphere (see Figure 1).

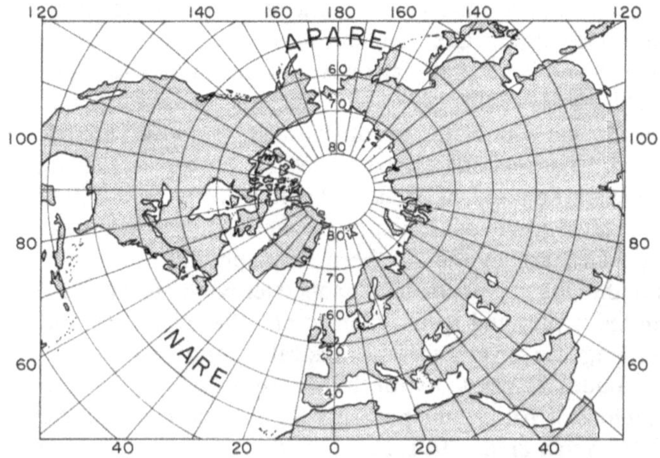

Figure 1. Map showing the proximity of the North Atlantic region to surrounding continents (land masses shaded gray) and to its sister IGAC marine activity, the East Asian/North Pacific Regional Experiment (APARE).

EVIDENCE FOR ANTHROPOGENIC PERTURBATIONS TO GLOBAL TROPOSPHERIC OZONE

The initial focus of NARE is to determine the impact of emissions of pollutants that lead to the production of tropospheric ozone. The research needed to meet this objective involves measurements of the distribution and trends of ozone and its precursors in the North Atlantic region, the determination of the sources of this ozone, and the elucidation of the processes responsible for its formation on a hemispheric scale.

Elevated ozone levels have been observed in summer over urban and large rural areas in western Europe (Cox *et al.*, 1975; Guicherit *et al.*, 1977; Grennfelt and Schjoldager, 1984), in central and eastern U.S. (Vukovich *et al.*, 1977, 1985), and in other industrialized countries. Observational and modeling studies show that high level rural ozone is the product of long-range transport of ozone precursors and multi-day photochemical production and accumulation of ozone (Vukovich *et al.*, 1977; Isaksen *et al.*, 1978). Figure 2 illustrates the latitudinal and seasonal distribution of the tropospheric residual amounts derived from measurements made from the TOMS and SAGE satellites of the total atmospheric column ozone and stratospheric ozone, respectively (Fishman *et al.*, 1990). These figures indicate that the ozone concentrations in the tropics are about a factor of 2 less than in the northern hemisphere midlatitudes and about a factor of 1.4 lower than in the southern hemisphere midlatitudes. This distribution suggests that there is probably greater stratospheric ozone injection and greater photochemical ozone production at midlatitudes, and greater photochemical ozone loss in the tropics. Integrated throughout the entire year, there is about 40% more ozone in northern hemisphere midlatitudes than at southern hemisphere midlatitudes. The asymmetry could be explained in two ways: the northern hemisphere stratospheric input flux is greater than the southern hemisphere flux, or there is a larger photochemical production of ozone in the northern hemisphere as a result of anthropogenic emissions of ozone precursors.

In addition there is a seasonal variation in the ozone distribution at midlatitudes in both hemispheres, with a significant increase in ozone column density in the respective springs in each hemisphere. In the northern hemisphere midlatitudes there is an additional increase in the summer that could be rationalized as a result of ozone production from ozone precursors of anthropogenic origin. The spring ozone increase has been observed over many remote stations and is generally attributed to increased stratospheric flux (Fabian *et al.*, 1968; Chatfield and Harrison, 1977). For example, seasonal variations of ozone have been observed to coincide with variations of ^{90}Sr and ^{7}Be, which are of stratospheric origin (Fabian *et al.*, 1968; Chatfield and Harrison, 1977). However, Penkett and Brice (1986) observed concurrent spring peaks of ozone and peroxyacetyl nitrate (PAN), leading them to suggest a photochemical origin for the spring ozone increase in the northern hemisphere. The spring ozone increase in the northern hemisphere could be associated with an accumulation of anthropogenic ozone precursors over winter (Penkett, 1987) and increased ozone lifetime in the winter and spring (Liu *et al.*, 1987). In addition, there is evidence supporting anthropogenic origin for the austral spring ozone maximum at southern hemisphere midlatitudes associated with the widespread biomass burning in the tropical and subtropical regions of Africa, and probably South America, during the dry season from July to October. (Fishman *et al.*, 1991; Fehsenfeld and Liu, 1993). However, a significant contribution from stratospheric ozone intrusion to the springtime ozone increase in both hemispheres is also likely.

Identification of the origin of the springtime ozone increase has profound implications for understanding the impact on tropospheric ozone of anthropogenic activities. If the origin is entirely natural, such as that due to stratospheric intrusion, then the only significant anthropogenic impact is the summertime increase at northern hemisphere midlatitudes. This would imply that photochemical production from anthropogenic precursors would account for only approximately 5% of the annual average ozone produced over the globe. On the other hand, if the origin is anthropogenic in both spring and summer, then the impact on tropospheric ozone due to human activities may be greater than 30% globally, and much more on a sub-hemispheric scale. Therefore, it is essential to design an effective research strategy to understand the seasonal variation of tropospheric ozone, and its cause.

Although most of the northern hemisphere appears to be significantly affected by anthropogenically produced ozone (Figure 2), the sparsity of observations prevents us from estimating the trends with certainty. However, there are some tantalizing indications in the data that are available. An interesting comparison can be seen in Figure 3, which shows measured ozone mixing ratios at Montsouris[1], a rural site in France where measurements were made in a period between 1876–1886 (cf., Volz and Kley, 1988), and modern-day observations at Arkona, a site near the Baltic coast (cf., ibid.). As the figure indicates, present-day ozone mixing ratios recorded at Arkona are a factor of 2 to 3 times higher than the measurements at Montsouris made in the late nineteenth century.

A further indication of an increase in the average tropospheric ozone mixing ratio in the northern hemisphere can be inferred from the data shown in Figure 4. This figure depicts the ozone trend derived from ozone sonde measurements between 2 to 8 km

---

[1]The Montsouris measurements and the potential interferences to the method used to made those measurements were critically evaluated and the results re-analysed (Volz and Kley, 1988). On this basis the results from these very valuable measurements are deemed to be reliable.

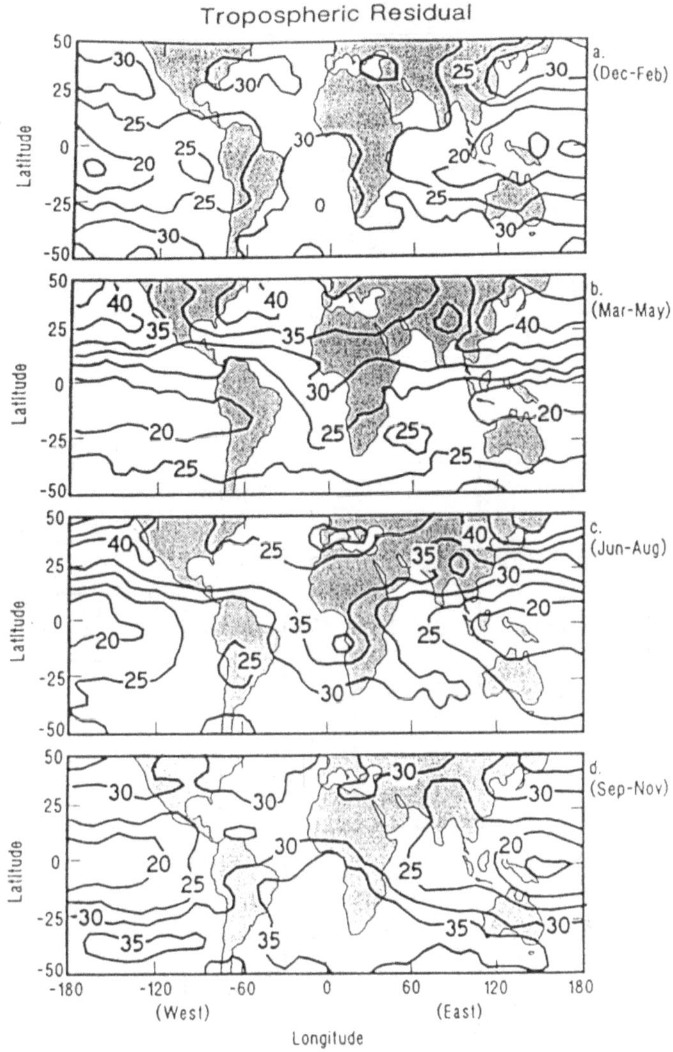

Figure 2. The climatological distribution of the integrated tropospheric residual amounts of ozone (Dobson units) divided into four periods (Fishman *et al.*, 1990).

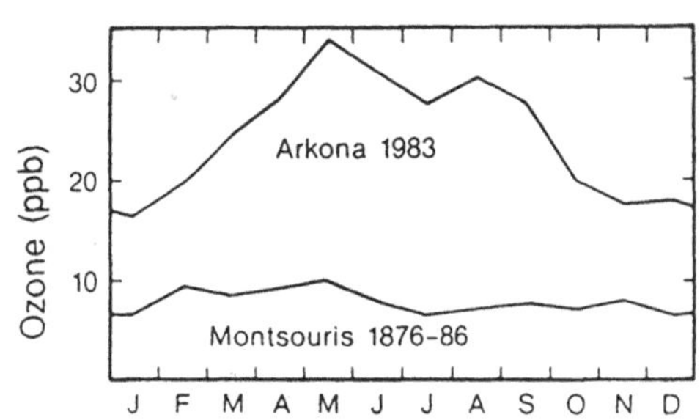

Figure 3. Average seasonal variation of ozone at Montsouris (1876–1986, southwest sector only) and at Arkona. Montsouris was located on the outskirts of Paris which could be characterized as a rural station at that time. Arkona is a remote station on the island of Rügen in the Baltic Sea (cf., Volz and Kley, 1988).

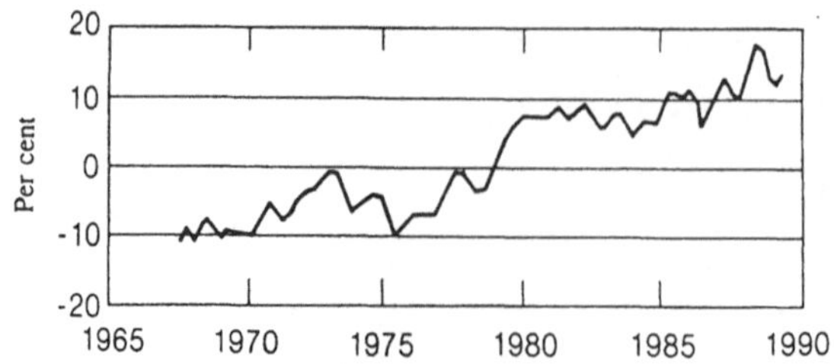

Figure 4. Deviation (in percent) from layer-mean ozone mixing ratios between 2 to 8 km altitudes in the north temperate latitudes from 1967-89 as derived from ozone sonde measurements (Fehsenfeld and Liu, 1993).

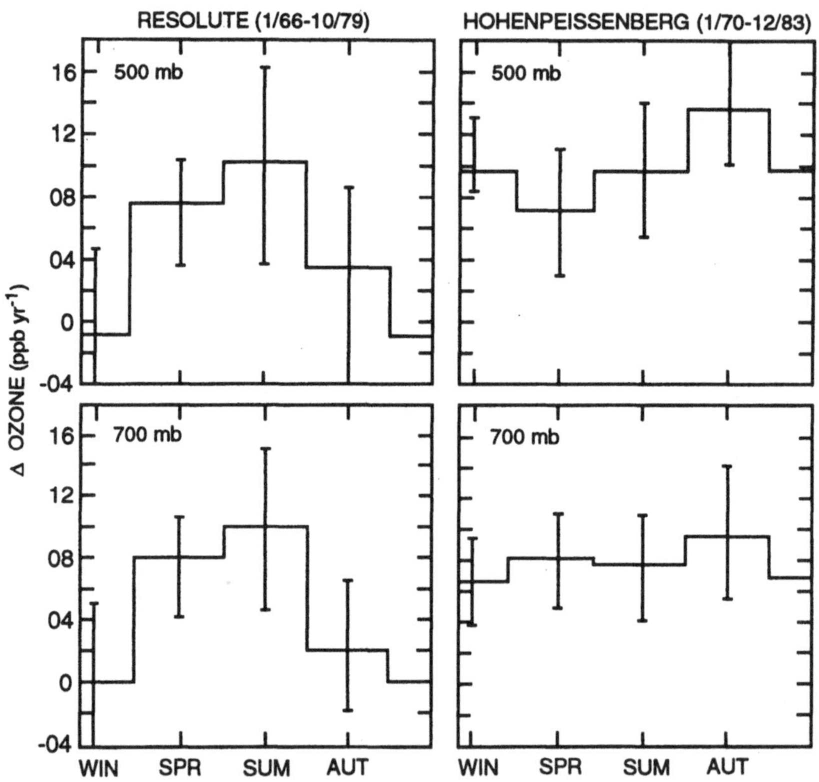

Figure 5. Vertical distribution of the trend in ozone (in percent per year) at Resolute and Hohenpeissenberg. The horizontal bars give the 90% confidence interval for the trend at each level. The shaded area shows the range of tropopause heights through the year.

altitudes in the north temperate latitudes from 1967 to 1989 (Angell, private communication). Likewise, Figure 5 shows the vertical distribution of the trend in ozone at Resolute, Canada and Hohenpeissenberg, Germany. Except near the surface, the two stations give a consistent picture of 1 to 2% per year increase in ozone in the troposphere and a gradual shift to a small decreasing trend in the lower stratosphere. This shift is, in general, consistent with current understanding of ozone trends in the troposphere and stratosphere.

INDICATIONS OF ANTHROPOGENIC POLLUTION OVER THE NORTH ATLANTIC FROM AEROSOL MEASUREMENTS

An objective of the Air-Ocean Chemistry Experiment (AEROCE) is to characterize the chemical climatology of a number of important chemical species over the North Atlantic, and to assess the relative importance of their various natural and anthropogenic sources. Other objectives focus on the study of the transport, transformation and removal processes, and on the possible impact of the deposition of these materials to the ocean. To these ends, a network of permanent coastal stations was established in the North Atlantic Ocean (Figure 6): Barbados; Bermuda; Mace Head, Ireland; Miami, Florida; and at Izania on Tenerife in the Canary Islands. A protocol of continuous measurements was begun in mid-1988 in a cooperative program involving scientists from universities and government laboratories in the U.S. and other countries. A wide range of chemical species and tracers are measured in the gas, aerosol and precipitation phases. Trends involving aerosols are of particular interest because recent model estimates suggest that anthropogenic aerosols, in particular those composed primarily of non-seasalt sulfate (nss-SO_4^{2-}), could have a significant impact on climate in the northern hemisphere, especially in the North Atlantic (Charlson *et al.*, 1991; Langner and Rodhe, 1991; Lelieveld and Heinzenberg, 1992).

Figure 6. Map of the North Atlantic showing the stations of the AEROCE network and those used in conjunction with and/or supported by the NOAA Climate and Global Change Program (NOAA CGCP).

The annual mean concentrations obtained at the AEROCE sites over a several year period are presented in Figure 7 along with means from remote islands in the Pacific and the coastal Antarctic where anthropogenic impacts should be minimal (Savoie *et al.*, 1989; 1992a). The North Atlantic values are clearly much higher. Nitrate concentrations are as much as 17 and 71 times greater than the American Samoa and Mawson means, respectively, while nss-SO_4^{2-} values are as much 7 and 28 times higher.

Pollution impacts are also indicated by the high concentrations of certain trace metals. At Bermuda, high nss-SO_4^{2-} concentrations are associated with sharp increases in Sb and Se (Arimoto *et al.*, 1992; Ellis *et al.*, 1993), elements that have major pollution sources; the concentrations and their ratios measured in Bermuda are consistent with those measured in the northeast U.S., suggesting fossil fuel combustion sources. It is notable that the average

Sb concentrations over the North Atlantic is 200 to 1000 times greater than at Samoa (Savoie *et al.*, 1993). This extremely large difference is indicative of the strong impact that the transport of material from continental sources has on the atmospheric chemistry over the entire North Atlantic Ocean. In fact, there was not even one sample from the entire AEROCE data set that had an Sb concentration as low as that at Samoa.

Although the pollution sources are clearly dominant in the samples from all sites, natural sources are also important, especially in the atmospheric sulfur cycle. At all sites, the concentration of aerosol methanesulfonate (MSA), an oxidation product of dimethylsulfide (DMS), shows a strong seasonal cycle that peaks in May–June and reaches a minimum in the December–January time period (Savoie *et al.*, 1994). This seasonal cycle conforms to the annual cycle of primary productivity in the oceans and, hence, reflects the changes in the emission of DMS to the atmosphere. Thus, the seasonal activity of the oceanic source of DMS is rather consistent over a large area of the tropical to midlatitude North Atlantic Ocean.

The AEROCE data can be used to estimate the relative contribution of natural and anthropogenic sources of sulfur over the North Atlantic on the basis of the concentration of MSA and Sb (Savoie *et al.*, 1994). At Barbados, the anthropogenic component is about 50% and it appears to be derived largely from Europe. At Bermuda, it is 70%, largely from North America. At Mace Head, the anthropogenic fraction is about 80–90%, mostly from Europe. Thus, the aerosol data suggest that the entire North Atlantic is impacted by continental anthropogenic sources for a major fraction of the year.

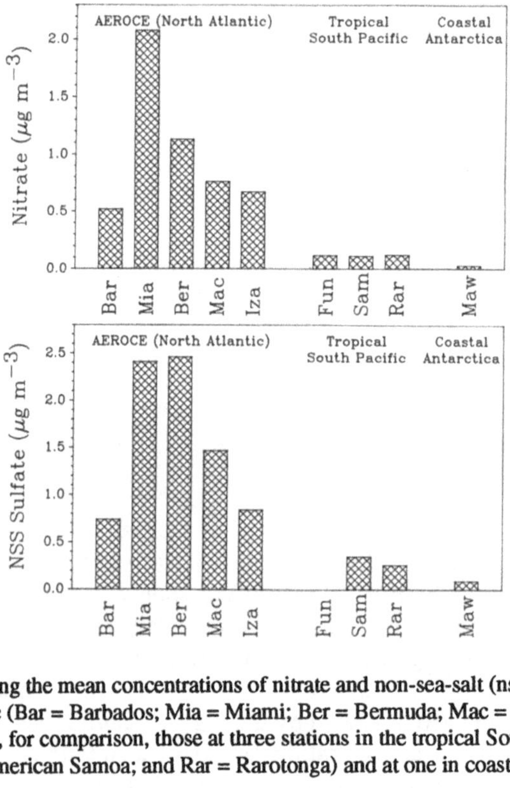

Figure 7. Bar chart showing the mean concentrations of nitrate and non-sea-salt (nss) sulfate at five stations in the in the North Atlantic (Bar = Barbados; Mia = Miami; Ber = Bermuda; Mac = Mace Head, Ireland; and Iza = Izania, Tenerife) and, for comparison, those at three stations in the tropical South Pacific (Fun = Funafuti; Sam = American Samoa; and Rar = Rarotonga) and at one in coastal Antarctica (Maw = Mawson).

OZONE EXPORT FROM NORTH AMERICA OVER THE NORTH ATLANTIC

Ozone has both natural and anthropogenic sources in the troposphere. The primary natural source is injection from the stratosphere. The anthropogenic source is photochemical production from precursors emitted by industrial and transportation combustion sources. To understand the budget, and hence the effects of ozone, it is important to quantify and to compare the magnitude of these two sources (Parrish, 1993; Parrish *et al.*, 1993). The heavily polluted eastern coast of North America is a particularly large source of ozone and its precursors, which can be transported to the temperate North Atlantic Ocean. During summer the atmospheric circulation in the temperate North Atlantic region is dominated by the Bermuda High, with flow to the northeast along the Atlantic Seaboard. To evaluate the significance of this transport, ozone and carbon monoxide levels were measured during 1991 and 1992 at sites on the Atlantic coast of Canada spaced at approximately 500 km intervals downwind from the northeastern urban corridor of the U.S. (see Figure 6).

Carbon monoxide is an anthropogenic pollutant that is relatively unreactive (lifetime about 1 month in the summer) and has been used as an effective tracer of anthropogenic pollution. The measured relations between the concentrations of ozone and carbon monoxide in the air advected over the North Atlantic, coupled with the inventoried emissions of carbon monoxide in eastern North America, provide the basis to estimate the amount of ozone exported from North America (Parrish *et al.*, 1993).

Measurements were made at three sites from mid-July to mid-September, 1991. Five minute average ozone and carbon monoxide concentrations were calculated. At each of the three sites the data (Figure 8) are characterized by generally constant, low levels of ozone and carbon monoxide interrupted frequently by episodes of correlated, elevated levels of both species. Linear regression analyses of the data yield relatively large correlation coefficients, which indicate that approximately one-half to three-quarters of the variance in the ozone levels at each site can be captured by a simple linear relation between ozone and carbon monoxide levels. The similarity of the derived slopes indicates that the relation between ozone and carbon monoxide levels is approximately constant over the 1000 km distance spanned by the sites.

The strong, positive correlation between ozone and carbon monoxide observed at Sable Island in August 1991 is reproduced in Figure 9 (Parrish, 1993). The linear correlation coefficient indicates that more than three-quarters of the variance in the measured ozone during this single summer month can be captured by a linear relationship between ozone and carbon monoxide. This suggests summertime photochemical production of ozone from anthropogenic ozone precursors emitted from the source regions on the North American continent. By contrast, in the winter at Sable Island, the data were characterized by high carbon monoxide levels and even lower ozone levels, which result in significant negative correlation between ozone and carbon monoxide. Figure 9 also presents the December data. For this month more than half of the variance in the measured ozone can be captured by a linear relationship. This indicates the chemical titration of ozone by anthropogenically emitted nitric oxides and volatile organic compounds (VOCs) in the absence of photochemical production.

Figure 10 summarizes the monthly linear correlations for the full data sets from the three sites (Parrish, 1993). Strong positive correlations were found at all three sites for the summer months. At Sable Island the correlation disappeared as fall progressed, and a negative correlation developed in the winter. This negative correlation in turn disappeared as spring progressed, followed by the reappearance of the positive correlation.

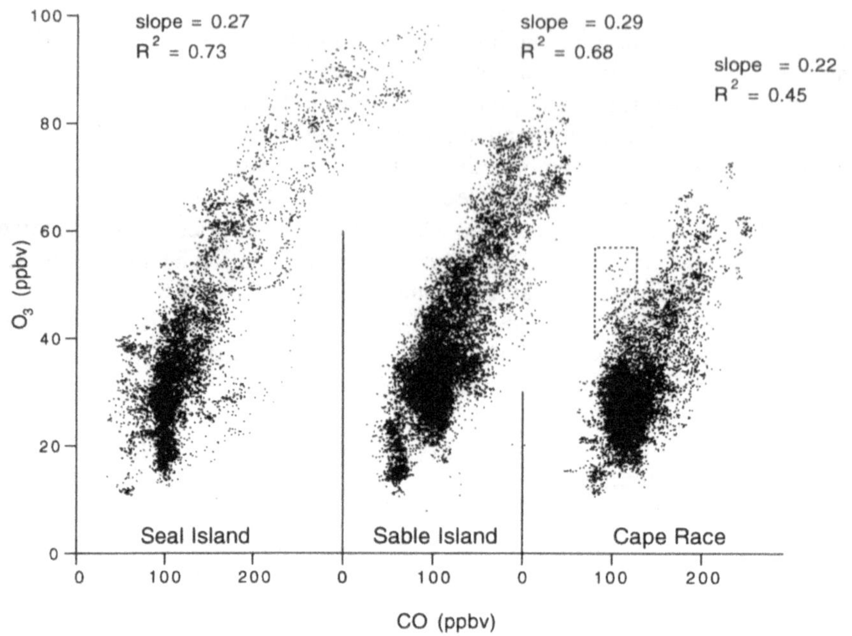

Figure 8. Relation between O_3 and CO levels at each of three sites. Each point indicates a five minute average. For each site, the slope and the square of the linear correlation coefficient (R^2) are given. At Seal Island, Sable Island and Cape Race, the measurement period was July 22 to September 22, July 8 to September 21, and July 13 to September 16, respectively, all in 1991 (Parrish, 1993).

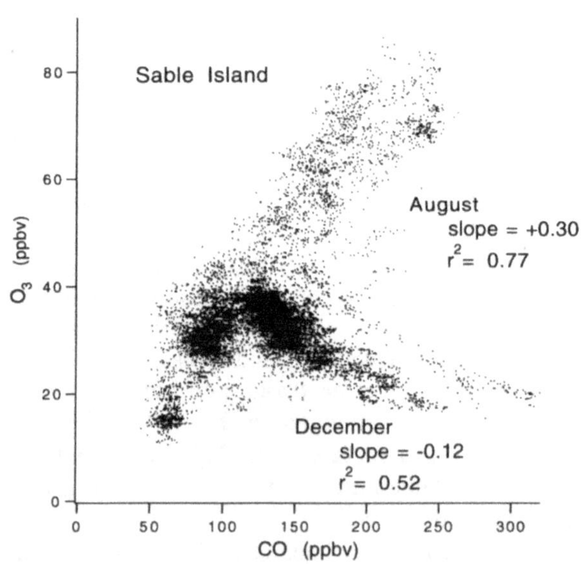

Figure 9. Relation between O_3 and CO at Sable Island in August and December. Each point indicates a five minute average. For each month, the slope from a linear, least squares fit and the square of the linear correlation coefficient (r^2) are given (Parrish, 1993).

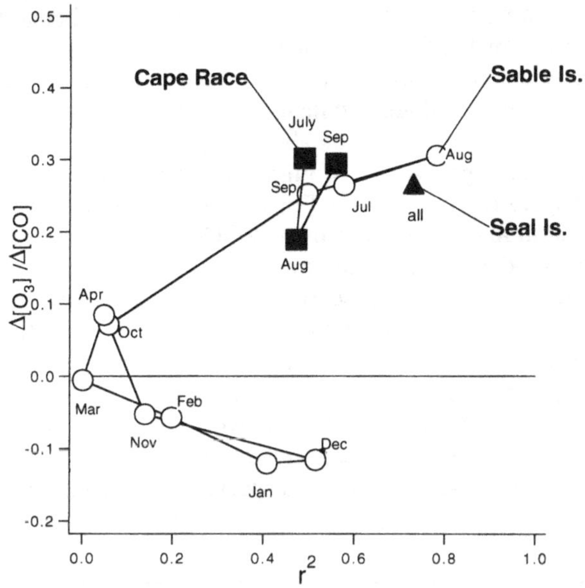

Figure 10. Summary of the linear regression analyses of the monthly data sets. The abscissa gives the square of the linear correlation coefficient, and the ordinate gives the derived slope. The different symbols indicate the different sites, and the months are labeled. The data from the three months at Seal Island were combined in one analysis, since the time coverage was relatively poor at this site.

The slopes of the linear regressions in Figure 8 and Figure 10 provide an indication of the amount of ozone photochemically produced from the precursors that were emitted into the air masses with the carbon monoxide. That is, the amounts of ozone and carbon monoxide in these transported air masses were at a molar ratio of about 0.3 to 1. Because the amount of carbon monoxide emitted over North America is approximately known from emission inventories, and because the fraction of this carbon monoxide that is transported to the North Atlantic can be estimated, it is possible to estimate the quantity of ozone photochemically produced from the precursors and transported to the North Atlantic: approximately 100 billion moles of ozone per summer (Parrish *et al.*, 1993). This amount is greater than that reaching the lower troposphere in this region from the stratosphere (WMO, 1986), the primary natural source of ozone. This conclusion supports the contention that ozone derived from anthropogenic pollution has a hemisphere-wide effect at northern temperate latitudes.

LARGE-SCALE OZONE TRANSPORT TO THE NORTH ATLANTIC

Some insights into the question of ozone transport on a larger scale over the Atlantic can be obtained by examining ozone data in the context of tracers for specific source regions or source types. In this regard, the data from the Spanish Meteorological observatory at Izania, Tenerife, are especially interesting. Izania serves as a Baseline Air Pollution Monitoring (BAPMoN) station and it hosts the activities of both the Tropospheric Ozone Research (TOR) program and AEROCE. The station is located at an elevation of 2367 m, which places it above the top of the marine inversion most of the time. Gas and aerosol sampling are carried out at night under downslope wind conditions and hence the data should be representative of free tropospheric air.

At Izania, ozone concentrations have a well-defined seasonal cycle with monthly means of about 40 ppbv in the winter and increasing rapidly in the spring to mean values of about 55 ppbv in April, May and June and about 50 ppbv in July (Schmitt *et al.*, 1988). It has been suggested that the persistence of high ozone concentrations in the summer could be due to the transport or ozone from Europe, which often experiences high ozone concentrations at that time of year. Recent AEROCE and TOR research (Prospero *et al.*, 1993) shows that day-to-day ozone concentrations can change substantially (Figure 11) and coherently during the summer in response to meteorological factors. However, as shown in Figure 11, ozone concentrations are highly correlated with ^{7}Be (half life 54 days), a radionuclide produced by cosmic rays in the stratosphere and upper troposphere (Brost, 1991). Conversely, ozone is negatively correlated with ^{210}Pb, a radionuclide that is produced from the decay of ^{222}Rn from soils and which is, hence, an excellent tracer for air parcels that have recently been in the continental boundary layer (Feichter *et al.*, 1991). These correlations (positive and negative) are especially strong during the summer months.

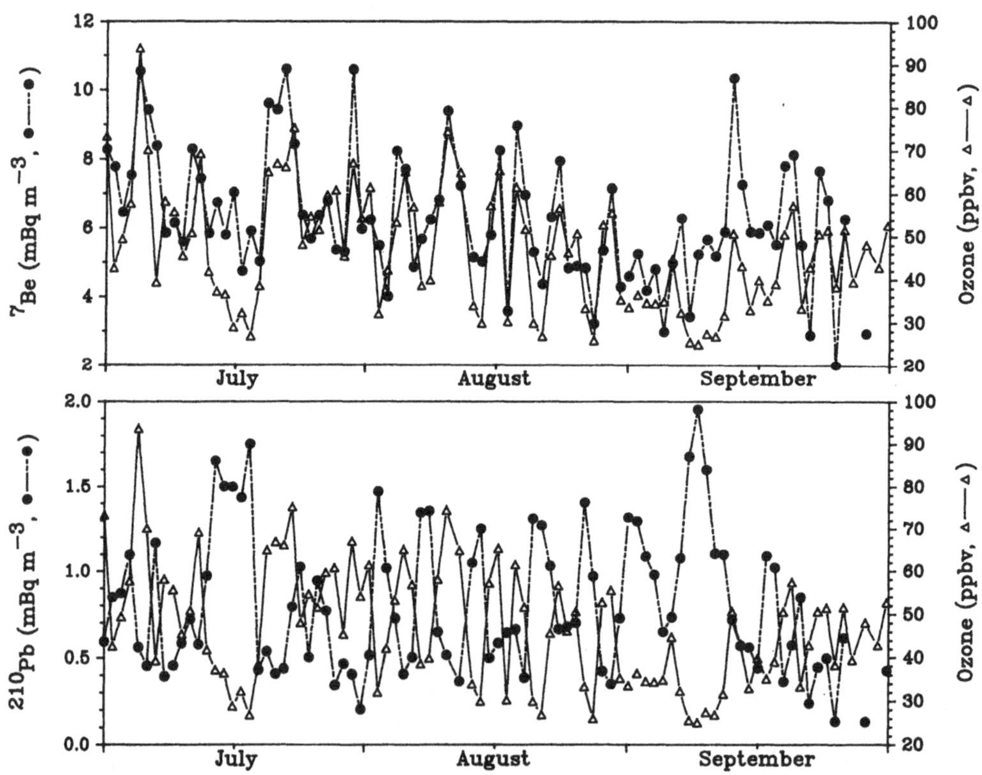

Figure 11. Comparative time series plots of ^{7}Be (top panel) and ozone and ^{210}Pb (bottom panel) and ozone at Izania, Tenerife, Canary Islands during July, August and September 1991 (Prospero *et al.*, 1993).

The seasonal cycle of ozone at Izania is similar to that for aerosol nss-SO_4^{2-} and NO_3^- (i.e., with maximum monthly means in the spring and summer). This similarity could be interpreted as supporting an anthropogenic source for ozone. However, on a day-to-day basis, ozone is strongly anti-correlated with aerosol NO_3^- and nss-SO_4^{2-} (Prospero *et al.*,

1993). Indeed, during periods of high ozone concentrations, the concentrations of nss-SO_4^{2-} are usually about a few tenths of a microgram per cubic meter of air, concentrations that are comparable to those in the remote South Pacific (Figure 7); during periods of low ozone concentration, nss-SO_4^{2-} values are typically about ten times higher.

The high coherence between the ozone time series and that for ^{7}Be, and the negative relationship with that for ^{210}Pb, suggests a strong free troposphere signature for the ozone source over the remote Atlantic. Because ozone concentrations tend to be low in the presence of high concentrations of NO_3^- and nss-SO_4^{2-} (and ^{210}Pb), it would appear that in many cases a significant fraction of the ozone, which might have been generated in polluted air masses, could have been destroyed during transport to Izania. Indeed, the isentropic trajectories for air samples having high concentrations of nss-SO_4^{2-}, NO_3^- and ^{210}Pb (and low concentrations of ozone and ^{7}Be) generally pass over North Africa and trend towards Europe. Conversely, events with high ozone and ^{7}Be (and otherwise low aerosol concentrations) tend to come from the direction of the high latitudes in the middle and upper troposphere. In this regard, the results that we obtained at Izania are similar to those obtained in the marine boundary layer at Barbados (Savoie *et al.*, 1992b), where high ozone events (and low aerosol concentrations) are associated with transport from the mid to high latitudes and the middle and upper troposphere; conversely, air masses with high concentrations of aerosols (including ^{210}Pb) had relatively low ozone concentrations and the trajectories went back to Africa (and possibly Europe).

The fact that we do not observe the direct transport of ozone from pollution sources during the summer at Izania does not necessarily mean that the ozone could not be derived from processes involving anthropogenic materials. Indeed, it is noteworthy that at Izania the concentration of aerosol NO_3^- during high ozone events is relatively high and that the ratio of NO_3^- to nss-SO_4^{2-} is also very high compared to that in polluted air masses. Thus it is conceivable that the ozone observed in the high ozone events could have been produced by reactions involving reservoir species in the free troposphere.

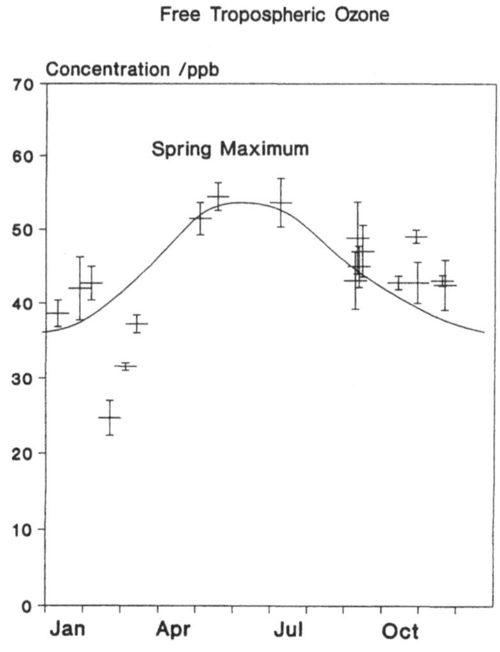

Figure 12. Seasonal variation of free tropospheric ozone over the North Atlantic Ocean.

HYDROCARBON MEASUREMENTS OVER THE NORTH ATLANTIC

The increase in ozone between January and May observed at Izania, and shown in Figure 12 for the Atlantic free troposphere at 56°N off the west coast of Scotland, is most intriguing. Its cause may be an increased flux of ozone from the large concentrations present in the lower stratosphere in the spring months, or possibly increased production from photochemical processes within the troposphere, which become more active as the light intensity increases (Penkett 1987; Liu *et al.*, 1987).

Evidence for this second explanation comes from aircraft measurements of many hydrocarbons in the free troposphere over the Atlantic made simultaneously with the ozone measurements shown in Figure 12 (Penkett *et al.*, 1993). Figure 13 shows the seasonal variation of total carbon in the form of reactive hydrocarbons, such as C_2H_2, C_2H_6, C_3H_8, C_4H_{10}, C_5H_{12}, C_6H_6, C_7H_8, etc., in free tropospheric air with a maritime polar origin reaching the west coast of Scotland. The amplitude of this variation in the reactive carbon is surprisingly constant from year to year, as demonstrated for C_2H_6 and C_2H_2. The data in Figure 14 show the seasonal variation between 1987 and 1989, but these values are very similar to ones collected over the whole period for which the experiment was conducted (1981 to 1990).

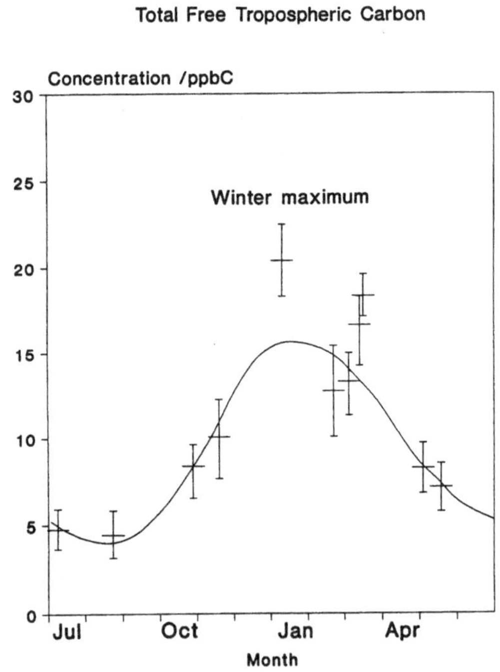

Figure 13. Seasonal variation of reactive carbon in the form of nonmethane hydrocarbons over the North Atlantic Ocean.

The latitudinal extent of this reservoir of about 20 ppbv of reactive carbon in the northern hemisphere (at least in the Atlantic region) is shown in Figure 15 for one hydrocarbon, ethane (Penkett, 1992). These measurements were made at sea level in November, 1982, and are typical for many hydrocarbons in that a large reduction to levels typical of the southern hemisphere occurs between 20°N and the equator. The contrast in the hydrocarbon concentration of air between latitudes in the region of 60°N, and in the region of 30°N, is also shown clearly in Table 1. Two aircraft flights, J124 and J125, were made in a similar location and altitude within four days of each other in January, 1989. The large contrast in concentration for free tropospheric hydrocarbons is entirely due to air mass origin. J124 took place in tropical maritime air with a temperature at 700 mb of 5°C, while J125 took place in maritime polar air with a temperature at 700 mb of −12°C. The respective free tropospheric ethane concentrations shown in Table 1 agree extremely well with those shown at the surface in Figure 15.

This large reservoir of reactive carbon is capable of generating an equivalent amount of ozone, on a mixing ratio basis, when it is oxidized in the atmosphere between the months of February and May, provided sufficient nitric oxide is co-present. Unfortunately, there is very little published data on nitrogen oxide concentrations in the Atlantic region for comparison. One of the objects of NARE is to fill that gap.

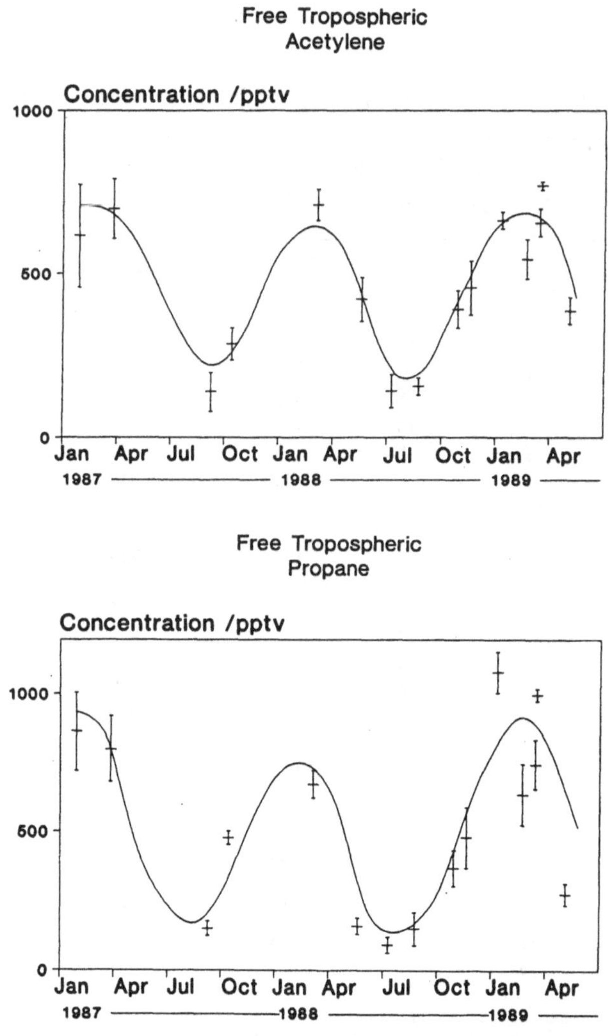

Figure 14. Seasonal variation of acetylene and propane over the North Atlantic Ocean.

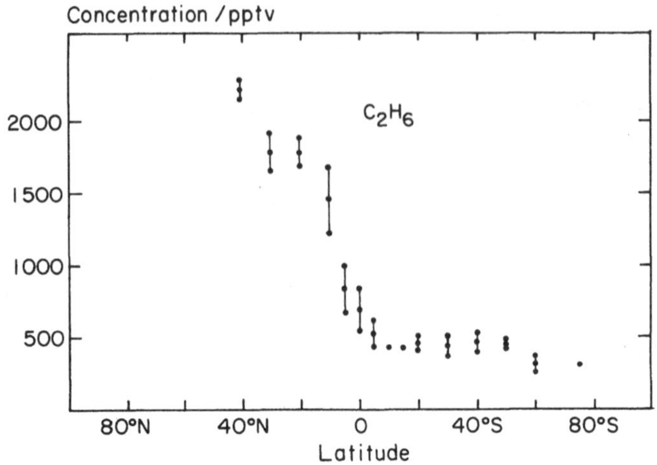

Figure 15. Latitudinal transect of ethane over the Atlantic from 40°N to 75°S (Penkett, 1992).

Table 1. Hydrocarbon concentrations measured in North Atlantic background flights

	J124, Jan. 6, 1989			J125, Jan. 10, 1989		
	Free Tropospheric		Low Alt	Free Tropospheric		Low Alt
Hydrocarbon Species	Average	S.D.	Average	Average	S.D.	Average
Ethane	1384	268	2253	2591	158	2442
Ethylene	607	396	542	543	149	454
Propane	336	107	790	1079	76	950
Propene	28	24	34	24	9	26
i-Butane	96	19	258	271	16	272
Acetylene	311	35	818	663	23	665
n-Butane	116	46	456	536	32	476
trans-2-Butene	9	5	7	10	2	11
1-Butene	8	5	11	10	4	4
i-Butene	96	54	83	5	7	70
cis-2-Butene	2	3	1	< d.l.		3
i-Pentane	39	11	171	211	10	193
n-Pentane	30	16	135	185	9	169
1-Pentene	5	2	5	5	3	4
n-Hexane	11	5	54	78	6	69
1-Hexene	5	1	5	4	2	3
n-Heptane	16	6	37	47	7	48
Benzene	90	8	231	263	8	262
n-Octane	8	3	12	19	4	15
Toluene	72	20	181	156	20	151
m+p-Xylene	16	7	56	46	9	44
o-Xylene	12	9	29	18	11	22
CFC-11	275	2	286	272	3	265
CFC 113	58	2	70	59	3	54
CH_3CCl_3	157	5	173	166	4	164
CCl_4	121	3	128	129	3	124
C_2Cl_4	74	20	167	17	4	15
Ozone, ppbv				38		31

In parts per 10^{12} by volume. S.D. = Standard Deviation; < d.l. = below detection limit.

A further feature of interest comes from a close examination of the hydrocarbon composition of air over the Atlantic Ocean when contrasted with air close to sources of major pollution. The relative abundance of the "iso" isomers (i.e. branched chain), for pentane, hexane, heptane and octane versus the "normal" isomers (i.e. straight chain) in the Atlantic samples is much lower than in air close to a major source, such as London. A possible explanation is the presence of a substantial source of these compounds other than motor vehicle emissions of typical western European or American origin. Alternatively the explanation could be removal from the atmosphere at differential rates for the "normal" and "iso" isomers of these hydrocarbons. This would be the case if the NO_3 radical played a significant role in the chemical removal of the atmospheric trace gases. Perhaps the most likely explanation is a combination of the two.

Resolving the many uncertainties concerning the source of the ozone observed in "clean" air over the North Atlantic, and by inference, the surrounding continents, and the nature of hydrocarbon oxidation processes, in addition to hydroxyl radical chemistry, are two of the major objectives of NARE.

The link between NARE objectives and IGAC-APARE (East Asian/North Pacific Regional Experiment) objectives is obvious from an inspection of the map shown in Figure 1. Given the nature of atmospheric circulation and the long lifetimes of ozone and its precursor molecules in winter, the same type and degree of chemical processing will probably occur in the North Pacific region and in the North Atlantic region.

FUTURE RESEARCH CAMPAIGNS

1993 Summer Intensive of NARE

Additional studies of ozone photochemistry in environments influenced by man-made pollution are required to understand better the key processes that control ozone production in the North Atlantic. Important insights into the chemistry of ozone production in the North Atlantic have been gained from the surface measurements conducted under NARE in 1991–1992. In 1993 a summer intensive is planned. The goals of the study are an improved understanding of the processes that determine the budget of ozone in the North Atlantic troposphere. Separate but coordinated measurements will be carried out in the western, mid, and eastern Atlantic areas. These studies will utilize both ground-based and airborne measurements. In addition, this activity will study the distribution of anthropogenic and co-transported secondary pollutants, in order to understand better the oxidizing capacity of the North Atlantic troposphere and the reactive nitrogen balance. Vertical profiles of ozone and meteorological parameters will be measured to determine if the surface measurements give an accurate picture of the transported species. These measurements will be supported by vertical wind profiles and additional surface ozone measurements associated with a variety of sites located throughout the North Atlantic region. The data base will be subjected to comparisons with model simulations by several coupled chemical-meteorological models that are being developed in North America and Europe to simulate the chemistry, meteorology, and transport that shape the ozone distribution of the region.

Winter Studies (1993–1994)

Detailed studies of the composition of air north and south of the polar front will be carried out over the Atlantic Ocean using a Hercules C130 aircraft. These studies will provide essential data on the concentration of reactive nitrogen compounds co-present with reactive carbon compounds. This information is needed to access the potential for ozone production during the transition from winter to spring or the transport of air from behind the polar front to more southerly latitudes where ozone production can proceed rapidly.

Winter/Spring Intensive (1995–1997)

During the past two decades considerable progress has been made toward identifying the principal features of gas phase photochemistry. However, nocturnal chemistry and heterogeneous chemistry have not been so fully developed. Oxidation of the ozone precursors, NO_x and most reactive carbon compounds, occurs rapidly (on the order of a day or less) during summer. During winter however, and particularly at high latitudes, the oxidation proceeds much more slowly, allowing NO_x to be transported over large distances. This is expected to give enhanced NO_x levels in the Arctic during winter months. Likewise, measurements of the other major ozone photochemical precursors, the nonmethane hydrocarbons, which have been made in polar maritime air, show a significant and systematic seasonal variation with pronounced winter maxima.

This seasonality is attributed to seasonal variations in the concentration of the OH radical, which is expected to be the major sink for these compounds in the troposphere. The decreasing solar flux over autumn and winter months lowers OH sufficiently to allow a large reservoir of these compounds to form in the troposphere. This reservoir, which covers a substantial fraction of the northern hemisphere, is calculated to hold significant reservoirs of reactive carbon and nitrogen at the January peak levels. The seasonality is also probably associated with the position of the polar front in winter; it can act as a barrier to dispersion of polluted air toward the tropics.

The large enhancement of these ozone precursors is expected to lead to efficient ozone production over large areas of the North Atlantic during spring. These nonmethane hydrocarbons are expected to play an important role in ozone production, and may partially explain the observed spring ozone maximum. Given the relatively long lifetime of ozone, the importance of this springtime northern latitude photochemical source must also be considered in connection with the traditional summer photochemical ozone production.

The importance of NO_3 as a primary oxidant must be determined and the role of heterogeneous chemistry investigated. In the northern polar regions during the winter copious quantities of NO_3 are probably formed. The cold temperatures attendant to this period promote association reactions and are conducive to the formation of aerosols. In the former case the reaction of NO_3 with hydrocarbons may tend to produce relatively large quantities of organic nitrates. In this regime the rate of conversion of SO_2 to sulfate is unknown. These processes will require both field measurements as well as laboratory investigation of the basic kinetic mechanisms.

REFERENCES

Angell, J.K., private communication, cited in F.C. Fehsenfeld and S.C. Liu, 1993, Tropospheric ozone: distribution and sources, *in:* "Global Atmospheric Chemical Change," C.N. Hewitt and W.T. Sturges (eds.), Elsevier Applied Science, New York.

Arimoto, R.A., R.A. Duce, D.L. Savoie and J.M. Prospero, 1992, Trace elements in aerosol particles from Bermuda and Barbados. Concentrations, sources and relationships to aerosol sulfate, *J. Atmos. Chem.*, 14:439-457.

Brost, R.A., J. Feichter and M. Heimann, 1991, Three-dimensional simulation of ^{7}Be in a global climate model, *J. Geophys. Res.*, 96:22,434-22,445.

Charlson, R.J., Langner, H. Rodhe, C.B. Leovy and S.G. Warren, 1991, Perturbation of the northern hemisphere radiative balance by backscattering from anthropogenic sulfate aerosols, *Tellus*, 43B:152.

Charlson, R.J., S.E. Schwartz, J.M. Hales, R.D. Cess, J.A. Coakley, Jr., J.E. Hansen, and D.J. Hofmann, 1992, Climate forcing by anthropogenic aerosols, *Science*, 255:423-430.

Chatfield, R. and H. Harrison, 1977, Tropospheric ozone II. Variations along a meridonal band, *J. Geophys. Res.*, 82:5969-5976.

Cox, R.A., A.E.J. Eggleton, R.G. Derwent, J.E. Lovelock, and D.E. Pack, 1975, Long-range transport of photochemical ozone in Norwestern Europe, *Nature*, 255:118-121.

Ellis, W.G., Jr., R. Arimoto, J.T. Merrill, D.L. Savoie, R.A. Duce and J.M. Prospero, 1993, Aerosol selenium at Bermuda and Barbados, *J. Geophys. Res.*, in press.

Fabian, P., W.F. Libby and C.E. Palmer, 1968, Stratospheric residence time and interhemispheric mixing of strontium 90 from fallout in rain, *J. Geophys. Res.*, 73:3611-3616.

Fehsenfeld, F.C. and S.C. Liu, 1993, Tropospheric ozone: distribution and sources, *in:* "Global Atmospheric Chemical Change," C.N. Hewitt and W.T. Sturges (eds.), Elsevier Applied Science, New York.

Feichter, J., R.A. Brost, and M. Heimann, 1991, Three-dimensional modeling of the concentration and deposition of ^{210}Pb aerosols, *J. Geophys. Res.*, 96:22,447-22,460.

Fishman, J., C.E. Watson, J.C. Larsen and J.A. Logan, 1990, Distribution of tropospheric ozone determined from satellite data, *J. Geophys. Res.*, 85:3599-3618.

Grennfelt, P. and J. Schjoldager, 1984, Photochemical oxidants in the troposphere: A mounting menace, *Ambio*, 13:61-67.

Guicherit, R. and H. Van Dop, 1977, Photochemical production of ozone in Western Europe (1971-1975) and its relation to meteorology, *Atmos. Environ.*, 11:145-156.

Isaksen, I.S.A., O. Hov and E. Hesstvedt, 1978, Ozone generation over rural areas, *Environ. Sci. & Tech.*, 12:1279-1284.

Langner, J. and H.Rodhe, 1991, A global three-dimensional model of the tropospheric sulfur cycle, *J. Atmos. Chem.*, 13:225-263.

Lelieveld, J. and J. Heinzenberg, 1992, Sulfate cooling effect on climate through in-cloud oxidation of anthropogenic SO_2, *Science*, 258:117-120.

Liu, S.C., M. Trainer, F.C. Fehsenfeld, D.D. Parrish, E.J. Williams, D.W. Fahey, G. Hübler and P.C. Murphy, 1987, Ozone production in the rural troposphere and the implications for regional and global ozone distributions, *J. Geophys. Res.*, 92:4191-4207

Parrish, D.D., 1993, "Carbon monoxide and light alkanes as tropospheric tracers of anthropogenic ozone," NATO ASI sub-series, Global Environmental Change, in press.

Parrish, D.D., J.S. Parrish, J.S. Holloway, M. Trainer, P.C. Murphy, G.L. Forbes and F.C. Fehsenfeld, 1993, Export of North American ozone pollution to the north Atlantic Ocean, *Science*, in press.

Penkett, S.A. and K.A. Brice, 1986, The spring maximum in photo-oxidants in the northern hemisphere troposphere, *Nature*, 319:655.

Penkett, S.A., 1987, Indications and causes of ozone increase in the troposphere, *in:* "The Changing Atmosphere," F.S. Rowland and I.S.A. Isaksen (eds.), pgs. 91-103, Dahlem Konferenzen, John Wiley & Sons Ltd.

Penkett, S.A., 1992, "Measurements of hydrocarbons in polar maritime air masses," NATO Meeting, Halifax, Nova Scotia, Canada.

Penkett, S.A., N.J. Blake, P. Lightman, A.R.W. Marsh, P. Anwyl and G. Butcher, 1993, The seasonal variation of nonmethane hydrocarbons in the free troposphere over the north Atlantic Ocean: Possible evidence for extensive reactions of hydrocarbons with the nitrate radical, *J. Geophys. Res.*, 98:2865-2885.

Prospero, J.M., R. Schmitt, E. Cuevas, D. Savoie, W.C. Graustein, S.J. Oltmans, H. Levy, II and A. Volz-Thomas, 1993, "Temporal variability of ozone and aerosols in the free troposphere over the eastern North Atlantic," paper presented at the Spring Meeting of the American Geophysical Union, Baltimore, Maryland, May 24 - 28, 1993.

Savoie, D.L., J.M. Prospero, and E.S. Saltzman, 1989, Nitrate, non-sea-salt sulfate and methane sulfonate over the Pacific Ocean, *in:* "Chemical Oceanography," Volume 10, J.P. Riley (ed.), Academic Press, New York, pgs. 219-250.

Savoie, D.L., J.M. Prospero, R.J. Larsen and E.S. Saltzman, 1992a, Nitrogen and sulfur species in aerosols at Mawson, Antarctica, and their relationship to natural radionuclides, *J. Atmos. Chem.*, 14:181-204.

Savoie, D.L., J.M. Prospero, S. Oltmans, W.C. Graustein, K.K. Turekian, J.T. Merrill and H. Levy, II, 1992b, Sources of nitrate and ozone in the marine boundary layer of the tropical North Atlantic, *J. Geophys. Res.*, 97:11,575-11,589.

Savoie, D.L., R. Arimoto, J.M. Prospero, R.A. Duce, W.C. Graustein, K.K. Turekian, J.N. Galloway and W.C. Keene, 1993, Oceanic and anthropogenic contributions to non-sea-salt sulfate in the marine boundary layer over the North Atlantic Ocean, *J. Geophys. Res.*, submitted.

Schmitt, R., B. Schreiber and I. Levin, 1988, Effects of long-range transport on atmospheric trace constituents at the baseline station Tenerife (Canary Islands), *J. Atmos. Chem.*, 7:335-351.

Volz, A. and D. Kley, 1988, Evaluation of the Montsouris series of ozone measurements made in the nineteenth century, *Nature*, 332:260-262.

Vukovich, F.M., W.D. Bach, Jr., B.W. Crissman and W.J. King, 1977, On the relationship between high ozone in the rural surface layer and high pressure systems, *Atmos. Environ.*, 11:967-984.

Vukovich, F.M., J. Fishman and E.V. Browell, 1985, The reservoir of ozone in the boundary layer of the eastern United States and its potential impact on the global tropospheric ozone budget, *J. Geophys. Res.*, 90:5687-5698.

WMO (World Meteorological Organization), 1986, "Atmospheric Ozone 1985: Assessment of our understanding of the processes controlling the present distribution and change," WMO Rep. 16, Global Change Research Monitoring Programme, Geneva, Switzerland.

MARINE AEROSOL AND GAS EXCHANGE AND GLOBAL ATMOSPHERIC EFFECTS

Barry J. Huebert[1], Timothy S. Bates[2] and Neil W. Tindale[3]

[1]Department of Oceanography
University of Hawaii
Honolulu, HI 96822 USA
[2]NOAA/PMEL
7600 Sand Point Way
Seattle, WA 98115 USA
[3]Departments of Meteorology and Oceanography
Texas A&M University
College Station, TX 77843 USA

ABSTRACT

It is becoming clear that the global climate system is controlled by numerous links between the biosphere and the atmosphere. The objective of IGAC's Marine Aerosol and Gas Exchange: Atmospheric Chemistry and Climate (MAGE) Activity is to quantify those links through interdisciplinary multinational research on air/sea exchange and its biological controls and impacts. We seek to bring together scientists from a variety of disciplines to study the interfaces between them. Wherever possible, we encourage collaborative work between marine scientists who look up at the interface from the water column and atmospheric chemists, whose work in the atmosphere has frequently treated the ocean's surface as a featureless source or sink.

Several problems require this interdisciplinary approach. Marine biological productivity in some areas is controlled by the supply of nutrients from the atmosphere. In certain nitrogen-rich regions, for instance, the supply of iron from atmospheric aerosols may limit productivity. In other areas, the wet and dry deposition of atmospheric nitrate and ammonium may be a significant source of fixed nitrogen to biological communities. MAGE helped to organize an international group of scientists who studied the effect of atmospheric iron on biological productivity, phytoplankton speciation, and DMS production as a part of the Equatorial Pacific JGOFS (Joint Global Ocean Flux Study) experiment in the spring of 1992. A second MAGE/JGOFS cruise studied the fluxes of biogenic gases through the air/sea interface in the same region.

Global Atmospheric-Biospheric Chemistry
Edited by R.G. Prinn, Plenum Press, New York, 1994

To quantitate the impact of marine biota on atmospheric aerosols, cloud properties, and climate, one must precisely measure (and then parameterize for use in models) the emission of trace gases from the ocean's surface. MAGE is seeking to develop new strategies for measuring some of these elusive fluxes. During the Atlantic Stratocumulus Transition Experiment (ASTEX) in June of 1992, MAGE organized scientists from five countries to study air/sea fluxes, their biological forcing, and their atmospheric effects. Three aircraft, two ships, two islands, and a dozen constant-density balloons were used to test a Lagrangian strategy for studying two airmasses. By repeatedly sampling the same air, we hoped to reduce the perennial problem of deconvoluting transport and chemistry so that we can clearly understand processes and fluxes. In this way we will improve our understanding of the marine nitrogen budget (including both ammonia emissions and nitrate deposition), as well as the climatically-important sulfur cycle and DMS emissions.

INTRODUCTION

The composition of the marine atmosphere is strongly influenced by sources and sinks at the ocean's surface. At present, these sources and sinks are poorly characterized relative to the needs of climate modelers. MAGE seeks to understand these interactions more fully, so that useful predictive descriptions of air/sea exchange can be included in climate and marine biological models.

There are numerous ways in which the air/sea exchange of gases and aerosols affects the earth's climate. MAGE is studying several:

1. Since atmospheric input of iron and fixed nitrogen may control primary productivity in certain marine ecosystems, the ocean's ability to produce biomass and bury atmospheric CO_2 is in part limited by the quantity of nutrients the atmosphere delivers to remote marine regions (Duce, 1986). Thus we have to understand atmospheric processes such as transport and surface deposition of non-carbon substances to model the CO_2 cycle and the greenhouse effect.
2. Numerous biogeochemical cycles are affected by surface fluxes: ammonia may be redistributed between regions by emission to the atmosphere and conversion to aerosol, followed by deposition to other locations (Liss and Galloway, 1993). Air/sea exchange is a critical part of balancing the budget of reduced nitrogen in surface water and using the Redfield ratio to predict the potential burial of atmospheric carbon (Duce, 1986).
3. The production of cloud condensation nuclei (CCN), which control cloud radiative properties (Albrecht, 1989), is thought to be the result of marine dimethyl sulfide (DMS) emissions (Andreae, 1986; Charlson *et al.*, 1987). Both this indirect impact of aerosols on the earth's radiation budget and the direct scattering of sunlight by particles (Charlson *et al.*, 1992) are thus strongly influenced by air/sea exchange.
4. The mechanisms that form and remove marine aerosols cannot be studied without knowing the rates of supply and removal of reactants and products (Langner *et al.*, 1993). The importance of SO_2, for instance, as a precursor to sulfate aerosol, depends heavily on knowing how much SO_2 is dry-deposited to the ocean's surface before it can be oxidized (Huebert *et al.*, 1993).

Existing methods for estimating surface fluxes are inadequate for many of the intended uses. In the case of DMS, for instance, some authors consider the uncertainties from thin-film flux estimates to be a factor of two or three (Andreae, 1986). But in view of the importance of marine aerosols to the earth's radiation budget, we need to know their source strength to at most tens of percent, not factors of several (Penner *et al.*, this volume).

MAGE has sponsored two field programs and is planning another. In the Equatorial Pacific Ocean in early 1992, MAGE conducted two shipboard studies alongside the U.S. JGOFS program. One group focused on gas emissions from the ocean and their effect on atmospheric chemistry, the other group focused on the deposition of aerosol particles to the ocean and their affect on ocean biological productivity. In June of the same year, MAGE used two ships and three aircraft to develop a Lagrangian observing strategy during the ASTEX/MAGE program near the Azores. Although it is still too early to report significant scientific results from those programs, some early insights are discussed below. MAGE has paid particular attention to the potential for doing Lagrangian experiments in the atmosphere, to isolate surface exchange processes from the confounding effects of horizontal advection. To further this goal MAGE is cosponsoring an Aerosol Characterization Experiment (ACE-1) in the vicinity of Cape Grim, Tasmania, in January and February of 1995.

JGOFS/MAGE

Many of today's most pressing problems cannot be understood without truly interdisciplinary research. MAGE helped organize two teams of atmospheric chemists and biologists, who took advantage of the U.S. Equatorial Pacific (EQPAC) JGOFS cruises in the spring of 1992 to work alongside the oceanographic community's best carbon-cycle researchers. The JGOFS and MAGE programs are quite complementary in that they share a common reservoir, the surface ocean. Whereas JGOFS focuses on the exchange of biogenic elements between the surface ocean and deep ocean, MAGE focuses on the exchange between the surface ocean and atmosphere. Both processes depend on the biology, chemistry and physics in the surface ocean. The biology and chemistry of the surface ocean in turn, depend on inputs of nutrients from the atmosphere and deep ocean. Together the two programs are beginning to quantify the important biogeochemical processes in the surface ocean that influence the earth's climate.

R/V Vickers Cruise

The first MAGE/JGOFS cruise took place during February and March of 1992 aboard the R/V Vickers. The main focus of this cruise was to:

1. Study the cycling of trace gases in the upper water column using measurements of key species as a function of depth and microbiological/photochemical rate studies.
2. Calculate the air-sea exchange of these trace gases using shipboard measurements of the important seawater and atmospheric species, air sea exchange models and micro-budget flux estimates.
3. Compare the atmospheric gas phase and aerosol phase chemistry over a wide variety of oceanic regimes.

The measurements included sulfur, halogen, nitrogen and carbon compounds in both biologically productive and oligotrophic waters along 140°W in the equatorial Pacific (Figure 1). Although the data from the cruise are still being analyzed there are already several interesting results.

Seawater dimethylsulfide (DMS) concentrations, and hence the flux of DMS to the atmosphere was much higher in the southern hemisphere than the northern hemisphere (Figure 2). The high concentrations of DMS were likely a result of lower microbial consumption rates. The entire seawater sulfur cycle in the equatorial pacific near 12°S was quite different from expected, based on previous measurements (Bates *et al.*, 1993) in temperate latitudes (48°N). At 12°S dimethyl-sulfoxide (DMSO), a photochemical

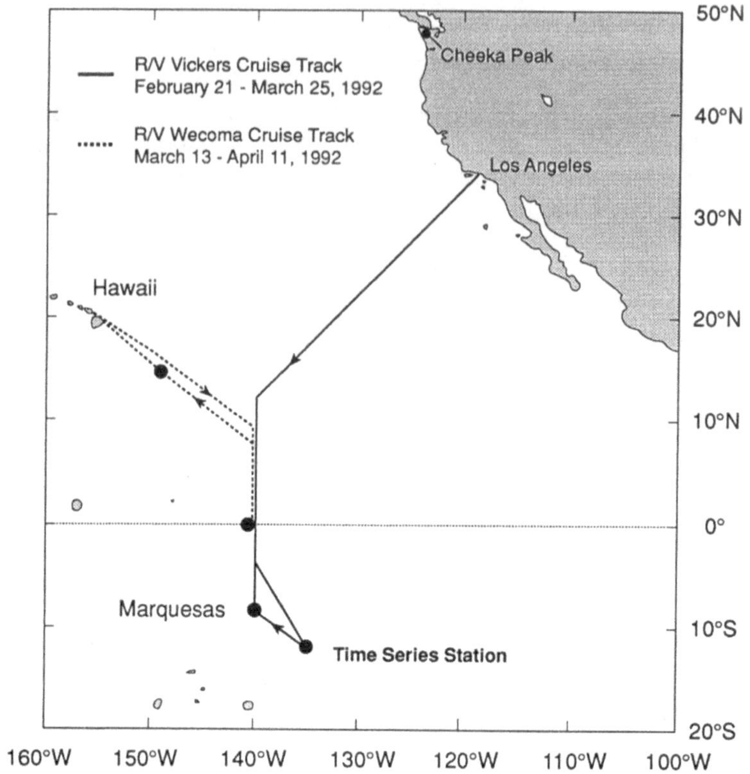

Figure 1. IGAC-MAGE Equatorial Pacific Field Study: Cruise tracks for the R/V Vickers and R/V Wecoma.

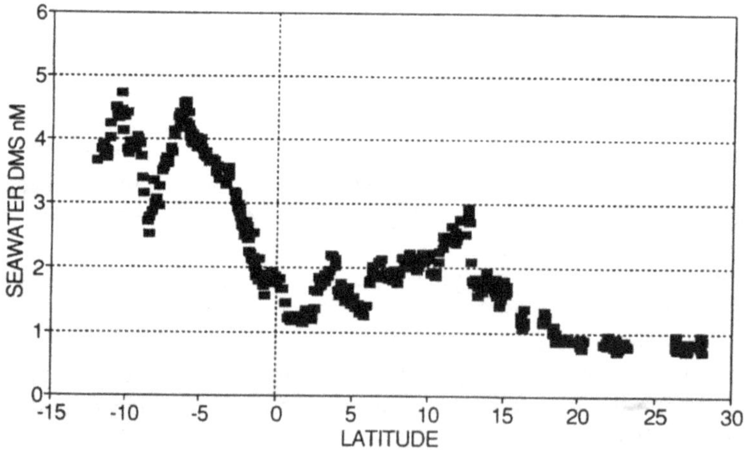

Figure 2. Seawater DMS concentrations along the Vickers cruise track. These data from the southbound leg were essentially identical to the values obtained during the return northbound leg. This implies that the features observed here were stable on the time-scale of one to two weeks.

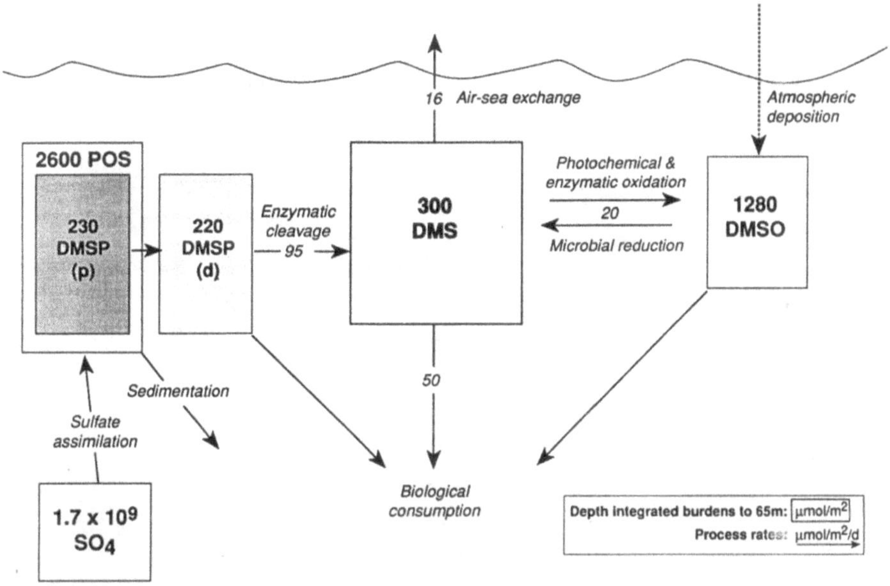

Figure 3. Seawater sulfur cycle at 12°S, 135°W. Note that the particulate DMSP burden was only about 10% of the total particulate organic sulfur burden and was approximately equal to the DMS burden. In this region the major sinks for DMS (air-sea exchange, biological consumption, and photochemical oxidation) are all of the same order of magnitude. (Data from R. Kiene, D. Kieber and P. Matrai, private communication.)

oxidation product of DMS, was the dominant sulfur species (Figure 3) and dimethylsulfoniopropionate (DMSP), the precursor of DMS, was present at concentrations similar to DMS. In contrast, DMSP concentrations were 25 times higher than DMS concentrations at the temperate latitude site. At 12°S air-sea exchange, photo-oxidation and biological consumption were all important sinks for DMS. Some of the extensive biological and chemical measurements from the JGOFS work may help explain why the biological consumption of DMS is so low in this region. These data will be described in more detail in a manuscript by R. Kiene and coworkers.

A time-series station at 12°S provided an opportunity to obtain a detailed picture of the atmospheric diurnal cycle of several trace gases. DMS concentrations, for example, markedly decreased during the day as a result of photochemical oxidation (Figure 4). This diurnal cycle can be coupled to an atmospheric photochemical model to estimate the DMS air-sea exchange rate needed to maintain the atmospheric DMS concentrations. This calculation, based on atmospheric DMS concentrations and an atmospheric photochemical model, is totally independent from the more traditional air-sea exchange calculations based on seawater DMS concentrations and a wind speed/transfer velocity relationship. At this 12°S station both methods produced nearly identical results with DMS fluxes in the range of 10 to 20 µmole/m^2/day. The details of this study will be described in a manuscript by S. Yvon and coworkers.

The gases that are emitted from the ocean affect the chemical composition of the atmosphere, the oxidative capacity of the atmosphere and the production of aerosol particles. DMS, for example, is the major natural source of sulfur to the atmosphere and is the primary precursor to aerosol particles in the remote marine atmosphere. The number and size distributions of aerosol particles are a functions of precursor source strength (DMS emissions), the existing aerosol size distribution, relative humidity, temperature and

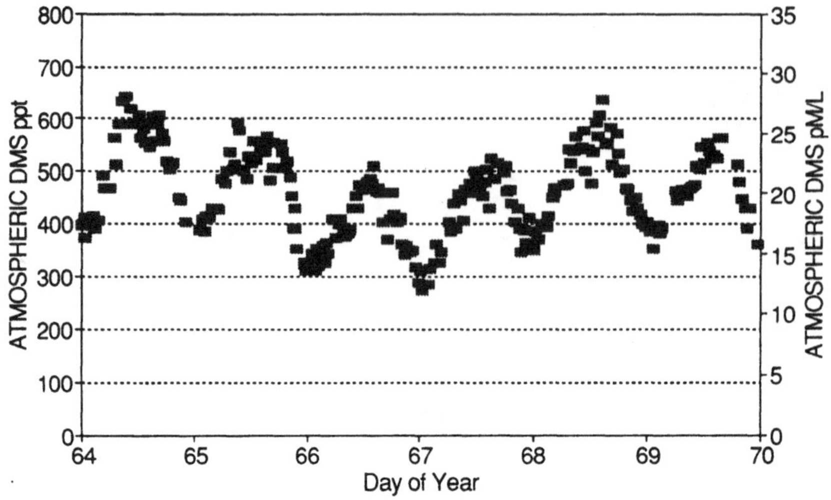

Figure 4. Atmospheric DMS concentrations at 12°S, 135°W. Time is shown in GMT. DMS concentrations decrease during the day as a result of photochemical oxidation. There were no significant changes in the marine boundary layer height during this time period.

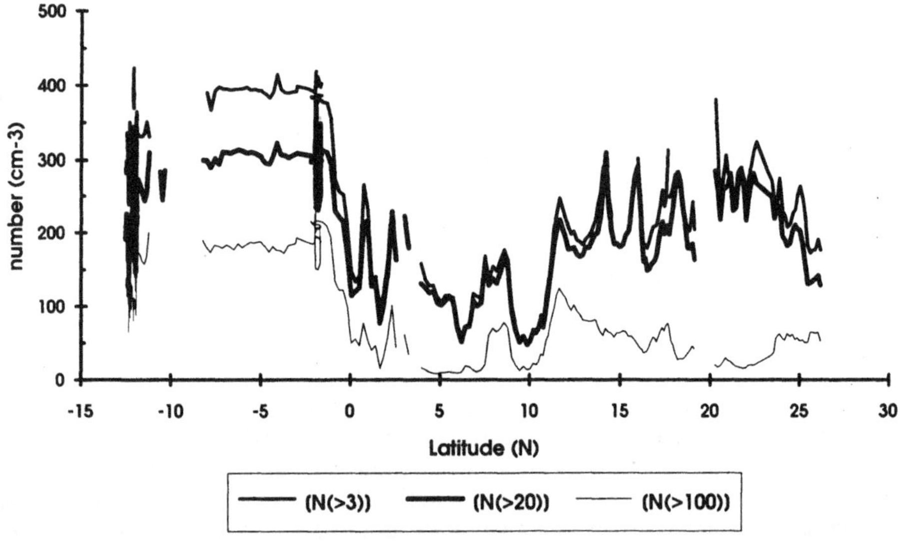

Figure 5. Aerosol concentrations on the northbound leg of the Vickers cruise track. The concentrations are given as number of particles per cm^3 with diameters greater than 3, 20 and 100 nm. Note that in the northern hemisphere there were few particles with diameters <20 or >100 nm. This single mode particle distribution implies that the particles were a few days old but yet had not had sufficient time to grow large enough to serve as cloud condensation nuclei (CCN). In the southern hemisphere the particle number size distribution was trimodal, suggesting that in addition to the nuclei mode there were also newly formed particles (an ultra-fine mode) and an aged accumulation mode. In this region most of the particles had diameters > 40nm diameter and were active CCN at 0.65% supersaturation. (Data from D. Covert and V. Kapustin, private communication.)

atmospheric circulation. The detailed aerosol chemical, physical and optical measurements collected aboard Vickers will provide a valuable data set for modeling aerosol formation and growth. Many of the processes important to aerosol formation and growth were evident along the Vicker's cruise track. For example, at the northern end of the cruise track along 140°W the aerosol particles were present in primarily a single mode with diameters around 20 to 30 nm (Figure 5). This size distribution is likely a result of subsidence from the free troposphere. Further south the aerosol size distribution was tri-modal suggesting a mixture of freshly formed aerosol and more aged boundary layer aerosol. The details of the aerosol physical measurements will be a manuscript by D. Covert and coworkers.

For up-to-date information on the above R/V Vickers manuscripts and others, contact Dr. Tim Bates, NOAA/PMEL, 7600 Sand Point Way, Seattle, WA 98115. (Phone: 206-526-6248, Internet: bates@pmel.noaa.gov.)

Wecoma Cruise: FeLINE II

The input of atmospheric material to the ocean has been suggested as a mechanism for limiting primary production in nutrient rich open ocean surface waters (Duce, 1986; Martin *et al.*, 1991). There is growing evidence, for instance, that eolian iron is particularly important as a limiting micronutrient (Young *et al.*, 1991). In three regions of the Pacific Ocean the input of eolian iron may regulate production in surface waters. These are the high latitude open ocean in the Gulf of Alaska and the Southern Ocean, and the band of nutrient rich surface water that exists in the central and eastern equatorial Pacific.

The equatorial Pacific (0, 140°W) was chosen as the study site for the 1992 MAGE FeLINE-II cruise (Figure 1). The main objectives of FeLINE-II included the study of: aerosol iron input, phytoplankton productivity and photosynthesis, dimethyl sulfide (DMS) production, dissolved organic carbon (DOC) production and consumption, photochemical cycling of iron and hydrogen peroxide in surface waters, and the oceanic vertical distribution of trace metals (primarily Fe and Zn).

Over a dozen shipboard incubation experiments (enrichment experiments) were completed during the cruise, evaluating the roles of iron (in several chemical forms), aerosol particles and leachate, nitrate, photosynthetically active radiation (PAR) and UV light, cobalt and vitamin B12 on phytoplankton productivity. Most of these factors were also measured in samples collected during conductivity-temperature-depth (CTD) casts and in trace metal clean samples collected using 30-liter GoFlo bottles deployed on Kevlar line.

Most of the experiments conducted during this cruise were collaborations between groups, which resulted in a much stronger and more flexible research program than would have resulted if the groups had worked independently. Highlights of the individual groups' efforts are summarized below.

The Moss Landing Marine Laboratories group examined the photochemistry of iron in surface waters and its implications to iron bioavailability. Their decktop incubation experiments showed enhanced growth from all of the iron additions, with the most dramatic enhancement in carboys enriched with atmospheric aerosol particles (Figure 6). They also observed clear diurnal patterns in iron levels, with a maximum at midday. This is consistent with a model involving the reductive dissolution of colloidal iron, its subsequent oxidation, and the uptake of dissolved iron (III). These results indicate that photochemical reactions can produce sufficient bioavailable iron (from what was previously unavailable colloidal iron) to support primary production in the euphotic zone in this area (Johnson *et al.*, 1993).

A University of Rhode Island (URI) team studied the influence of ultraviolet radiation, photo-chemical cycling, and rain events on the distribution of iron and hydrogen peroxide, and the speciation and bioavailability of iron. They:

1. quantified the air-sea fluxes of peroxides and cathodic-stripping voltammetric (CSV) labile iron during a rain squall (Figure 7), and documented the influence of such a rain event on the horizontal and vertical distribution of labile iron and peroxides in the surface waters of the region (Figure 8),
2. compared the effects of iron-containing rainwater additions on phytoplankton growth in nutrient-rich and nutrient-depleted surface waters,
3. determined CSV labile iron vertical distributions in euphotic and deeper waters,
4. characterized the temporal and latitudinal variability in the spectral distribution of downwelling ultraviolet and photosynthetically active radiation in euphotic waters,
5. examined the importance of natural UV levels to iron availability, phytoplankton growth and the photo-redox cycling of iron and peroxides (Hanson *et al.*, 1993).

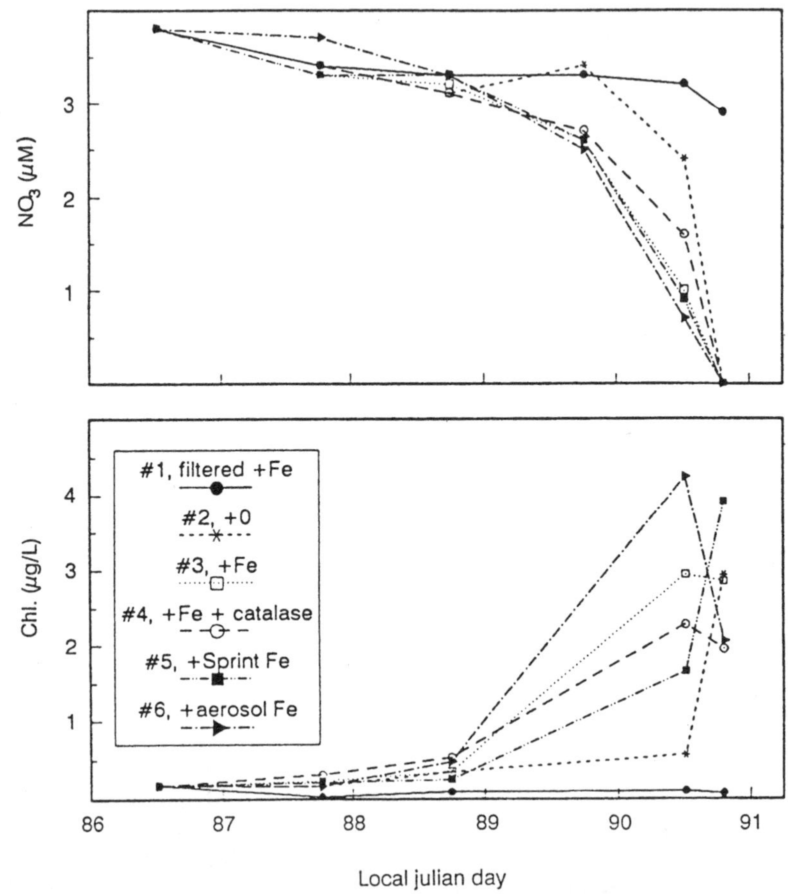

Figure 6. Nitrate **(top panel)** and chlorophyll **(bottom panel)** concentrations measured over time in carboys containing seawater collected at 0°N, 140°W. Seawater in Carboy #1 was filtered and 0.22 μm $FeNH_4(SO_4)_2$ was added to a final concentration of 10 nM. Carboy #3 contained unfiltered seawater with 10 nM $FeNH_4(SO_4)_2$. Carboy #4 was similar to #3, except that the enzyme catalase was added every day to keep H_2O_2 concentrations less than 5 nM as compared to ambient concentrations near 25 to 30 nM. Seawater in Carboy #5 was unfiltered and contained 10 nM Fe added as sodium ferric ethylenediamine di-(o-hydroxyphenyl) acetate. Seawater in Carboy #6 was unfiltered and approximately 10 nM iron was added in the form of aerosol particles collected on Oahu, Hawaii. (Data from Johnson *et al.*, 1993.)

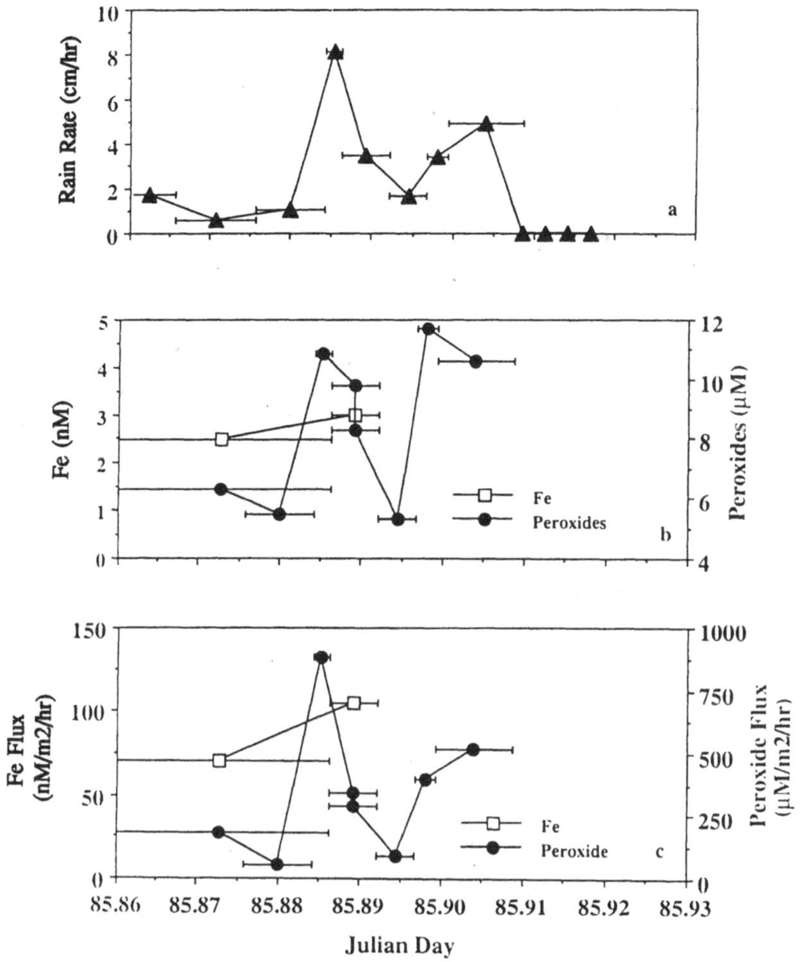

Figure 7. Time course of precipitation observations during the squall. **(top panel)** Average rain rate in cm per hour for the time period of collection (horizontal bars); **(middle panel)** concentrations of dissolved iron (open squares) and peroxides (open circles) in the precipitation; and **(bottom panel)** calculated wet depositional fluxes of dissolved iron (open squares) and peroxides (closed circles). (Data from Hanson *et al.*, 1993.)

The University of East Anglia group (G. Malin, S. Turner and P. Liss) studied the transfer of volatile DMS from the surface waters to the atmosphere. This link between the atmosphere and ocean is made through the action of phytoplankton which synthesize DMSP (the dissolved, non-volatile precursor to DMS) as an osmolyte and cryoprotectant.

Another URI effort involved deckboard incubation experiments in which surface waters were enriched with aerosol leachate, inorganic iron, cobalt, nitrate and vitamin B12. The results indicated that the addition of cobalt alone does not influence phytoplankton growth. However, the addition of cobalt and iron together may have allowed more rapid growth (as indicated by the increase in chlorophyll) than was observed in a solely iron stimulated population. For iron stimulation to be observed, nitrate had to be present (in situ, or added). Again, enhanced growth was seen in deckboard enrichment experiments with the addition of natural aerosol material (D. Swift, private communication).

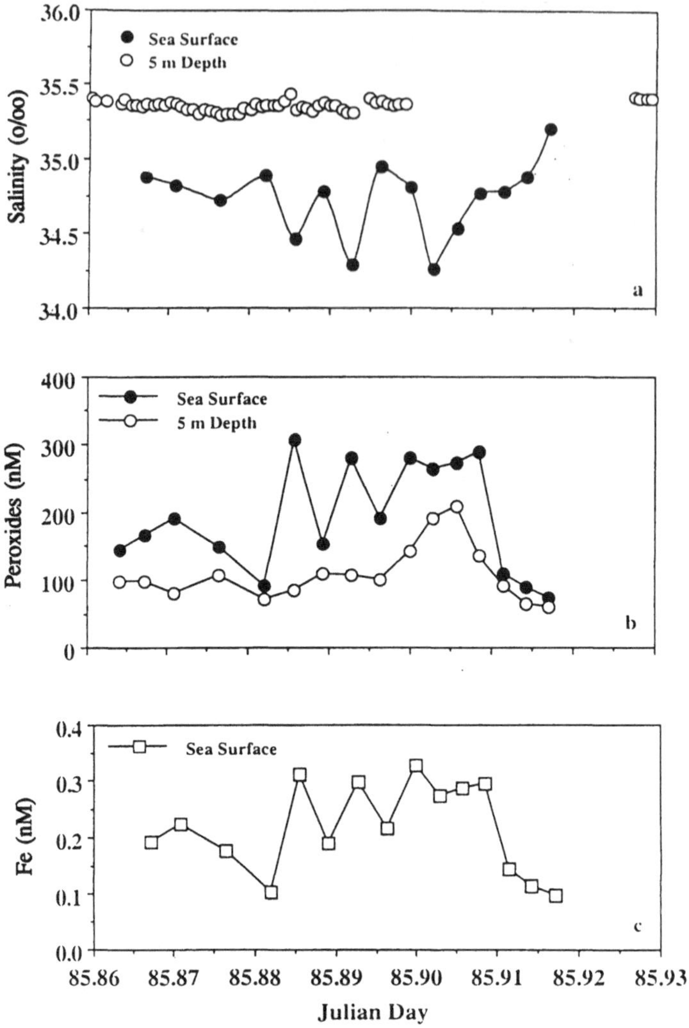

Figure 8. Time course observations of seawater during the squall. **(top panel)** Salinity of surface (closed circles) and water at 5 m depth (open circles); **(middle panel)** concentration of total peroxides in seawater samples collected from the surface (closed circles) and 5 m depth (open circles); and **(bottom panel)** the concentration of dissolved iron as determined by ACSV in sea surface samples. (Data from Hanson *et al.*, 1993.)

The persistence of high nutrient, low biomass waters along the equatorial Pacific at 140°W is thought to result from some form of resource sub-utilization. The two hypotheses put forth to explain this phenomenon involve either resource limitation or grazing pressure. R. Greene, Z. Kolber, and P. Falkowski (Brookhaven National Laboratory) examined the variability in the quantum yield of chlorophyll fluorescence using a fast repetition rate fluorometer. The parameter derived (denoted ~hm) is a quantitative measure of photochemical energy conversion efficiency in photosystem II. They found that ~hm was relatively low throughout the nutrient-rich equatorial waters (and higher in more oligotrophic waters) indicating greater physiological limitation of the photochemical energy conversion efficiency in the natural phytoplankton populations at the equator. Shipboard enrichment studies (Figure 9) showed that ~hm increased following addition of nanomolar concentrations of inorganic iron and aerosol dust leachate. This response suggests that iron (rather than relaxed grazing pressure) increased the photosynthetic efficiency of the phytoplankton, resulting in increased productivity in the shipboard enrichments. The Brookhaven results indicate that low iron availability limits photochemical energy conversion efficiency and is the principle mechanism controlling rates of photosynthesis in these nutrient-rich equatorial Pacific waters.

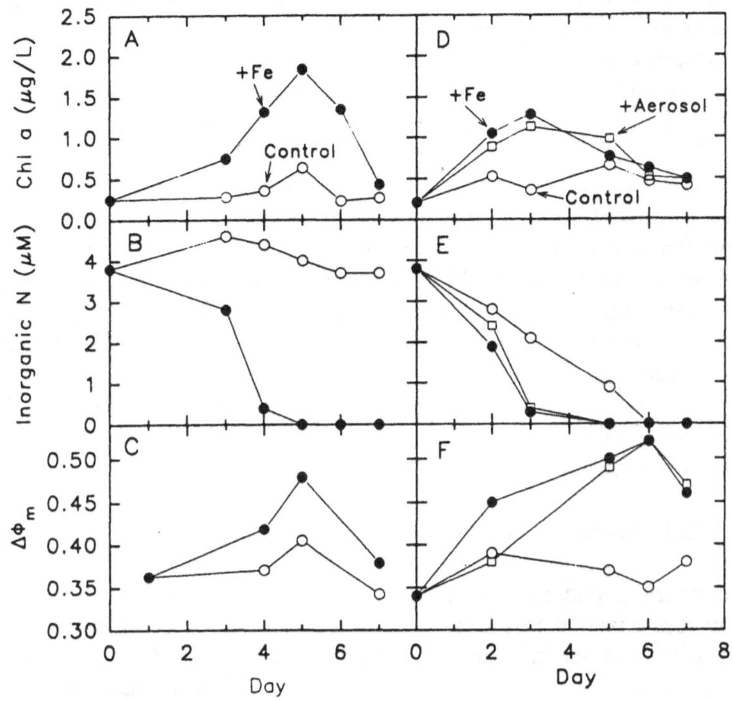

Figure 9. Changes in Chl a (μg L^{-1}, panels A and D), inorganic nitrogen (μM, panels B and E), and the quantum yield of fluorescence ($\Delta\phi_m$, panels C and F) in shipboard incubations. The panels on the left show results from experiments initiated on March 23, 1992 with addition of inorganic Fe to a final concentration of 4.5 nM Fe (depicted as +Fe). The panels on the right show results from experiments initiated on March 28, 1992 with addition of inorganic $FeSO_4$ (4.5 nM Fe final concentration) and Fe as aerosol dust leachate (2 nM Fe final concentration, depicted as +aerosol). Control bottles were natural samples without enrichment. (Data from Greene *et al.*, 1993.)

Several links between the marine sulfur cycle, atmospheric iron availability and non-seasalt sulfur in ice cores emphasize the climatic significance of DMS. Preliminary results indicate that the production of both DMS and DMSP, in the dissolved and particulate phase, is enhanced in the presence of added iron. Although preliminary in nature, there is also strong evidence that the production of chloroiodomethane is also dramatically increased in the presence of iron.

J. Martin's group (Moss Landing Marine Laboratory) conducted iron enrichment experiments on both the FeLINE-II and JGOFS-EQPAC cruises. Dissolved iron concentrations in the equatorial Pacific were extremely low (less that 0.03 nM) in the surface waters, in contrast to the 10 nM concentrations found in coastal waters. Results from their enrichment experiments indicated enhanced growth in both addition treatments and controls although the treatments usually showed greater growth more quickly. The subsequent growth in the controls remains a mystery. Several explanations may account for this observation: (1) an increased Fe source, (2) contamination of the controls, or (3) reduced grazing pressure as a result of enclosure. Although the evidence for increased Fe deposition during this time is positive, it is highly tentative and requires further examination (see below). Recently several investigators have suggested that phytoplankton growth may be controlled by a balance of factors including nutrients, Fe limitation, and grazing. It is possible that containment alone was responsible for some of the observed effects under the conditions of this oceanographic regime. This points to the need for an uncontained iron enrichment experiment.

The atmospheric group of N. Tindale and B. Duce from Texas A&M University collected both air and rain samples on station and underway, in order to measure the concentration and flux of aerosol particles during a transect through the Intertropical Convergence Zone (ITCZ). Portions of aerosol and rain water samples were provided to the other groups for use in their enrichment experiments on both the FeLINE and JGOFS cruises. Aerosol concentrations in this region were extremely low, especially south of the ITCZ in southern hemisphere air. Preliminary results indicate there was a higher atmospheric concentration of aerosol iron in the EQPAC region in 1992 than in an earlier cruise in 1990, yet these differences are small and are not yet completely resolved. It is unclear whether the difference in atmospheric concentrations would lead to a significant difference in the flux of iron from the atmosphere to the ocean surface.

For up-to-date information on the R/V Wecoma cruise results, contact Dr. Neil Tindale at the Department of Meteorology, Texas A&M University, College Station, TX 77843 USA. (Phone: 1-409-862-4401, Internet: tindale@ariel.tamu.edu.)

ASTEX/MAGE: Lagrangian Measurements

To quantitate the impact of marine biota on atmospheric aerosols, cloud properties, and climate, one must precisely measure (and then parameterize for use in models) the emission of trace gases from the ocean's surface (Bates and Cline, 1985; Langner *et al.*, 1993). MAGE is trying to develop new strategies for measuring these fluxes. During the Atlantic Stratocumulus Transition Experiment (ASTEX) in June of 1992, MAGE organized scientists from five countries to study air/sea fluxes and their atmospheric effects. ASTEX/MAGE was a large boundary-layer, cloud, and radiation experiment conducted in the region of the Azores.

Surface fluxes and atmospheric reaction rates are often very difficult to derive from Eulerian observations (at a fixed point). Following one parcel of air and observing its changes over time (the Lagrangian reference frame) would make the rates much easier to deduce. The difficulty is that one must first know where an air parcel is moving and then follow it with a large suite of analytical instruments. It took the combined efforts of two research ships (the Woods Hole R/V Oceanus and NOAA's R/V Malcolm Baldrige), three aircraft (NCAR's Electra, the UK C-130, and the Univ. of Washington's C-131), 12 GPS-tracked balloons, real-time trajectory predictions, and the cooperation of literally hundreds of scientists from several countries to test a Lagrangian strategy during ASTEX/MAGE.

The analysis of Lagrangian data is based on a continuity equation, which has previously been used in various forms for evaluating fluxes of heat, momentum, and ozone in the marine boundary layer (Lenschow *et al.*, 1981; Kawa and Pearson, 1989). For a substance S in a Lagrangian parcel,

$$\frac{d}{dt}\langle S\rangle = J_o(S) - J_h(S) + F(S) - D(S) \tag{1}$$

Here, J_o is its surface flux, J_h is the flux at top of the mixed layer (i.e. entrainment of free tropospheric air), F is the column formation rate (from chemical reactions), D is the column destruction rate, and <S> is the column concentration of S. Implicit in this equation is the Lagrangian assumption that transfer through the sides of the parcel is not causing significant concentration change. Although neither advection nor dispersion appears in the budget equation above, they will contribute to the uncertainty in quantities derived from Equation 1, and must be explicitly included in a propagation of errors analysis.

To track the air's movements, we tagged our airmasses with constant-density balloons and inert perflourocarbon tracers. After being launched from the Oceanus (upwind of Santa Maria Island) the balloons floated with the air and radioed their GPS-derived positions to the aircraft, which were flying nearby. By using a continuous relay of the three aircraft, the airmass was under continual observation for about 42 hours. The Malcolm Baldrige, positioned to intercept the air as the balloons floated out of aircraft range downstream, did a detailed characterization of its gases and aerosols for comparison with the upstream values.

Two Lagrangian experiments were attempted during ASTEX/MAGE, with varying degrees of success. The first took place in extremely clean marine air. Cloud condensation nuclei (CCN) counts of less than 50 cm^{-1} ensured that cloud formation would generate a small number of large droplets, with the result that drizzle was frequently observed. While air which maintains its low aerosol concentrations by drizzling is an interesting case, the addition of just 0.1 mm of water on each balloon's top was enough to drop it to the surface, drowning the transmitter.

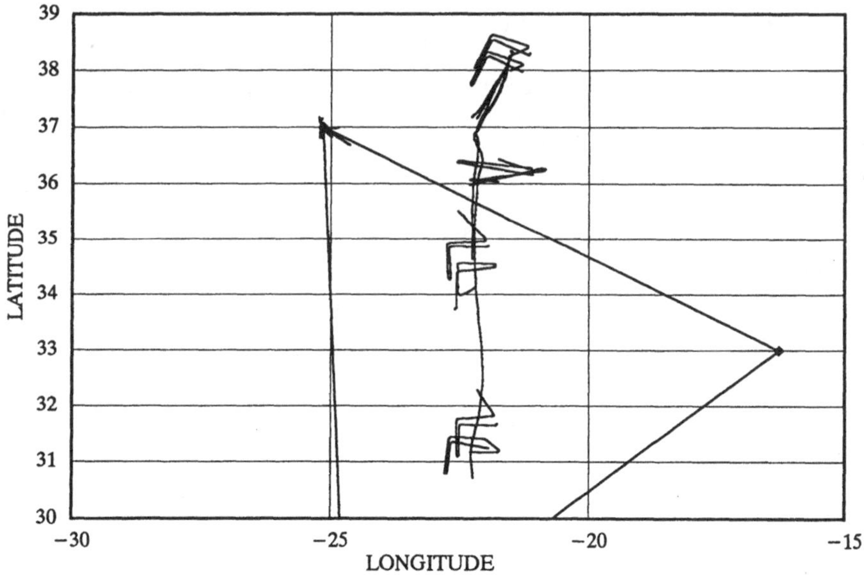

Figure 10. ASTEX/MAGE Lagrangian #2. The upper left corner of the triangle is Santa Maria, Azores; the right apex is Porto Santo, Madeira; and a ship occupied the lower apex. The wavy vertical lines are the tracks of five balloons, starting from the top of the figure where they were launched by the R/V Oceanus and ending by passing over the R/V Malcolm Baldrige at the bottom (ships not shown). Only one balloon survived to the end of the experiment. The back-and-forth patterns are Electra boundary layer sampling legs from flights 11–14 (one every 12 hours).

The second Lagrangian was a perfect contrast to the first: the air was heavily polluted by European aerosols, which suppressed precipitation and maintained high particle concentrations. We followed one balloon for the entire 42 hour observational period. Seven coordinated aircraft flights (four of which are diagrammed in Figure 10) generated an unmatched set of observations on how boundary layers decouple and develop over time.

We learned important lessons about how to conduct a Lagrangian experiment:

1. High-quality long range communications between all platforms is essential.
2. The balloons must be robust, so that they can withstand drizzle and modest downdrafts without loss. Protocols need to be established for how each platform would respond to the loss of one or more balloons.
3. The launch point of the balloons needs to be selected with some knowledge of the homogeneity of the airmass. Working on a boundary between airmasses (Figure 11) makes it harder to isolate the chemistry in either airmass from the effect of dynamics.
4. The best possible meteorological data and predicted trajectories need to be used to position ships and aircraft, to ensure that the aircraft can reach the starting and ending points of the tagged trajectory.
5. Even though advective effects are minimized, several techniques should be used to characterize the remaining atmospheric transport processes (entrainment, surface fluxes, and dispersion), so that chemical conversions can be isolated from these (inescapable) dynamic effects.
6. Natural and manufactured tracers should be used to confirm the extent to which the original airmass is still intact at the end of the experiment. The ability to analyze tracer samples in real-time (preferably on the aircraft) is essential for optimizing experimental strategies.
7. Extreme care must be used to avoid sampling contaminants from the ships and aircraft used in the study. The experimental plan should include measurements of the exhaust impact.
8. It is essential that 10 to 20% of the flight hours and ship days be devoted to instrument intercomparisons between platforms, to ensure that observed differences can be interpreted.

During most of the ASTEX/MAGE experiment, the deep marine boundary-layer was decoupled, meaning that a secondary inversion divided the moist air into a surface layer and one largely isolated from the ocean. Significant differences in chemistry between the layers created concentration gradients within the boundary layer (Figure 12). Some of the variability in the measurements of NSS vs. time (Figure 13) was due to the fact that the filter samplers collected material from altitudes where the concentrations differed.

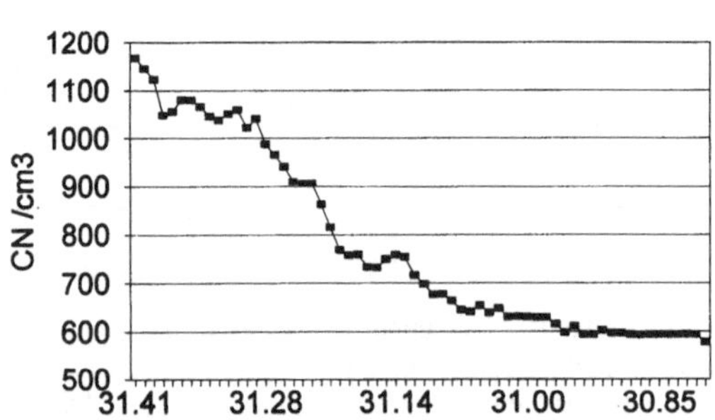

Figure 11. Condensation nuclei data along one boundary layer sampling leg. The CN gradient results from our sampling on the boundary between clean marine and polluted continental air. (Data from T. Clarke, private communication.)

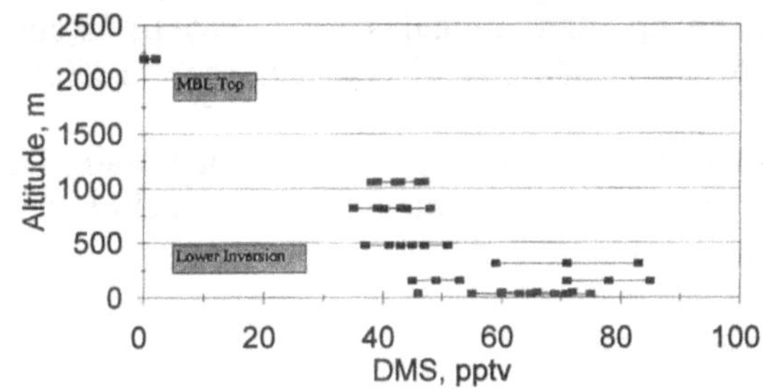

Figure 12. DMS concentrations versus altitude for Electra flight 7. A secondary inversion between 400 and 800 m decoupled the surface mixed layer from the upper boundary layer. This stratification produced vertical gradients of many chemical species. (Data from B. Blomquist, A. Bandy and D. Thornton, Drexel University, private communication.)

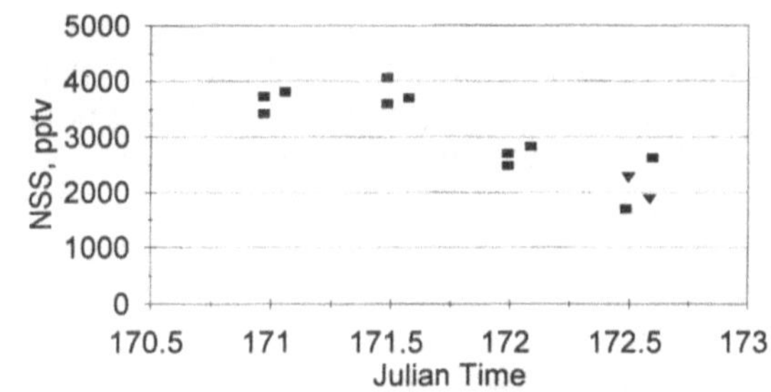

Figure 13. Evolution of non-seasalt sulfate concentrations during Lagrangian #2. The squares represent Electra samples. On each of the first three flights, one pair of samples was collected simultaneously with parallel samplers. The triangles indicate measurements from the R/V Malcolm Baldrige as the Lagrangian parcel passed over. (Malcolm Baldridge data from A. Pszenny, private communication.)

We could have modified our sampling strategies to account for the vertical gradients, had we been sufficiently aware of their impact. This is one of many lessons learned during ASTEX/MAGE that will improve our ability to separate dynamics and chemistry on subsequent Lagrangian experiments. We learned that the balloons can be ballasted differently to reduce their chances of being dunked, and that we should use a few "smart" balloons (which can adjust their buoyancy in flight) to ensure balloon survival. We also learned that satellite data and a pre-Lagrangian survey flight should be used to avoid starting the experiment in a region of sharp horizontal concentration gradients. ASTEX/MAGE brought us much closer to implementing a practical Lagrangian strategy.

Although the data from ASTEX/MAGE are still being analyzed, preliminary results suggest that the Atlantic Ocean near the Azores was a net source of ammonia vapor during the second Lagrangian experiment. To maintain the (relatively constant) ammonium aerosol concentrations in the presence of rapid dilution by clean free-tropospheric air, the ocean's surface had to be emitting tens of micrograms of ammonia nitrogen per square meter per day. This calls into question the idea that precipitation ammonium is a new (as opposed to recycled) fixed-nitrogen source to the oceans (Duce, 1986). Similar analyses of other species budgets are now underway.

Lagrangian observations can play an important role in the study of atmospheric chemistry, marine aerosol dynamics, boundary layer dynamics, and cloud/climate interactions. They offer a unique way of constraining fluxes and reaction rates by observing the effect of those processes on a parcel of air. Their implementation, however, requires cooperation on an international scale: MAGE is involved because such experiments are beyond the capabilities of any one group, discipline, or nation working alone.

For further information about ASTEX/MAGE, contact Barry Huebert, Department of Oceanography, University of Hawaii, Honolulu, HI 96822 USA (Phone: 1-808-956-6896, Internet: huebert@okika.soest.hawaii.edu).

Future MAGE Experiments: ACE-1

Under the umbrella of the International Global Atmospheric Chemistry Program (IGAC), MAGE and a sister project Multiphase Atmospheric Chemistry (MAC) are planning to employ a Lagrangian experiment to study DMS, aerosols, clouds, and climate during the Aerosol Characterization Experiment (ACE-1) in the vicinity of Cape Grim, Tasmania in early 1995. The new NCAR C-130 will support an even comprehensive chemical experiment, and one or two ships will be used to study the surface water forcing of air/sea exchange.

Anyone interested in learning more about ACE-1 should contact either Barry Huebert (the MAGE Convenor), or the MAC Co-convenors, Tim Bates or John Gras (Division of Atmospheric Research, CSIRO, Private Bag 1, Mordialloc, Vic. 3195, Australia. Phone: 61-3-586-7666, Internet: jlg@dar.csiro.au).

REFERENCES

Albrecht, B.A., 1989, Aerosols, cloud microphysics, and fractional cloudiness, *Science*, 245:1227-1230.

Andreae, M.O., 1986, The ocean as a source of atmospheric sulfur compounds, *in:* "The Role of Air-Sea Exchange in Geochemical Cycling," P. Buat-Menard (ed.), D. Reidel, Hingham, MA, U.S., pgs. 331-362.

Bates, T.S. and J.D. Cline, 1985, The role of the ocean in a regional sulfur cycle, *J. Geophys. Res.*, 90:9168-9172.

Bates, T.S., R.P. Kiene, G.V. Wolfe, P.A. Matrai, F.P. Chavez, K.R. Buck, B.W. Blomquist and R.L. Cuhel, 1993, The cycling of sulfur in surface seawater of the northeast Pacific, *J. Geophys. Res*., in press.

Charlson, R.J., J.E. Lovelock, M.O. Andreae and S.G. Warren, 1987, Oceanic phytoplankton, atmospheric sulfur, cloud albedo, and climate, *Nature*, 326:655-661.

Duce, R.A., 1986, The impact of atmospheric nitrogen, phosphorus, and iron species on marine biological productivity, *in:* "The Role of Air-Sea Exchange in Geochemical Cycling," P. Buat-Menard (ed.), pgs. 497-529, D. Reidel, Hingham, MA, U.S.

Greene, R.M., Z.S. Kolber, D.S. Swift, N.W. Tindale and P.G. Falkowski, 1993, Physiological limitation of phytoplankton photosynthesis in the eastern equatorial Pacific determined from natural variability in the quantum yield of fluorescence, *Limnol. Oceanogr*., in press.

Hanson, A.K., N.W. Tindale, M.A.R. Abdel-Moati, J.E. Prentice, D.W. O'Sullivan, 1993, Influence of a rain event on iron, peroxide, and phytoplankton in the surface waters of the equatorial Pacific ocean, in preparation.

Huebert, B.J., T.S. Bates, A. Bandy, S. Larsen and R.A. Duce, 1990, IGAC/MAGE: International Planning for Air/Sea Exchange Research, *EOS,* 71(35).

Huebert, B.J., S.G. Howell, P. Laj, J.E. Johnson, T.S. Bates, P.K. Quinn, S.A. Yvon, E.S. Saltzman, V. Yegorov, A.D. Clarke and J.N. Porter, 1993, Observations of the atmospheric sulfur cycle on SAGA-3, *J. Geophys. Res*., 98:16,985-16,996.

Johnson, K.S., K.H. Coale, V.A. Elrod and N.W. Tindale, 1993, Iron photochemistry and bioavailability in equatorial Pacific waters, *Marine Chem.*, in press.

Kawa, S.R. and R. Pearson, Jr., 1989, Ozone budgets from the dynamics and chemistry of marine stratocumulus experiment, *J. Geophys. Res.*, 94:9809-9817.

Langner, J., T.S. Bates, R.J. Charlson, A.D. Clarke, P.A. Durkee, J. Gras, J. Heintzenberg, D.J. Hoffman, B. Huebert, C. Leck, J. Lelieveld, J.A. Ogren, J. Prospero, P.K. Quinn, H. Rodhe and A.G. Ryaboshapko, 1993, "The global atmospheric sulfur cycle: An evaluation of model predictions and observations," Report No. CM-81, ISSN 0280-445X, Dept. of Meteorology, Stockholm Univ.

Liss, P.S., J.N. Galloway, 1993, Air-Sea exchange of sulphur and nitrogen and their interaction in the marine atmosphere, *in:* "Interactions of C, N, P and S Biogeochemical Cycles and Global Change," R. Wollast, F. T. Mackenzie and L. Chou (eds.), Springer-Verlag, pgs. 259-281.

Lenschow, D.H., R.J. Pearson and B.B. Stankov, 1981, Estimating the ozone budget in the boundary layer by use of aircraft measurements of ozone eddy flux and mean concentration, *J. Geophys. Res.*, 86:7291-7297.

Martin, J.H., R.M. Gordon, and S.E. Fitzwater, 1991, The case for iron, *Limnol. Oceanogr.*, 36:1793-1802.

Yin, F., D. Grosjean and J.H. Seinfeld, 1990, Photooxidation of DMS and DMDS. I: Mechanism development, *J. Atmos. Chem.*, 11:309-364.

Young, R.W., K.L. Carder, P.R. Betzer, D.K. Costello, R.A. Duce, G.R. Ditullio, N.W. Tindale, E.A. Laws, M. Uematsu, J.T. Merrill and R.A. Feely, 1991, Atmospheric iron inputs, primary productivity, and phytoplankton responses in the north Pacific, *Global Biogeochem. Cycles*, 5:119-134.

MULTIPHASE ATMOSPHERIC CHEMISTRY: IMPLICATIONS FOR CLIMATE

Robert J. Charlson[1] and Jos Lelieveld[2]

[1]Department of Atmospheric Sciences
University of Washington/BG-10
Seattle, WA 98195 USA
[2]Max-Planck-Institut for Chemistry
Atmospheric Chemistry Division
P.O.B. 3060, D-6500 Germany

ABSTRACT

While it has been known for over a century that atmospheric aerosol particles are an important factor governing the interaction of solar radiation with the earth, both through direct influences on solar radiation and indirectly as cloud condensation nuclei, the large degree of variability in both aerosol composition and concentration precluded all but crude global appraisals of the actual climate forcing (W m^{-2}). The advent of global chemical reaction/transport/removal models, including parameterizations of heterogeneous processes, has made it possible to estimate the direct climatic forcing for sulfate aerosol and for condensed organic materials from biomass combustion. Thus the study of multiphase atmospheric chemistry has made possible a species-by-species and mechanism-by-mechanism approach to assessing these physical effects. Examination of the generalized heat balance equation for the earth suggests that there must be numerous other important aerosol species and several more mechanisms by which climate is affected. To date, the assessment of effects by anthropogenic sulfate and smoke from biomass combustion indicate that these aerosols cause a climatic forcing that, when averaged over the northern hemisphere, is comparable in magnitude but opposite in sign to the "greenhouse" forcing by CO_2, CH_4, chlorofluorocarbons, etc. Three steps are involved in these coupled chemical/radiative transfer models: simulation of the geographically, time and height dependent aerosol formation process, prediction of the microphysical properties as functions of the source processes or species characteristics, and coupling of these to geographically dependent radiative transfer calculations. We first present a description of the main aerosol production mechanisms and resultant physical properties. Subsequently we address the *direct* and *indirect* (CCN-cloud albedo) radiative forcings, leaving open the important questions of influence of aerosol particles on cloud amount and cloud-droplet longevity. Furthermore, we discuss multiphase chemical processes that affect the abundance of tropospheric O_3, the latter being a potent greenhouse gas. We will emphasize the importance of the open scientific questions, concluding that it is not yet possible to quantify climate forcing by anthropogenic or natural aerosols fully. Finally, we describe the approach of the IGAC Multiphase Atmospheric Chemistry (MAC) Activity to improve the understanding of these issues.

INTRODUCTION

It has long been known that aerosol particles and clouds must play important roles in the radiative balance and climate of the earth. Tyndall (1861) made observations of both aerosol light scattering and infrared light absorption by the atmosphere. Bergeron (1928) identified air masses by their content of "opalescent haze." Ångström (1929) began measurement of the amount of solar radiation reaching the ground under clear skies and defined the "turbidity coefficient" as a quantitative measure of the clear-sky attenuation of the solar beam. More recently Flowers *et al.* (1969) and Ball and Robinson (1982) showed that the non-cloud aerosol over the industrialized eastern U.S. caused an average loss of solar irradiance of several percent. However, in all instances, the large (factor of ten) temporal variability of atmospheric extinction of sunlight, coupled with large (also factor of ten) geographic variability precluded any extrapolation to the larger hemispheric to global spatial scales. Bolin and Charlson (1976) attempted a global extrapolation from regional-scale turbidity data, incorrectly concluding that industrial haze, dominated by sulfate, was of marginal importance to global heat balance but was only important regionally (here the underestimate was due to a lack of geographical coverage by the measurement network).

It also has been known since the time of Aitken (1880–1881) and Köhler (1926) that the formation of cloud droplets requires the presence of a nucleus around which water could condense at low supersaturation (less than a few percent); it was implicitly recognized that the entire hydrologic cycle depends on clouds and thus on these aerosol particles acting as cloud condensation nuclei (CCN). More recently, Twomey (1977) presented calculations suggesting that the population of CCN (cm^{-3}) actually influenced cloud albedo and, hence, the heat balance and climate as well. Charlson *et al.* (1987) estimated that a change of CCN of ±30%, acting only on marine stratus clouds would change the heat balance of the entire planet by ~1 W m^{-2}. Because this quantity approaches the magnitude of today's greenhouse gas forcing (~2.1 to 2.4 W m^{-2}), even relatively small changes in CCN could be important to climate. However, just like the optical effects of aerosol, the CCN concentrations are enormously variable both in space and time, making it difficult to ascertain the factors controlling them.

Most importantly, over the course of these historical developments, it has been increasingly recognized that multiphase atmospheric chemical processes are an important controlling factor influencing all atmospheric aerosol particles, and are the very source of most CCN and of much of the world light scattering aerosol. Currently, increased attention is being paid to these multiphase atmospheric chemical processes, due to improved understanding and the increasing role of anthropogenic aerosol, both optically and as CCN. In particular, anthropogenic perturbation of the global sulfur cycle and of organic aerosol substances from biomass combustion are calculated to have altered radically the totality of atmospheric aerosol and its effects over perhaps the past one to two centuries (Charlson *et al.*, 1991, 1992; Penner *et al.*, 1992, 1993). Substantial evidence exists to suggest that the growing abundances of these particles, taken together, may result in a northern hemisphere average forcing of −1 to −2 W m^{-2}, possibly counteracting greenhouse gas forcing on a regional scale. This may have masked the detection of global warming due to the changes in other gas phase constituents of the atmosphere. Indeed, this juxtaposition of different climate forcing factors provides a starting point for an objective survey of climatic implications of multiphase atmospheric chemistry in the context of the numerous identifiable forcings.

Q = radiative flux

$$\frac{dQ}{dt} = \frac{\partial Q}{\partial GG}\frac{dGG}{dt} \quad \text{"Greenhouse Gases"}$$

$$+ \frac{\partial Q}{\partial H_2O(v)}\frac{dH_2O(v)}{dt} \quad H_2O \text{ Vapor}$$

$$+ \frac{\partial Q}{\partial LCQ}\frac{dLCQ}{dt} \quad \text{Long wave effect of Cloud Fraction}$$

$$+ \frac{\partial Q}{\partial SCQ}\frac{dSCQ}{dt} \quad \text{Short wave effect of Cloud Fraction}$$

$$+ \frac{\partial Q}{\partial LWP}\frac{dLWP}{dt} \quad \text{Liquid Water Path}$$

$$+ \frac{\partial Q}{\partial CCN}\frac{dCCN}{dt} \quad \text{Cloud Condensation Nuclei}$$

$$+ \frac{\partial Q}{\partial LAP}\frac{dLAP}{dt} \quad \text{Long wave Absorption by Particles}$$

$$+ \frac{\partial Q}{\partial SSP}\frac{dSSP}{dt} \quad \text{Short wave effect Stratospheric Particles}$$

$$+ \frac{\partial Q}{\partial STP}\frac{dSTP}{dt} \quad \text{Short wave effect Tropospheric Particles}$$

$$+ \frac{\partial Q}{\partial SUN}\frac{dSUN}{dt} \quad \text{SUN Solar Flux}$$

$$+ \quad ...\ ?$$

Figure 1. Major terms in the equation for the changing heat (Q) balance of the earth as a function of time.

MECHANISMS OF CLIMATE FORCING

Figure 1 presents a generalized equation for the time rate of change of the radiative balance of the surface of the globe in the form of a chain-rule expansion of its time derivative, dQ/dz, including the major terms known to depend on atmospheric composition. If we take the thermal emission from the surface to be a purely dependent variable in the climate system, and disregard solar variability as being relatively small, all the remaining terms can in principle be dependent in some way on the chemical composition of the atmosphere, and all but the greenhouse gas terms are expected to demonstrate a connection to multiphase atmospheric chemical processes and properties. It is even likely that one of the potent greenhouse gases, ozone, depends to some extent on reactions on the surfaces of clouds and aerosol particles and on liquid phase chemistry.

Table 1 summarizes the mechanisms by which aerosol particles directly or indirectly influence the heat balance of earth, including the aerosol sources and types that are relevant to each mechanism, the known or possible anthropogenic forcing and the status of scientific understanding of each. Several points emerge from this list and recent literature:

- There are many chemical species in aerosol form that are known to be involved.

Table 1. The influence of aerosol particles on the heat balance of the earth.

Mechanism	Relevant Aerosol Sources	Global Anthropogenic Forcing	Status of Scientific Knowledge
A. DIRECT:			
Reflection and absorption of solar radiation by tropospheric aerosol			
	$(CH_3)_2S$ (DMS) $\rightarrow SO_4^=$	—	small effect
	Industrial $SO_2 \rightarrow SO_4^=$	−0.3 to −1.2 W m^{-2}	amenable to reduction of uncertainty
	Volcanic $SO_2 \rightarrow SO_4$	—	sporadic large events of short duration
	Soot from combustion	—	approachable in principle
	Biomass combustion	−0.2 to −1 W m^{-2}	amenable to observation and reduced uncertainty
Reflection of solar radiation by stratospheric aerosol			
	OCS $\rightarrow SO_4^=$	—	
	Volcanic $SO_2 \rightarrow SO_4^=$	—	large events, ca −2 to −3 W m^{-2}
Absorption of terrestrial radiation by tropospheric aerosol			
	Wet seasalt aerosol	—	
	Mineral dust from soils	?	difficult due to sporadic nature of dust storms
B. INDIRECT:			
Physical			
Dependence of cloud albedo on CCN number concentration			
	Marine DMS $\rightarrow SO_4^=$	—	substantial supporting evidence
	Industrial $SO_2 \rightarrow SO_4^=$	possibly −1W m^{-2}	potentially large regional effects
	organic condensates	?	
	soil dust	—	small effects due to small
	seasalt	—	number concentration of coarse particles
Dependence of cloud droplet lifetime or cloud cover on CCN			
	same	potentially large	totally open question
Dependence of water vapor lifetime on CCN			
	same	potentially large	H_2O vapor is the main greenhouse gas
Nucleation of cirrus clouds			
	? aircraft exhaust	potentially large	amenable to direct observation
Chemical			
Loss of O_3 on aerosol and in clouds			
	all	?	amenable to exploratory observation and modeling
Loss of NO_x, change in O_3			

- Anthropogenic perturbation is known to exist or is likely for many of these aerosols.
- The magnitude of anthropogenic climate forcing currently can only be estimated for sulfates and biomass combustion aerosol due to a lack of adequate data on other aerosol types.
- The magnitude of forcing by anthropogenic sulfate particles and biomass combustion aerosols is comparable, at least regionally in the source areas, to forcing by greenhouse gases, but it is opposite in sign.
- The uncertainty of these estimated forcings is large but not large enough to allow the possibility that these forcings are negligible.

- There is great variability in the level of understanding of the different aerosol types, but several of them are amenable to study with known methods and instruments. Sulfate and biomass burning smoke offer the possibility of local, column and regional to global closure (internal consistency) experiments.
- Indirect forcing via anthropogenic changes in cloud microphysics is potentially large, as are changes from the natural variability of CCN.
- Volcanic and anthropogenic indirect forcing may result from changes in ozone due to chemical reactions on the surfaces of aerosol and/or in cloud particles.

This general background, along with the fact that much of the aerosol mass and most of the aerosol number concentration are produced by chemical reactions in the atmosphere, also suggest that it is necessary to understand the role of multiphase atmospheric chemical processes as determinants of the climatically relevant properties of the aerosol particles. Included are the formation mechanisms, rates of reaction, physical transformations (e.g., phase changes), removal processes and biogeochemical fluxes due to multiphase processes.

As noted in Table 1, there is great variation in the status of understanding of various aspects of these processes. In general, processes related to the direct radiative forcing of climate by the products of gas-to-particle conversion reaction have been studied much more extensively and successfully than the processes controlling the number concentration of particles. Investigations on direct radiative forcing of climate have provided many well posed questions and opportunities for quantifying and minimizing uncertainties. In contrast, the indirect forcing is only approachable in a limited way, and in some cases, there is no approach yet evident, for example in studying the possibility that surface-active organic species may alter the lifetimes of droplets and hence cloud longevity, cloud cover, the lifetime of water vapor in the atmosphere or the exchange of gases with droplets.

CHARACTERISTICS OF THE ATMOSPHERIC AEROSOL

While the atmospheric aerosol is chemically very inhomogeneous and has particles ranging in size from several nm to over 10 μm, some general statements can be made based on the mechanism of controlling processes:

- *Coarse particles* with diameters, d, greater than about 1 μm are in general produced mechanically; e.g., wind-blown soil dust or seasalt.
- These particle surfaces are often basic and may be the site of reactions with gases, e.g., SO_2.
- *Fine particles*, $d \leq 1$ μm, are in general produced by either low or high temperature condensation of materials from the gas phase, e.g., sulfates from the oxidation products of SO_2, organic species from material volatilized in biomass combustion or soot (impure elemental carbon) from certain types of combustion, especially of fossil fuels. Figure 2 illustrates some examples of aerosol-forming chemical reactions.
- Transformation processes occur during the residence time of the particles in the air. This includes mixing of particles by Brownian motions and coagulation, growth by the addition of condensable species into or onto the surfaces of particles, modification while the particles are dissolved into the liquid phase of cloud droplets, and, possibly, particle to gas conversion if OH removes organic matter from the condensed phase.
- *Removal* of aerosol particles occurs to a small extent by sedimentation and Brownian deposition to the earth's surface and to a much larger extent by particles acting as CCN and coalescing into falling hydrometeors.

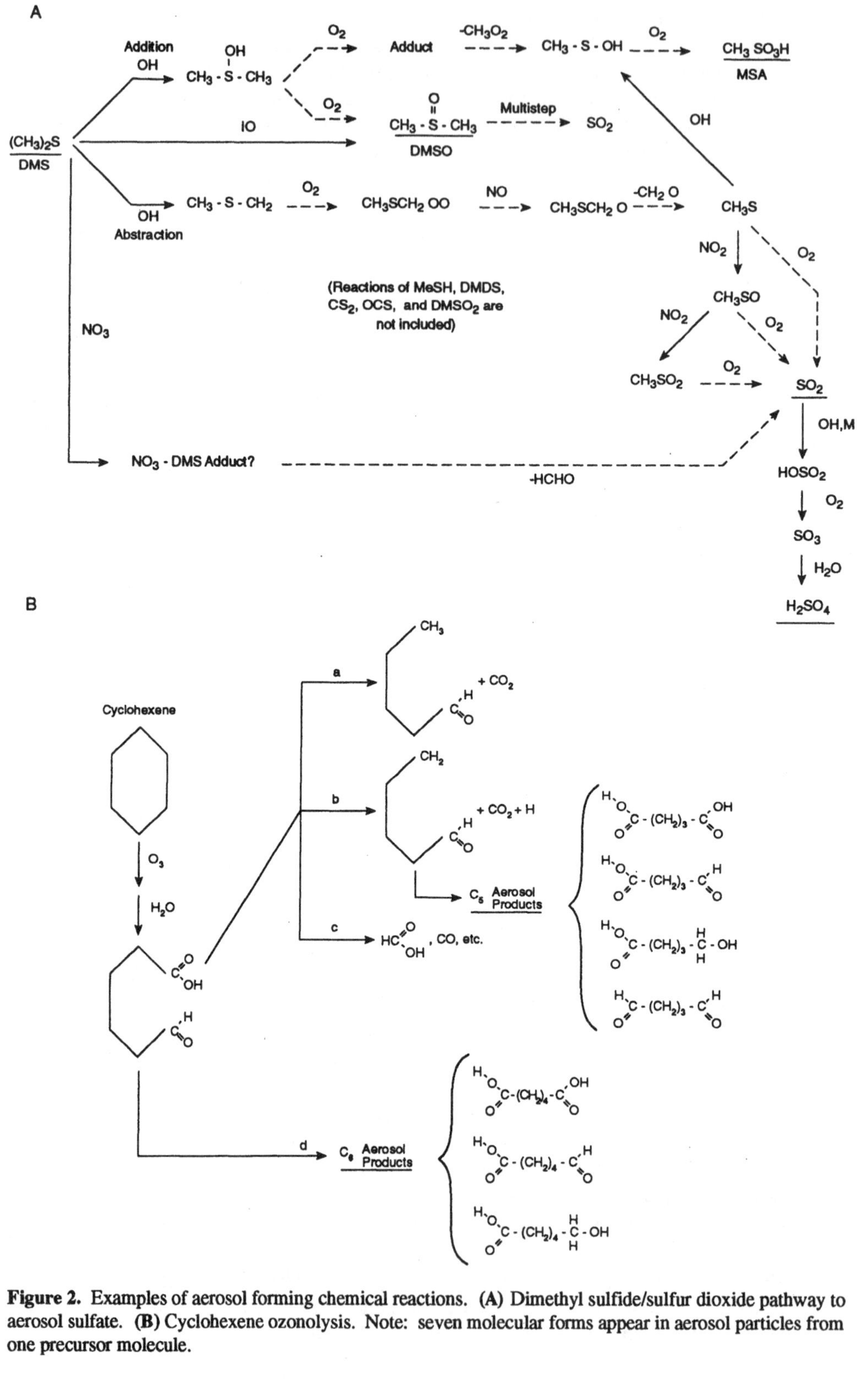

Figure 2. Examples of aerosol forming chemical reactions. **(A)** Dimethyl sulfide/sulfur dioxide pathway to aerosol sulfate. **(B)** Cyclohexene ozonolysis. Note: seven molecular forms appear in aerosol particles from one precursor molecule.

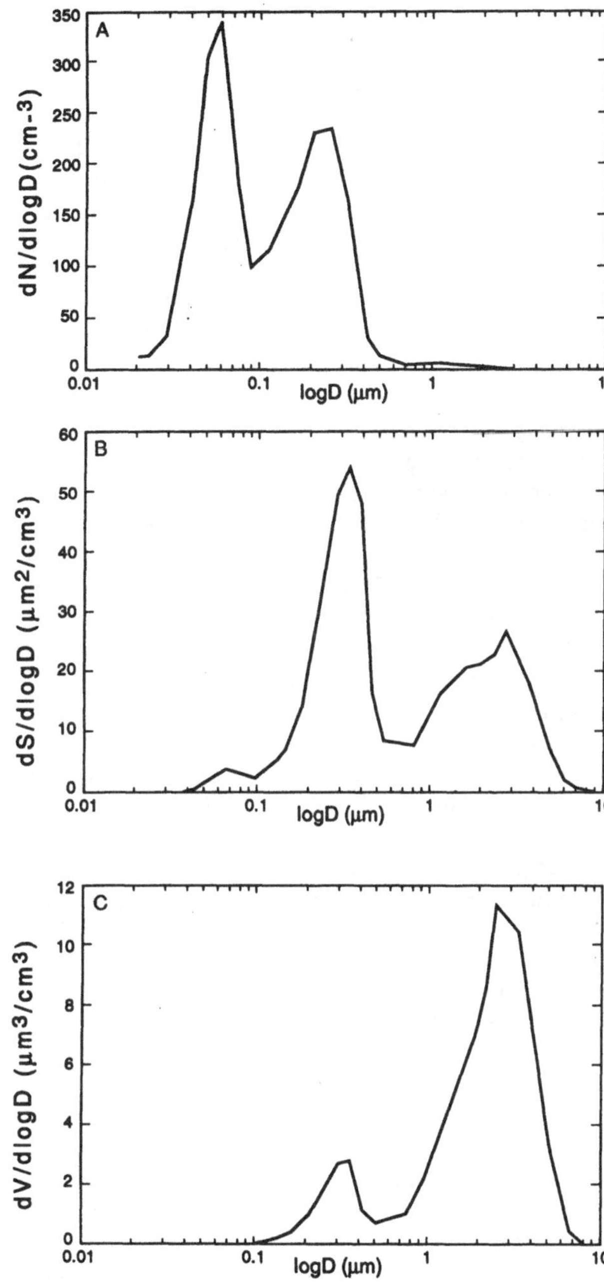

Figure 3. Example marine aerosol size distribution. **(A)** Number distribution. **(B)** Surface distribution. **(C)** Volume (mass) distribution.

This complex set of processes results in distribution of particle number concentration, area and mass concentration being distributed over a large range of sizes. Figure 3 (Covert, 1993) shows an example of measured number, surface and volume distribution functions, and illustrates the usual systematic difference of particle morphology above and below $d \approx 1\ \mu m$. It is known that different chemical species dominate these three different moments of the size distribution.

MULTIPHASE CHEMICAL PROCESSES

Perhaps the best understood aerosol constituents, and often the ones dominating mass, surface area and number concentration below ~1 μm diameter, are sulfates ranging in composition from H_2SO_4 to $(NH_4)_2SO_4$. A small fraction of this $SO_4^=$ is produced during new particle formation. However, most oxidation of sulfur dioxide into sulfate in the troposphere takes place in the liquid phase in cloud and rain droplets. On a global scale, about 80% of the sulfate formation occurs within clouds, mainly through the rapid reaction between SO_2 and H_2O_2 in solution (Hegg, 1985; Langner and Rodhe, 1991). When the H_2O_2 is depleted, or within clouds with a relatively high pH, the aqueous phase reaction of SO_2 with O_3 can become important, a reaction that may also account for some SO_2 oxidation in seasalt aerosol in the marine boundary layer (Sievering *et al.*, 1992). Heterogeneous oxidation into sulfate adds to the mass of pre-existing particles, notably those that have previously served as CCN, with the implication that only the relatively small remaining amount of SO_2 that is oxidized by OH in the gas phase yields H_2SO_4 molecules that can cluster into new nuclei. The low available aerosol particle surface in the upper troposphere makes it possible that new particle formation is enhanced at high altitude (Raes, 1992). Thus, only relatively small fractions of the anthropogenically emitted SO_2 and the naturally produced SO_2 from $(CH_3)_2S$ (DMS) contribute to new particle formation, which is relevant for potential changes in CCN number concentrations. It should be stressed, however, that gas phase SO_2 concentrations are a direct function of SO_2 and DMS emissions. In fact, SO_2 concentrations in the anthropogenically perturbed lower troposphere may be enhanced by several orders of magnitude compared to the natural atmosphere, which greatly increases the potential for new particle nucleation (see Figure 4).

It may be important to distinguish between SO_2 oxidation in stratiform and convective clouds. In convective clouds relatively high water vapor supersaturations occur, so that not only large but also relatively small particles are activated as CCN (Twomey, 1977). Aqueous phase SO_2 oxidation adds to the aerosol mass, as larger and more efficient CCN are released after convective cloud evaporation, thus effectively enhancing the number of CCN for subsequent stratiform cloud formation, which occurs at much lower supersaturations. Note that most clouds evaporate rather than precipitate and that the reflective properties of stratiform clouds are calculated to be most sensitive to changes in CCN number (Charlson *et al.*, 1987). Note also that we do not argue that clouds are sources of new particles. The potential importance of different cloud types can be estimated by using the global data sets in the NCAR cloud atlases (see Lelieveld, 1993). Stratiform clouds cover a volume fraction of about 5% of the global troposphere, while cumuliform clouds cover only about 0.3%. On the other hand, the relatively short turnover times of convective clouds cause them to "process" air at an approximately 5 to 10 times higher rate than stratiform clouds. If we assume, to a first approximation, that all SO_2 is converted into sulfate in all cloud types, it follows that the amount of SO_2 to sulfate oxidation in stratiform clouds is only about twice as large as in convective clouds, so that in the latter up to about 70 Tg sulfate is formed annually, of which most occurs in the continental northern hemisphere (Lelieveld, 1993). This amount of sulfate, of which only a small fraction (≈10%) is directly removed by precipitation (Pruppacher and Klett, 1980), plus some liquid water mass collected by deliquescence, could contribute significantly to the growth of CCN in the relatively small geographical areas where most sulfur pollution accumulates.

Aqueous phase sulfate formation and the subsequent release of the dry aerosol by cloud evaporation appear to also influence the particle scattering properties (Lelieveld and Heintzenberg, 1992). Through in-cloud SO_2 oxidation the sulfate is deposited on the particles that have been activated as CCN. The preferential growth of particles in the size range 0.05 to 0.2 μm, which coincides with the minimum radii for CCN (Twomey, 1977), strongly increases the aerosol scattering efficiency. This effect is particularly pronounced for sulfate pollution of the order of 0.5 to 2 $\mu g\ m^{-3}$, which may occur on regional to hemispheric scales (Langner and Rodhe, 1991), enhancing the particle scattering efficiency by a factor of 2 to 4 compared to sulfate formation in cloud-free environments. One of the difficulties in quantifying this anthropogenic aerosol perturbation, required to estimate the climate cooling forcing, is to define the properties of the unperturbed aerosol. Time series of sulfate concentrations in southern Greenland derived from ice core analyses confirm that sulfate pollution occurs on a hemispheric scale (Mayewski *et al.*, 1990). Studies of the pristine aerosol should therefore concentrate on the southern hemisphere. Nevertheless, the relatively high scattering efficiency of sulfate aerosol as observed in the contemporary atmosphere in rural parts of Europe and the U.S. (Charlson *et al.*, 1992) can be explained by model calculations that account for the cloud "processing" of aerosol particles in the anthropogenically sulfur enriched lower troposphere (Lelieveld and Heintzenberg, 1992).

Heterogeneous cloud and aerosol related processes also affect tropospheric ozone and, indirectly, the atmosphere's oxidation efficiency. For example, photochemical O_3 formation, which occurs in nitrogen oxide (NO) containing cloud-free air is reversed into O_3 destruction within clouds (Lelieveld and Crutzen, 1990). Dissolution of soluble peroxy radicals (notably HO_2) in the cloud droplets arrests the most important gas phase O_3 production reaction with NO, as the latter is very insoluble; thus the reactants are separated. Moreover, O_2^-, the dissociated form of HO_2, destroys O_3 in the aqueous phase. Furthermore, the methane and higher hydrocarbon oxidation reaction intermediate formaldehyde, an important source of radicals in the atmosphere through its photodissociation, is quite reactive in the aqueous phase, contributing to O_3 breakdown. As a consequence of the reduction of O_3 and radical abundances the atmospheric oxidation efficiency is decreased. Cloud droplets also provide a reactive surface in tropospheric nighttime chemistry. In the absence of photodissociation reactions nitrogen oxides and ozone react into NO_3 and subsequently N_2O_5. The latter is effectively converted into nitric acid within clouds (Platt *et al.*, 1984), which is an important sink for reactive nitrogen (NO_x), in turn reducing daytime photochemical ozone formation.

Heterogeneous removal of NO_3 and N_2O_5 on sulfate or seasalt aerosol particles or on cloud droplets, mostly during nighttime, is an additional NO_x sink, in particular in the aerosol-rich anthropogenically influenced part of the atmosphere (Dentener and Crutzen, 1993). The N_2O_5 reactions both on clouds and aerosols, the former particularly in the relatively clean background atmosphere (southern hemisphere), the latter more importantly in the continental northern hemisphere, account for most of the NO_x removal in the winter. Photochemical NO_x destruction by hydroxyl radicals (OH) predominates in the summer and tropics. In fact, owing to these two complementary NO_x oxidation mechanisms, nitrate deposition rates are generally quite uniform throughout the year (Dentener and Crutzen, 1993). Recently, laboratory work has indicated that significant nighttime aqueous-aerosol chemistry may also be initiated by NO_3 radicals (Herrmann *et al.*, 1992), which could add to the loss of NO_x and O_3. Furthermore, metal-ion catalyzed reactions may play an important role in heterogeneous daytime radical chemistry (Behra and Sigg, 1990). We emphasize that the study of heterogeneous tropospheric chemistry is still in its infancy. Many potentially important reactions need to be investigated in the laboratory and evaluated for their atmospheric significance in numerical models. Also, the representations of clouds and aerosols in these models require considerable improvements (Ogren and Charlson, 1992).

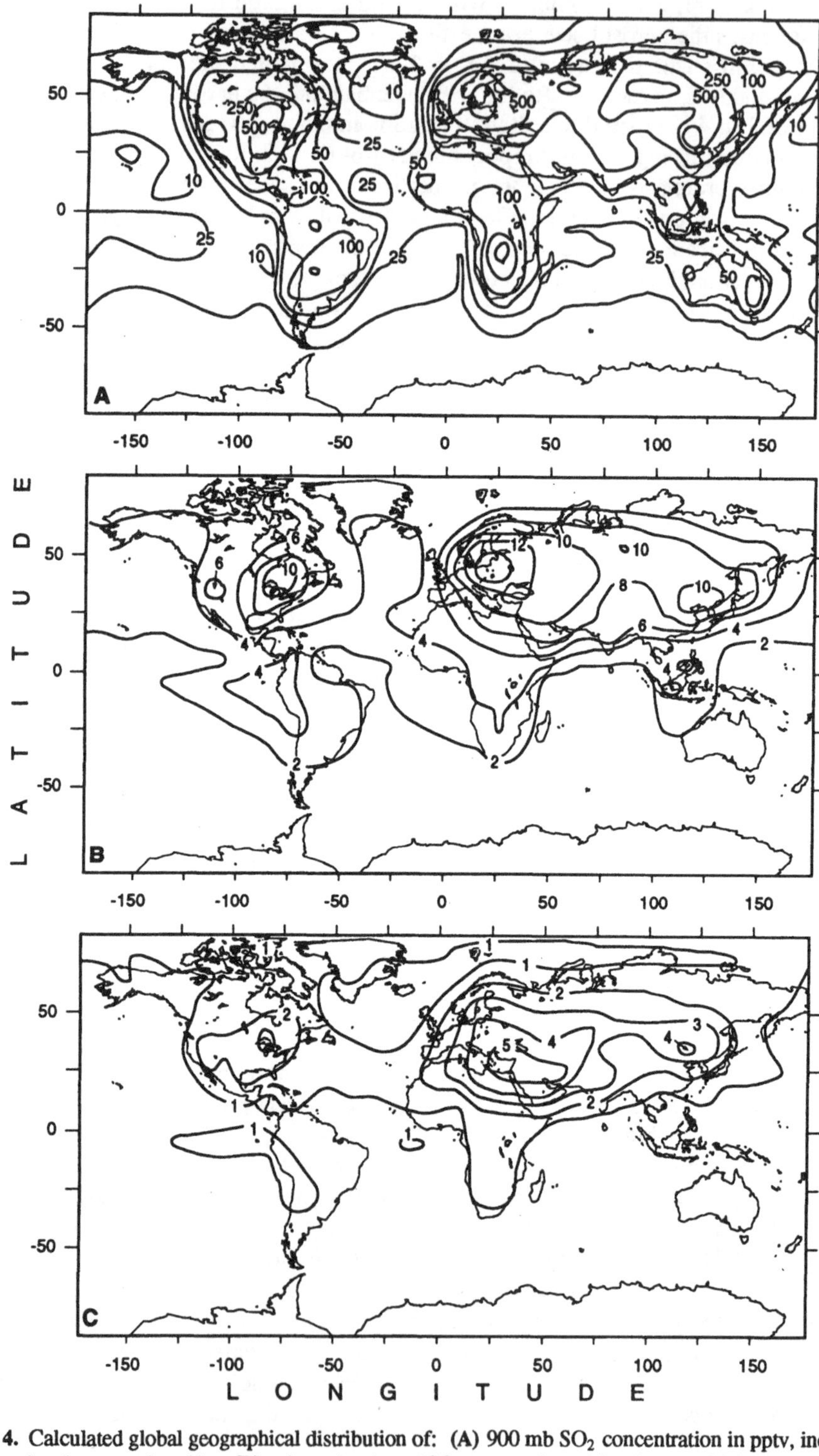

Figure 4. Calculated global geographical distribution of: **(A)** 900 mb SO_2 concentration in pptv, including natural and anthropogenic sulfur; **(B)** column burden of $SO_4^=$ aerosol resulting from multiphase atmospheric chemical processes (mg m^{-2}); and **(C)** resultant forcing (W m^{-2}).

QUANTITATIVE EXAMPLE: DIRECT FORCING BY AEROSOL SULFATE

In order to demonstrate that climate forcing indeed does derive from multiphase atmospheric chemical processes, the example of direct forcing by anthropogenic sulfate can be given. Briefly summarizing the numerous published calculations of the direct effect of sulfate, it is critically important to note that this is only part of one term in Figure 1. It does not include other chemical forms of tropospheric aerosol, nor does it include stratospheric aerosol or the longwave (IR) effects of particles.

Figure 4 shows the results of model calculations based on the work of Langner and Rodhe (1991) for the gaseous aerosol precursor SO_2 (anthropogenic plus natural), column burden of sulfate aerosol, and radiative forcing due to the anthropogenic SO_2-derived $SO_4^{=}$ alone. Several features of this system deserve to be mentioned. First, there is substantial anthropogenic enhancement of the radiative forcing, with regional maxima around a few Watts per square meter. Second, the SO_2, $SO_4^{=}$ and forcing fields are geographically complex with the anthropogenic influence appearing almost everywhere over the industrial regions of the northern hemisphere. Third, the horizontal spatial scale of the calculated geographical variation is of magnitude comparable to the spatial scale of synoptic weather systems.

CONCLUSION: THE ROLE OF IGAC

Research over the past decade, some of which is reflected in the examples in this paper, has significantly altered the singular view of dominance of the climatic role of atmospheric composition by infrared absorbing gases. Additionally, the view of the atmospheric aerosol as a static "background" has changed to one dominated by multiphase chemical processes. Indeed, the role of aerosol particles in directly and indirectly influencing radiative balance poses one of the largest uncertainties in the forecast of global climatic changes. The consequence of these new developments is clear; it is necessary to develop a thorough and workable understanding of the source, transport, transformation and sink processes that control the properties and effects of the atmospheric aerosol.

IGAC has developed the Multiphase Atmospheric Chemistry Activity (MAC) in order to integrate the chemical and physical studies that are needed to gain the understanding of factors that control the influence of aerosols on climate. Here the focus is indeed general, but there is a clear emphasis on delineating the anthropogenic departures from a natural system. The research focus of MAC is the role of multiphase atmospheric chemical processes as determinants of those aerosol properties which are relevant for climate. The goals of MAC are:

- Identification: To develop a chemical and physical characterization of aerosol particles in the key air masses (i.e., clean vs. polluted marine air; clean vs. polluted continental air; and upper tropospheric air).
- Dynamics: To understand the physical and chemical factors that produce aerosol particles and control their evolution and relevant physical properties.
- Spatial extrapolation: To evaluate the means to utilize and interpret remote sensing to extend local aerosol concentration measurements to larger geographical areas.
- Modeling: To develop improved aerosol and cloud parameterizations in order to relate the factors controlling the radiative and cloud nucleating properties of aerosol particles to the sources of aerosols and aerosol precursors and to the large scale processes and variables addressed in climate model calculations.

It is fully intended that this research will be carefully integrated with purely physical radiation and meteorological research on the one hand, and with modeling on the other.

The first experiments under MAC, the Aerosol Characterization Experiments (ACE), are intended to provide an opportunity for the international community to bring together their measurement and modeling capabilities. ACE-1 will be conducted near Tasmania in early 1995 with a focus on the relatively unperturbed marine aerosol system. ACE-2 will follow as an investigation of anthropogenic perturbation of this same system over the North Atlantic region.

REFERENCES

Aitken, J., 1880-1, On dust, fogs and clouds, *Trans. Roy. Soc. Edinb.*, 30:337.

Ångström, A., 1929, On the atmospheric transmission of sun radiation and dust in the air, *Geograph. Annal.*, 11:156-166.

Ball, R.J. and G.D. Robinson, 1982, The origin of haze in the central U.S.and its effects on solar radiation, *J. Appl. Meteorol.*, 21:171-188.

Behra, P. and L. Sigg, 1990, Evidence for redox cycling of iron in atmospheric water droplets, *Nature*, 344:419-421.

Bergeron, T., 1928, "Über die dreidimensional verknüpfende Wetter analyse," Geof. Pub, V, No. 16, Oslo.

Bolin, B. and R.J. Charlson, 1976, On the role of the tropospheric sulfur cycle in the short wave radiative climate of the earth, *Ambio*, 2:4754.

Charlson, R.J., J. Langner, H. Rodhe, C.B. Leovy and S.G. Warren, 1991, Perturbation of the northern hemisphere radiative balance by backscattering from anthropogenic sulfate aerosols, *Tellus*, 43AB:152-163.

Charlson, R.J., S.E. Schwartz, J.M. Hales, R.D. Cess, J.A. Coakley, Jr., J.E. Hansen and D.J. Hofmann, 1992, Climate forcing by anthropogenic aerosols, *Science*, 255:423-430.

Crutzen, P.J. and P.H. Zimmermann, 1991, The changing photochemistry of the troposphere, *Tellus*, 43AB:136-151.

Dentener, F. and P.J. Crutzen, 1993, Reaction of N_2O_5 on tropospheric aerosols: impact on the global distributions of NO_x, O_3 and OH, *J. Geophys. Res.*, in press.

Flowers, E.C., R.A. McCormick and K.R. Kurtis, 1969, Turbidity over the United States 1961-1966, *J. Appl. Meteorol.*, 8:955-962.

Hegg, D.A., 1985, The importance of liquid-phase oxidation of SO_2 in the troposphere, *J. Geophys. Res.*, 90:3773-3779.

Herrmann, H., M. Esner and H. Zellner, 1992, Laser based studies of reactions of the nitrate radical in aqueous solution, *Ber. Bunsenges. Phys. Chem.*, 96:470-485.

Köhler, H., 1926, Zur Kondensation des Wasser dampfes in der Atmosphäre, *Geofys. Publ.*, 2, 3, 6, 26.

Langner, J and H. Rodhe, 1991, A global three-dimensional model of the tropospheric sulfur cycle, *J. Atm. Chem.*, 13:225-263.

Lelieveld, J., 1993, Multi-phase processes in the atmospheric sulfur cycle, *in:* "Interactions of C, N, P and S biogeochemical cycles and global change," H. Wollast, F. T. Mackenzie and L. Chou (eds.), NATO ASI Series I, Vol. 4, Springer Verlag, Berlin, pgs. 305-331.

Lelieveld, J. and P.J. Crutzen, 1990, Influences of cloud photochemical processes on tropospheric ozone, *Nature*, 343:227-233.

Lelieveld, J. and J. Heintzenberg, 1992, Sulfate cooling effect on climate through in-cloud oxidation of anthropogenic $SO_4^{=}$, *Science*, 258:117-120.

Mayewski, P.A., W.B. Lyons, M.J. Spencer, M.S. Twickler, C.F. Buck and S. Whitlow, 1990, An ice-core record of atmospheric response to anthropogenic sulfate and nitrate, *Nature*, 346: 554-556.

Ogren, J.A. and R.J. Charlson, 1992, Implications for models and measurements of chemical inhomogeneities among cloud droplets, *Tellus*, 44B:208-225.

Platt, U.F., A.M. Winer, H.W. Biermann, R. Atkinson and J.N. Pitts Jr., 1984, Measurement of nitrate radical concentrations in continental air, *Environ. Sci. Technol.*, 18:356-369.

Pruppacher, H.R. and J.D. Klett, 1980, "Microphysics of Clouds and Precipitation," Reidel, Dordrecht, 714 pp.

Raes, F., 1992, private communication.

Sievering, H., J. Boatman, E. Gorman, Y. Kim, L. Anderson, G. Ennis, M. Luria and S. Pandis, 1992, Removal of sulfur from the marine boundary layer by ozone oxidation in sea-salt aerosols, *Nature*, 360:571-573.

Twomey, S., 1977, "Atmospheric Aerosols," Elsevier, Amsterdam, 302 pp.

Tyndall, J. (1861) On the absorption and radiation of heat by gases and vapors, and on the physical connexion of radiation, absorption and conduction, *Phil. Mag. S.*, 4, 22, No. 446.

ATMOSPHERIC CHEMISTRY OF THE EAST ASIAN NORTHWEST PACIFIC REGION

Hajime Akimoto[1], Douglas D. Davis[2], Shaw C. Liu[3]
and PEM-West (A) Science Team[4]

[1]Research Center for Advanced Science and Technology
University of Tokyo
4-6-1 Komaba, Meguroku, Tokyo 153, Japan
[2]School of Earth and Atmospheric Sciences
Georgia Institute of Technology
Atlanta, Georgia 30332-0340, U.S.A.
[3]Aeronomy Laboratory
325 Broadway, Boulder, Colorado 80303, U.S.A.
[4]See acknowledgment section for complete listing of participants

ABSTRACT

The East Asian Continental Rim region is characterized by high and rapidly growing anthropogenic emissions of NO_x, SO_2, hydrocarbons, and other air pollutants due to its high and growing population density and, in some regions, intensive industrial development. Emissions of NO_x and SO_2 from the East Asian region have been growing at a rate of about 4% per year in the last decade, which is in marked contrast to the trends in Western Europe and North America where those emissions have been stabilized or started to decline after 1980. The main focus of the IGAC-APARE (East Asian/North Pacific Regional Experiment) is to understand the increasing human impact on the chemical processes occurring in the atmosphere of the East Asian Northwest Pacific Ocean. Scientists from Pacific Rim countries (China, Taiwan, Hong Kong, Japan, South Korea and Russia) as well as the U.S. and Australia have joined to steer the APARE activity.

The first intensive APARE field study was completed in September–October 1991, by the cooperative leadership of the U.S. National Aeronautics and Space Administration (NASA) and the Japanese National Institute of Environmental Studies (NIES). NASA's initiative, named PEM-West (Pacific Exploratory Mission), consisted of airborne measurements using the NASA DC-8 and ground-based studies. The DC-8 flew across the Pacific ocean and intensive data flights were conducted from operational bases established in Japan, Hong Kong and Guam. Data were obtained over a wide range of selected geographical areas and meteorological conditions. NIES organized a study called

PEACAMPOT (Perturbatioxn by East Asian Continental Air Mass to the Pacific Oceanic Troposphere), which consisted of airborne measurements using a CESSNA-404 and ground-based studies. The CESSNA-404 sampled the air in a selected area of the East China Sea, Yellow Sea and Japan Sea. Ground stations were situated on the eastern coast of China, and in Japan, Korea, Taiwan and Hawaii, as well as on several remote Pacific islands, and scientists from these involved countries joined the study.

The PEM-West measurements provided an extensive characterization of air masses over the western North Pacific Ocean including clean oceanic air, continental outflow, stratospheric air, equatorial marine air and southern hemispherical marine air. The major greenhouse gases, ozone and its precursors, sulfur species, and continental and marine aerosols were measured, and the airborne remote lidar system provided a two-dimensional mapping of the atmospheric ozone and aerosol distributions along the entire set of flight tracks. Unique results were obtained concerning the transport of trace gases in a major typhoon. The PEACAMPOT measurements characterized the air masses near the continental rim of East Asia.

INTRODUCTION

The main focus of IGAC's East Asian/North Pacific Regional Experiment (APARE) is to study chemical processes and long-range transport over the Northwest Pacific Ocean and to estimate the magnitude of impact of anthropogenic emissions (e.g., NO_x, SO_2, hydrocarbons), and other air pollutants on the oceanic atmosphere over this area.

The study of seasonal variations of aerosol composition at several remote islands in the Pacific revealed that sulfate, nitrate and mineral dust of Asian continental origin are transported as far as 5,000 km to Midway during the spring (Prospero and Savoie, 1989). Meanwhile, significant increase of tropospheric ozone between 1968 and 1990 have been observed at three Japanese island sites near the Asian Continent (Akimoto *et al.*, 1991). The latter authors have suggested that the ozone increase might be due to an increase in the emissions of NO_x and other species from anthropogenic sources in East Asia. If confirmed, these observations would indicate that the impact of human activity has already influenced the chemistry of the oceanic atmosphere in the Northwestern Pacific region.

Despite the obvious urgent need to make a systematic survey of the characteristics of the atmosphere in this area, there have been no intensive field experiments conducted in this region up to the time of the APARE program initiative. Clearly, the task of quantifying emissions, chemical transformations, and the transport of atmospheric constituents in this region necessitates international cooperation of scientists from the Pacific Rim Nations including those from China, Taiwan, Hong Kong, Japan, South Korea, U.S., Australia and Russia. All of the above countries have now joined in a common effort in pursuing the goals of the APARE program.

Specific tasks that have been defined to achieve the goals of APARE include:

1. develop an emission inventory of the region,
2. conduct surface measurements of environmentally important trace gases,
3. organize and carry out intensive field sampling programs, and
4. establish a ground monitoring network.

Concerning activity (1), emission inventories of SO_2 and NO_x derived from fuel combustion and industrial activities in Asia on a country-by-country base have been reported by Kato and Akimoto (1992). This is an ongoing study which is now focused on developing a latitude-longitude mesh data base with 1° × 1° resolution.

The first intensive APARE field study was completed in September–October 1991, under the cooperative leadership of the U.S. National Aeronautics and Space

Administration (NASA) and the Japanese National Institute for Environmental Studies (NIES). NASA's initiative, named PEM-West (A) (Pacific Exploratory Mission), consisted of both an airborne measurement component using the NASA DC-8 and ground-based study. A second field study organized by NIES called PEACAMPOT (Perturbation by East Asian Continental Air Mass to the Pacific Oceanic Troposphere) utilized a CESSNA-404 to make airborne measurements in conjunction with ground based studies. The CESSNA-404 sampled air in the East China Sea, Yellow Sea and Japan Sea.

This paper discusses emissions of SO_2 and NO_x from East Asian countries, and presents selected preliminary results from PEM-WEST (A).

ANTHROPOGENIC EMISSIONS OF SO_2 AND NO_x IN EAST ASIA

Table 1 shows the emissions of SO_2 and NO_x in East Asian Northwest Pacific Rim regions in 1980 and 1987. Briefly, the emissions were calculated by an energy balance method based on: (1) the fuel consumption data taken from the energy balance tables of the Organization of Economic Cooperation and Development/International Energy Agency (OECD/IEA) (IEA, 1990), (2) the production of non-ferrous metals, (3) the estimated sulfur content in fuels in each country, (4) the estimated retention factor of sulfur, and (5) selected emission factors of NO_x in each economic sector. A more detailed description of the methodology used in this analysis has been describe by Kato and Akimoto (1992). As shown in Table 1, total emissions of SO_2 and NO_x have increased from 9.10 to 11.68 Tg S/yr, and from 2.46 to 3.28 Tg N/yr respectively, from 1980 to 1987. Thus, both have increased by nearly a factor of 1.3 in seven years. It should be noted, however, that the emission rate of SO_2 has decreased in Taiwan, Hong Kong, Japan and South Korea during this period of time; whereas the emission rate of NO_x has continued to increase except for Japan. Emissions of SO_2 and NO_x in China have increased by a factor of ~1.5 during the same period. The molar ratio of sulfur to nitrogen varies between 4.46 in China to 0.84 in Japan. While these differences might affect the atmospheric composition and chemistry in different areas of the Northwest Pacific Rim Region, the overwhelming emission rate from China may be the major factor which defines the chemical composition of air transported into the northwest Pacific basin.

Table 1. Emissions of SO_2 and NO_x in East Asian Northwest Pacific Rim region.

	SO_2 (Gg S/yr)		NO_x (Gg N/yr)	
	1980	1987	1980	1987
China	6686	9995	1493	2243
Taiwan	518	302	68	99
Hong Kong	83	75	21	41
Japan	720	493	649	589
North Korea	136	166	117	143
South Korea	959	647	111	169
TOTAL	9102	11678	2459	3284

Table 2 presents province-based emissions of SO_2 and NO_x in China in 1980 and 1987. From these data it can be seen that the highest emission rates of SO_2 are centralized in inner land, Sichuan, and in the southeastern and coastal region along the Yellow Sea and East China Sea, Shangdong, Jiangsu, Liaoning, Hebei and Henan. High emissions of NO_x

are seen in the same coastal regions and in the south eastern part of China. The high SO_2 emission rate relative to NO_x in Sichuan province appears to be due to the use of coal with a very high sulfur content. The relatively high and rapidly growing SO_2 and NO_x emissions along the coastal areas of China now appear to be one of the most important factors influencing the atmospheric chemistry in the oceanic troposphere in the Northwest Pacific Rim region.

Table 2. Province-based emissions of SO_2, and NO_x in China.

	SO_2 (Gg S/yr)		NO_x (Gg N/yr)	
	1980	1987	1980	1987
1. Beijing	198	208	69	81
2. Tianjing	139	147	43	50
3. Hebei	375	533	107	157
4. Shanxi	268	419	71	118
5. Neimenggu	196	325	39	67
6. Liaoning	400	499	136	177
7. Jilin	145	176	61	80
8. Heilonjiang	220	291	84	122
9. Shanghai	294	326	78	91
10. Jiangsu	492	804	86	149
11. Zhejiang	116	207	31	60
12. Anhui	156	251	43	71
13. Jujian	75	123	19	34
14. Jiangxi	131	227	29	48
15. Shangdong	611	987	104	172
16. Henan	350	491	91	130
17. Habei	196	287	55	83
18. Hunan	200	275	56	82
19. Guandong	162	264	44	80
20. Guanxi	142	229	24	37
21. Sichuan	848	1289	81	120
22. Guizhou	218	376	22	35
23. Yunnan	219	374	23	40
24. Xizang	0.6	1.0	0.5	0.8
25. Shaanxi	306	509	37	60
26. Gansu	88	135	23	37
27. Qinghai	19	30	7.5	12
28. Ningxia	47	107	7.1	17
29. Xinjiang	73	105	21	32
TOTAL	6685	9995	1492	2243

The accuracy and consistency of emission inventory data is of great concern to a good understanding of atmospheric chemistry of the Pacific basin as well as to the transboundary air pollution problem. Table 3 compares the results of this study with other recent reports. From these data it can be seen that the SO_2 emission in China reported by Wang *et al.* (1993) is about 20% lower than our value. This difference most likely can be ascribed to the difference in the assumed average sulfur content in coal; they assumed 1.12% in weight as compared to our assumption of 1.35%. If we take into account this difference, the agreement is excellent. For Taiwan, our value is in excellent agreement with that reported by Liu *et al.* (1993) in spite of the latter authors using a totally different methodology, i.e. the former is based on an energy balance method and the latter on an individual source

Table 3. Comparisons of the anthropogenic emissions of SO_2 and NO_x as reported in literature sources.

	This Study	Literature Value (Year)	Reference
	SO_2 (Tg S/yr)		
China	10.00	9.22 (1988)	Wang *et al.* (1993)
		9.00 (1989)	Shen *et al.* (1992)
		9.45 (1986)	Foell and Green (1990)
Taiwan	0.30	0.30 (1988)	Liu *et al.* (1993)
		0.43 (1986)	Foell and Green (1990)
Hong Kong	0.075	0.14 (1986)	Foell and Green (1990)
Japan	0.49	0.47 (1986)	FEA (1987)
North Korea	0.1	0.30 (1986)	Foell and Green (1990)
South Korea	0.65	0.80 (1991)	Park (1993)
		0.61 (1986)	Foell and Green (1990)
	NO_x (Tg N/yr)		
China	2.24	2.39 (1988)	Wang *et al.* (1993)
		2.33 (1986)	Foell and Green (1990)
Taiwan	0.10	0.12 (1988)	Liu *et al.* (1993)
		0.10 (1986)	Foell and Green (1990)
Hong Kong	0.021	0.033 (1986)	Foell and Green (1990)
Japan	0.59	0.40 (1986)	JEA (1987)
North Korea	0.14	0.19 (1989)	Foell and Green (1990)
South Korea	0.17	0.27 (1991)	Park (1993)
		0.20 (1986)	Foell and Green (1990)

method. Agreement between our value for South Korea and that reported by Park (1993) is also satisfactory if we take into account the difference in the year of estimation.

Concerning NO_x, the level of agreement for China and Taiwan is satisfactory, whereas the agreement between our value for Korea and that by Park (1993) is not satisfactory and cannot be reconciled even taking into account the difference in the year of estimation. The reason for the discrepancy is not clear presently. The disagreement between the value reported in this study and by JEA (1987) for Japan is mainly due to different methodologies used in estimating emissions from automobiles. Data reported by Foell and Green (1990) are in general consistent with our values for both SO_2 and NO_x although a large discrepancy exists for values of Hong Kong and North Korea. For these regions, there is no detailed report available and the accuracy of this estimation cannot be evaluated at this time.

PEM-West (A): Overview of Field Campaign and Initial Results

The DC-8 flight component of the PEM-West (A) program was initiated on September 16, 1991. The primary and secondary scientific objectives of this program were:

- To investigate the atmospheric chemistry of O_3 and its precursors (NO_x, CO, CH_4, and NMHCs) over the Western Pacific, and to examine the natural budgets of these species and the impact of anthropogenic sources.
- To investigate some important aspects of the atmospheric sulfur cycle over the Western Pacific with emphasis on the identification of the relative importance of continental and oceanic sources.

As shown in Figure 1 the mission originated from NASA Ames Research Flight Center in San Francisco, California. Over the time period of September 16 to October 21, 1991, a total of 18 missions were flown, which involved 119 flight hours. Eight of these missions were "survey" type missions in which the aircraft was moving from one base of operation to another. The remaining ten flights were "intensive" missions in which specific aspects of the aforementioned scientific objectives were examined. The central staging sites for these "intensive" missions were: Yokota, Japan, Hong Kong, Guam, and Hawaii.

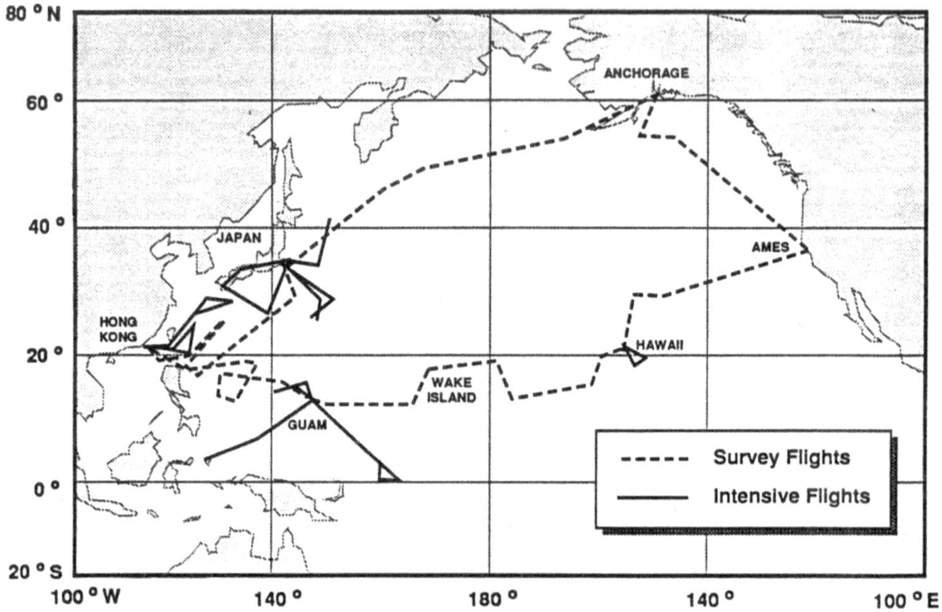

Figure 1. Flight tracks for the DC-8 aircraft during the PEM-West (A) mission.

To better understand the influence of anthropogenic emissions during the time of the airborne phase of PEM-West (A) and to get an improved picture of the "representativeness" of the airborne data, a series of ground-based stations were also operated before, during, and after the DC-8 flight operation. The ground based component, shown in Figure 2, utilized both the existing APARE/PEACAMPOT sampling network as well as several new stations that helped develop a more complete emissions picture for the PEM-West (A) overflight region.

Shown in Tables 4 and 5 (in the Acknowledgment section) is a listing of the chemical and meteorological parameters recorded on the DC-8 as well as those at several of the ground-based stations operated during PEM-West (A). Also presented is a listing of principal investigators (PIs) and their respective institutions. Of considerable importance in regard to the aircraft program are the indicated instrument time resolutions. Most of these were ≤5 min and many were operated at 1 Hz. Even in the case of the hydrocarbon grab samples, typically more than 140 samples were taken per flight. These data recording rates proved especially useful on the DC-8 in performing detailed horizontal and vertical profiling of trace gas species. In the case of ozone and aerosol scattering measurements, the in situ aircraft data were augmented by an on-board differential absorption lidar (DIAL), which extended the measurements of these species/parameters vertically upward to the stratosphere and downward to the ocean surface. The latter technique provided an extremely valuable assessment of mixing/transport processes occurring near the atmospheric boundary layer and near the tropopause.

From a meteorological perspective, PEM-West (A) provided an opportunity to examine the chemical characteristics of several different air mass types. Included in these were: midlatitude stratospheric air, midlatitude maritime, subtropical maritime, tropical maritime, subtropical typhoon, midlatitude modified continental outflow, subtropical continental outflow, and subtropical "convective" continental outflow. As expected during the PEM-West flight series, clear cases were found of both clean marine as well as polluted continental air. However, in many cases the meteorological picture was complex because of the widespread incidence of convection over much of the sampling region. Nevertheless, considerable consistency was found in most cases between chemical signatures and the likely path of the large-scale air mass motion based on isentropic trajectory analysis.

Although the analysis of the PEM-West (A) data base is still actively under way, those analyses completed to date already have led to several significant provisional findings. These are summarized below under several sub-topical headings.

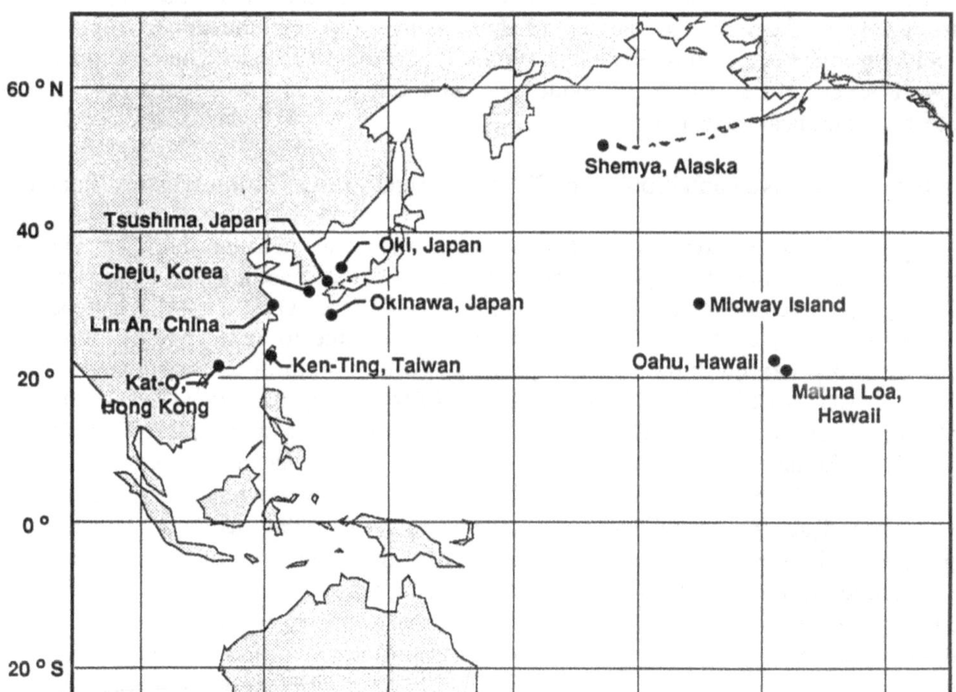

Figure 2. Locations of collaborative ground-based sites during APARE/PEM-West (A) mission.

Characterization of background Western Pacific air

By combining trajectory analysis with a photochemical signature derived from hydrocarbons and other species, the characteristics of background marine air have been contrasted with relative fresh outflow from the Asian continent. As expected, the contrast in the concentration levels of species observed in one to two day old continental air versus well-aged background Pacific air was quite dramatic in some cases. For CO the median value was 97 ppbv for the former and 72 ppbv for the latter; whereas, for C_2H_2, having a much shorter lifetime, the corresponding median ranged from 150 to 25 pptv. Similarly, the change in the median value for the long-lived species CH_4 was 1755 to 1685 ppbv; by contrast, C_6H_6 was found to range from a median value of 35 to 7 pptv. Compounds that had both significant photochemical sources and sinks (e.g., O_3, NO_x and H_2O_2) were not found to exhibit any clear tendency after a time period of ~2 days. For background free-tropospheric Pacific air NO_x ($NO + NO_2$) and O_3 mixing ratios averaged 40 pptv and 27 ppbv, respectively. By comparison, the median mixing ratio values for these same species for 1 to 2 day old air were significantly higher at 80 pptv and 55 pptv, respectively.

Evaluate the impact of a major typhoon on the transport and redistribution of trace gases

On September 27, 1991 the DC-8 aircraft made a flight to study the role of Mireille, a level IV typhoon, in the transport of trace gases. Observations made during this flight indicated that very substantial amounts of DMS, H_2O, and other trace species were entrained from the boundary layer and then transported into the upper troposphere, particularly in the eye wall region of the typhoon. On the other hand, contrary to a recent study (Rodgers *et al.*, 1990), no evidence was found for any downward intrusion of stratospheric air near the eye region.

Evaluate the relative contributions of different ozone sources in the Western Pacific

Using large-scale distributions of ozone and aerosols, as measured remotely via lidar techniques and augmented by in situ measurements of trace gases, seven major air mass types were observed and characterized during the PEM-West (A) field mission as related to O_3. For example, air that contained enhanced ozone due to stratosphere-troposphere exchange was observed in the upper troposphere (within 5 km of the tropopause) nearly 45% of the time for the latitude region of 20° to 40°N and approximately 70% of the time for the region of 40° to 60°N. Thus, the downward transport of stratospheric air observed at mid and highlatitudes was found to be a major source of O_3 over the Western Pacific. At lower latitudes, O_3 enhancements due to this exchange process occurred infrequently.

A second source of enhanced ozone was related to the eastward advection of continental air over the Western Pacific. In this case pollution sources with photochemically enhanced ozone were observed in association with aerosol-enhanced plumes directly advected over the ocean in the lower troposphere (typically below 6 km) and with low aerosol plumes in the mid to upper troposphere. Overall, continental outflow was found to be a major source of O_3 below 5 km in nearly 37% of the observations in the region 20° to 40°N and in approximately 21% of the observation in the region 0° to 20°N. Collectively, these results showed that the contributions of different air mass types to the tropospheric O_3 budget is a very strong function of latitude and altitude in the Western Pacific.

Estimate the regional sources and sinks of the major greenhouse gases: CO_2, CH_4, and N_2O

The high data collection rates for CO_2, CH_4, N_2O, and CO permitted a statistically meaningful analysis of the correlations among these species. Positive correlations were found among CO, CH_4, and N_2O, suggesting sources from generally common land areas. By contrast, CO_2 was anti-correlated with these gases, suggesting a terrestrial biogenic sink for this gas during the fall time period. The slopes of these correlation profiles were found to be proportional to the relative source strengths and sinks of these gases.

Estimate the impact of Pinatubo eruption on upper tropospheric levels of SO_2 and $SO_4{}^{2-}$ via stratospheric intrusions

Several observations involving air masses showing a significant influence of stratospheric air were found to have elevated levels of both SO_2 and $SO_4{}^{2-}$. These sulfur species were found to be correlated with stratospheric tracers such as O_3 and anticorrelated with tropospheric tracers such as CH_4, N_2O and H_2O. These observations show that three months after the eruption of Mt. Pinatubo, the amount of downward transported sulfur was approximately equivalent to that of NO_y transported from the stratosphere. This suggests that following a major volcanic eruption, the stratospheric flux of sulfur into the upper troposphere may have a significant influence on the levels of sulfur species observed at high altitudes.

Evaluate the effect of convective processes on the vertical profiles of trace gases

Observed vertical profiles of trace species strongly associated with land sources were frequently (more than 50% probability) found to exhibited higher mixing ratios in the upper troposphere (above 6 km) than the middle troposphere (2 to 6 km). Species showing these trends include CO, CH_4, N_2O, and most of the hydrocarbons and halocarbons. It was concluded that the observed vertical profiles are the result of major convective events over land followed by transport by strong westerly flow into the Western Pacific basin. Both O_3 and NO_y exhibited vertical profiles similar to the above trace gases; however, their gradients were significantly greater, indicating an additional source or sources in the upper troposphere or a greater sink in the lower troposphere.

Estimate the oxidizing capacity of the Western Pacific

Based on the results from approximately one thousand independent observations, photochemical box modeling calculations were used to explore the oxidizing characteristics of the Western Pacific. By filtering the observations according to a fixed zenith angle range (e.g., 30° to 55°), meaningful latitudinal and altitudinal trends were assessed for model estimated levels of OH as well as values for the ozone tendency, $P(O_3)$. These results showed that the highest levels of OH occurred at high latitudes (20° to 40°N) and low altitudes (0 to 3 km). By comparison, OH levels from 0° to 20°N, for the same altitude range, were nearly a factor of 3 times lower. Also, at all latitudes for the highest altitudes sampled (10 to 12 km), the average OH level was nearly a factor of 3 lower than those in the 0 to 3 km range. The trends in OH tend to reflect the distribution of the critical photochemical species O_3, NO and H_2O.

A similar analysis of the $P(O_3)$, showed a somewhat similar trend to OH as related to latitude; but, with the exception of a small sub-set of data collected in the polluted marine boundary layer near the coast of Japan and China, the altitudinal trend in $P(O_3)$ was found to be completely out of phase with that for OH. For example, with the exception of the coastal data, $P(O_3)$ was on average negative at all latitudes for altitudes less than 6 km. However, for the latitude range 20° to 40°N and altitudes above 6 km, $P(O_3)$ was nearly always positive. Between 0° to 20°N and above 6 km, $P(O_3)$ was found to be much lower in magnitude and to oscillate between negative and positive values. Overall, one of the most significant $P(O_3)$ results was the observed trend that positive values of this calculated parameter increased with latitude and altitude. Correlations with other photochemical species/parameters revealed that the mixing ratio for NO was the single most important factor controlling the sign and magnitude of $P(O_3)$. This finding is in general consistent with previous studies (Liu *et al.*, 1980; Chameides *et al.*, 1987; Carroll *et al.*, 1990).

Table 4. PEM-West (A) DC-8 Measurements.

Species or Parameter	Technique	Sample Frequency	Principal Investigator, Co-P.I.	Institution
NO	LIF	3 min.	John Bradshaw, Scott Sandholm	Georgia Tech.
NO	Chemi Lum.	~1 min.	Yutaka Kondo	Nagoya Univ.
NO_2	PF/LIF	3 min.	John Bradshaw, Scott Sandholm	Georgia Tech
NO_y	CC/LIF	3 min.	John Bradshaw, Scott Sandholm	Georgia Tech
NO_y	CC/Chemi Lum.	~1 min.	Yutaka Kondo	Nagoya Univ.
O_3	Chemi Lum.	<1 min.	Gerald Gregory, Brace Anderson	NASA (LaRC)
CO, N_2O, CH_4	TDL	<1 min.	Glen Sachse	NASA (LaRC)
O_3 (Remote)	UV Dial	<1 min.	Edward Browell	NASA (LaRC)
H_2O	Frost Pt. Hygrometer	<1 min.	DC-8 Support	NASA (ARC)
H_2O	Lyman Flour.	<1 min.	Kenneth Kelly	NOAA
H_2O_2	Enzyme Lum.	1 min.	Brian Heikes	U. of Rhode Island
Aerosol (Size, No., Distrib.)	Nollenberg P.	<1 min.	Gerald Gregory, Brace Anderson	NASA (ARC)
Aerosol (Remote)	Lidar	<1 min.	Edward Browell	NASA (ARC)
RO_2H	Enzyme Lum.	10 min.	Brian Heikes	U. of Rhode Island
DMS, COS, SO_2, CS_2	GC/QMS	5 min.	Alan Bandy, Donald Thornton	Drexel Univ.
PAN	GC/FID	5-10 min.	Hanawant Singh	NASA (ARC)
HNO_3, RCO_2H	Mist Chamber	~20 min.	Robert Talbot	U. New Hampshire
Aerosol (Comp.)	High Vol. Filter	~20 min.	Robert Talbot	U. New Hampshire
NMHC	Grab Sample	1-20 min.	Fumio Sakamaki	NIES
NMHC	Grab Sample	1-20 min.	Sherry Rowland, Donald Blake	U. of Cal., Irvine
CO_2	IR Absorption	<1 min.	Bruce Anderson	NASA (LaRC)
UV Radiation	Eppley Radiometer	~1 min.	John Barrick	NASA (LaRC)

ACKNOWLEDGEMENTS

The authors would like to thank the principal investigators listed in Table 4 and 5 for their contributions to the airborne and ground-based field measurement components of PEM-West (A) as well as several other science team members for their significant contributions to the meteorological and modeling analysis of these data. Included in the latter category are: Reginald Newell, John Merrill, Mark Shipman, Scott Bachmeir, Nien Dak Sze, Jose Rodriguez, Stuart McKeen, and Gregory Carmichael.

Table 5. Collaborating Ground-based Measurements during APARE/PEM-West (A).

Species or Parameter	Location	Principal Investigator	Institution
CLIMATE AND AIR QUALITY TAIWAN STATION (CATS)			
Aerosol Composition: SO_4, NO_3, Cl	Ken-Ting, Taiwan	J.C. Wu Gaston	Nat'l Taiwan Normal U.
C_2H_4, C_2H_6, C_3H_8, C_4H_{10}, C_3H_6, C_4H_8, C_5H_{12}, CO	Ken Ting, Taiwan	Tai-Ly Tso	Nat'l Tsing Hua Univ.
C_6H_{14}, C_6H_6, C_7H_{16}, C_7H_8, C_8H_{18}, C_8H_{10}, SO_2	Ken-Ting, Taiwan	Jiunn-Guann Lo	Nat'l Tsing Hua Univ.
O_3, NO_x	Ken-Ting, Taiwan	Martin Buhr Gloria Hsu	NOAA Nat'l Tawan Univ.
Surface Meteorology	Ken-Ting, Taiwan	Chung-Ming Liu	Nat'l Taiwan Univ.
PEACAMPOT			
CO, O_3	Oki, Japan	Hajime Akimoto	NIES
O_3	Okinawa, Japan	Hajime Akimoto	
O_3	Tsushima, Japan	Hajime Akimoto	
NO, NO_y	Oki, Japan	Daniel Jaffe	Univ. of Alaska
CHINA STATION			
O_3, SO_2, NO, NO_x, NO_y, HNO_3, NO_3, SO_4	Lin An, China	Michael Rodgers	Georgia Tech.
SO_2, O_3, NO, NO_2, NO_x, Black Carbon, Surface Met.	Lin An, China	Zhou Xiuji	Academy of Met. Sci.
PEM-WEST (A)			
Aerosol Composistion Al, Br, Ca, Cl, Cu, I Mg, Mn, Na, V	Cheju, Korea Kat-O, Hong Kong Ken-Ting, Taiwan	Richard Arimoto	Univ. of Rhode Island
Aerosol Composition Cl, NO_3, SO_4, Na, NSS, SO_4, MSA, NH_4	Midway Island Okinawa, Japan Shemya, Alaska	Joseph Prospero	Univ. of Miami
MLOPEX-II			
Halocarbons/$RONO_2$	Mauna Loa, Hawaii	Elliot Atlas	NCAR
CN, Nephelometer	Mauna Loa, Hawaii	Barry Bodhaine	NOAA/CMDL
CH_2O	Mauna Loa, Hawaii	Teresa Campos	NCAR
RO_2	Mauna Loa, Hawaii	Chris Catrell, Jack Calvert	NCAR
CH_4	Mauna Loa, Hawaii	Ed Dlugokencky	NOAA/CMDL
OH	Mauna Loa, Hawaii	Fred Eisele, David Tanner	Georgia Tech.
CCl_4, CFC-11, CFC-12, CH_3Cl_3, N_2O	Mauna Loa, Hawaii	Tim Elkins	NOAA/CMDL
CO, NMHC	Mauna Loa, Hawaii	J. Greenberg, S. Schauffler	NCAR
Formaldehyde, Peroxides	Mauna Loa, Hawaii	Brian Heikes, Brian McCully	Univ. of Rhode Island
GC/MS	Mauna Loa, Hawaii	Deiter Helmig, Walt Pollock	NCAR
Dpt., T, Winds	Mauna Loa, Hawaii	Gary Hebert	NOAA/CMDL
Trajectories	Mauna Loa, Hawaii	P.Hess, K.Mosher, S.Madronich	NCAR
NO_y, SO_2	Mauna Loa, Hawaii	Gerhard Hzübler	NOAA/CIRES
Peroxides (cont. HPLC)	Mauna Loa, Hawaii	Greg Kok, Tom Staffelbech	NCAR
HNO_3, NO_3	Mauna Loa, Hawaii	John Lind, Teresa Campos	NCAR
Dpt., T, Winds	Mauna Loa, Hawaii	Denise Montzka, Jim Walega	NCAR
CH_2O coil method	Mauna Loa, Hawaii	X. Zhou	Brookhaven Nat'l Lab.
Aldehyde catridge method	Mauna Loa, Hawaii	Y.-N. Lee	Brookhaven Nat'l Lab.
Carbonyls in precipitation	Mauna Loa, Hawaii	K. Mopper	Washington State U.

(continued on next page)

Table 5. (continued)

Species/Parameter	Location	Principal Investigator	Institution
HNO_3, NO_3, Org. Acids	Mauna Loa, Hawaii	Dick Norton	NOAA/Aeronomy Lab.
CO	Mauna Loa, Hawaii	Paul Novelli	NOAA/CMDL
O_3	Mauna Loa, Hawaii	Sam Oltmans	NOAA/CMDL
MN, PAN	Mauna Loa, Hawaii	Brian Ridley	NCAR
MO, NO_2, O_3	Mauna Loa, Hawaii	Brian Ridley, Jim Walega	NCAR
j(NO_2) actinometer		Rick Shetter	NCAR
j(NO_2) radiometer		Chris Cantrell	NCAR
j(O_3) actinometer			
UV biometer			
CO_2	Mauna Loa, Hawaii	Pieter Tans	NOAA/CMDL
CH_2O, HOOH	Mauna Loa, Hawaii	Gervase Mackay, D. Karecki	Unisearch
Eppley UV Radiometer	Mauna Loa, Hawaii	Jim Walega	NCAR
Radon, Thoron	Mauna Loa, Hawaii	Stewart Whittleston	ANSTO

NCAR = National Center for Atmospheric Research (U.S.); ANSTO = Australian Nuclear Science and Technology Organization; NOAA = National Oceanic and Atmospheric Administration (U.S.); CMDL = Climate Modeling and Diagnostics Laboratory (NOAA); CIRES = Cooperative Institute for Research in Environmental Sciences (NOAA); NIES = National Institute for Environmental Studies (Japan)

REFERENCES

Akimoto, H., H. Nakane and Y. Matsumoto, 1991, "Tropospheric ozone trend at selected sites in Japan," The Chemistry of the Atmosphere: Its Impact on Global Change, CHEMRAWN VII. December, 1991, Baltimore, U.S.

Carroll, M.A., et al., 1990, Aircraft measurements of NO_x over the eastern Pacific and continental United States and implications for ozone production, *J. Geophys. Res.*, 95:10,205-10,234.

Chameides, W.L., D.D. Davis, M.O. Rodgers, J. Bradshaw, S. Sandholm, G. Sachse, G. Hill, G. Gregory and R. Rasmussen, 1987, Net ozone photochemical production over the eastern and central North Pacific as inferred from GTE/CITE 1 observations during fall 1983, *J. Geophys. Res.*, 92:2131-2152.

Foell, W.K. and C.W. Green, 1990, "Acid Rain in Asia: An Economic, Energy and Emission Overview," Proceedings of the Second Annual Workshop on Acid Rain in Asia, November, 1990, Bangkok, Thailand.

JEA (Japan Environment Agency), 1987, "Comprehensive survey on the emission of air pollutants," Air Quality Bureau, Tokyo.

Kato, N. and H. Akimoto, 1992, Anthropogenic emissions of SO_2 and NO_x in Asia: emission inventories, *Atmos. Environ.*, 26A:2997-3017.

Liu, C.-M., H.W. Chen and S.-H. Fang, 1993, "Emission inventory of SO_2 and NO_x in Taiwan," International Workshop on Harmonization of Monitoring Techniques of Acid Deposition and Methodology of Emission Inventories of SO_2 and NO_x in East Asia, March 1993, Tsukuba, Japan.

Liu, S.C., D. Kley, M. McFarland, J.D. Mahlman and H. Levy, II, 1980, On the origin of tropospheric ozone, *J. Geophys. Res.*, 85:7546-7552.

Park, I.-S., 1993, "Emission inventories of SO_2 and NO_x in Korea," International Workshop on Harmonization of Monitoring Techniques of Acid Deposition and Methodology of Emission Inventories of SO_2 and NO_x in East Asia, March 1993, Tsukuba, Japan.

Prospero, J.M. and D.L. Savoie, 1989, Effect of continental sources on nitrate concentrations over the Pacific Ocean, *Nature*, 339:687-689.

Rodgers, E.B., J. Stout, J. Steranka and S. Chang, 1990, Tropical cyclone-upper atmospheric interaction as inferred from satellite total ozone observations, *J. Appl. Meteorol.*, 29:934-954.

Shen, D.-X. and D.-W. Zhao, 1992, "Emission of trace gases and reduction strategy in China," International Symposium on Emission Inventory and Prevention Technology for the Atmospheric Environments, March, 1992, Tsukuba, Japan.

Wang, W.-X., S.-X. Hong and T. Wang, 1993, "Anthropogenic emission inventories of SO_2 and NO_x in China," International Workshop on Harmonization of Monitoring Techniques of Acid Deposition and Methodology of Emission Inventories of SO_2 and NO_x in East Asia, March 1993, Tsukuba, Japan.

BIOMASS BURNING IN THE GLOBAL ENVIRONMENT: FIRST RESULTS FROM THE IGAC/BIBEX FIELD CAMPAIGN STARE/TRACE-A/SAFARI-92

M.O. Andreae[1], J. Fishman[2], M. Garstang[3], J.G. Goldammer[1], C.O. Justice[4], J.S. Levine[2], R.J. Scholes[5], B.J. Stocks[5], A.M. Thompson[4], B. van Wilgen[6] and the STARE/TRACE-A SAFARI-92 Science Team[7]

[1]Biogeochemistry Department
Max Planck Institute for Chemistry
P.O. Box 3060, D-55020 Mainz, Germany
[2]NASA Langley Research Center
Hampton, Virginia 23665, USA
[3]Department of Environmental Sciences
University of Virginia
Charlottesville, Virginia 22903, USA
[4]NASA Goddard Space Flight Center
Greenbelt, Maryland, USA
[5]Forestry Canada
Great Lakes Forestry Center, Sault St. Marie, Ontario Canada
[6]Forestek/CSIR, Stellenbosch, South Africa
[7]See acknowledgment section for listing of participants

ABSTRACT

Biomass burning is now recognized as a major source of important trace gases, including CO_2, NO_2, CO and CH_4, and of aerosol particles. It takes on many forms: burning of forested areas for land clearing, extensive burning of grasslands and savannas to sustain grazing lands, burning of harvest debris, and use of biomass fuel for heating.

The emissions from biomass burning represent a large perturbation to global atmospheric chemistry, especially in the tropics. Here, satellite observations have shown high levels of O_3 and CO over vast areas of Africa, South America, and the tropical Atlantic and Indian Oceans. Recent studies have linked this phenomenon to biomass burning plumes, and demonstrate that pyrogenic emissions affect regional ozone concentrations and the oxidative characteristics of the tropical atmosphere. The particulates affect regional global radiation budgets by their light-scattering effects and their influence on cloud microphysical processes.

Fire also has both short- and long-term effects on trace gas emissions from affected ecosystems which, for instance in the case of CO_2 and N_2O, may be more significant than their immediate release during the fire. Fire also alters the long-term dynamics of the cycling and storage of elements within terrestrial ecosystems, thereby altering their significance as sources or sinks of various trace gases. Finally, deposition of compounds produced by biomass burning on pristine tropical ecosystems may affect their composition and dynamics.

The importance of biomass burning is well recognized in IGAC, and one of the initial IGAC Activities, Biomass Burning Experiment: Impact on the Atmosphere and Biosphere (BIBEX), is focused on this phenomenon. It is aimed at quantifying the extent of the temporal and spatial distribution, dynamics, species emissions, and atmospheric consequences of biomass burning.

The first large, coordinated program being conducted under BIBEX is the Southern Tropical Atlantic Regional Experiment (STARE), an aircraft and ground-based measurement program to investigate the sources of trace gases, their atmospheric transport, and the chemical processes in the atmosphere, which lead to elevated levels of ozone, CO, and other trace gases over the southern tropical Atlantic Ocean. During August–October 1992, the first STARE field campaign took place successfully. Approximately 160 researchers from 14 nations participated in the field experiments, using eight aircraft and numerous ground-based measurement systems. Some measurement sites have remained in operation in order to investigate the seasonal and longer-term trends in the distribution of pyrogenic emissions and ecosystem behavior. High levels of O_3, CO, NO_x, and other trace gases, as well as aerosol particles were found in the entire region from South America to southern Africa, and could be related to forest and savanna fires on the two continents. Initial results are presented in this paper, together with a discussion of plans for future BIBEX experiments.

INTRODUCTION

Research conducted mostly during the last decade has shown that large-scale vegetation burning has considerable effects on the characteristics of the atmosphere and consequently on regional and global climate (Crutzen and Andreae, 1990; Andreae, 1993; Crutzen and Goldammer, 1993; Lacaux *et al.*, 1993). As a result, environmental scientists have become increasingly interested in the effects of biomass burning on the global environment. Since the second half of the 1980s, the threat to tropical forests has been brought to the public's attention, and has become increasingly a subject of international environmental politics. Fire is used primarily in the conversion of tropical forests into other types of land-use, whether it is used for the traditional but rapidly increasing practice of clearing with fire employed by farmers in preparing new fields when their old ones have been exhausted, or whether it is for large scale clearing to create pasture for cattle ranching or fields for plantations.

Besides using fire to clear forested areas, local populations set fires regularly and over large areas in several vegetation types. These are primarily the savannas (tree, bush, and grass) of the tropics and subtropics that comprise an area worldwide of 2.6 billion hectares (~16 million km^2), and which are burned every 1 to 5 years on average (Table 1). Fire is also used extensively in agriculture, especially in the tropics, for example to burn rice straw and sugar cane residue.

Within IGAC's Effects of Biomass Burning on the Global Atmosphere (BIBEX) Activity, an international team of scientists is currently investigating the impact of biomass burning on the atmosphere and climate through integrated field measurement campaigns.

Starting in 1990, burning in the tropics, especially Africa, was targeted, and beginning in 1993, fires in the boreal forests of Siberia are also being investigated. Results from the 1990 FOS/DECAFE (Fire of Savannas/Dynamicque et Chimie Atmosphérique en Forêt Équatoriale) experiment, which took place in Côte d'Iviore, have been published recently (Lacaux *et al.*, 1993a, b). In 1993, the measurement campaign FIRESCAN (FIre RESearch Campaign Asia-North) focused on forest fires in Siberia. Thirty participants from eight nations studied fire behavior and emissions from a fire in a Siberian pine forest (FIRESCAN Science Team, 1993).

Table 1. Global estimates of annual amounts of biomass burning and of the resulting release of carbon to the atmosphere (Andreae, 1993).

Burning Source	Biomass Burned (Tg dm/yr) Tropical	Extratropical	Total	Carbon Released (Tg/a)
Forest	1260	1150	2410	1080
Savanna	3690	—	3690	1660
Biomass Fuel	1720	220	1940	880
Charcoal	20	1	21	82
Agricultural Waste in Fields	420	420	850	380
World Total	7110	1790	8910	4080

The goal of these international measurement campaigns is to conduct research on the effects of fire in the various ecosystems, and simultaneously to determine the composition, amount, and fate of emissions from these fire events. Sampling and measurement of emissions takes place on the ground, as well as from research aircraft flying at various altitudes above the fires and at various distances from the fires. This strategy allows not only the amount and chemical composition of the gaseous and particulate emissions to be measured, but also the transport and potentially their further fate, such as chemical reactions within the atmosphere. A long term goal is to be able to determine the area and type of vegetation burned through remote sensing measurements. This is essential, because, while the areas of forest and scrub land burned in Europe and North America are relatively accurately reported, the reports from the rest of the world are only projected estimates that require verification. Consequently, the estimates of savanna burned yearly range from about one half to one billion ha per year. Data from the former Soviet Union are also very inaccurate, ranging from 1 to 10 million ha per year.

From August to October of 1992, after two years of preparation, the first field measurements were conducted in South America and southern Africa in the context of the BIBEX project Southern Tropical Atlantic Region Experiment (STARE). This project had two subcomponents, TRACE-A (Transport and Atmospheric Chemistry near the Equator-Atlantic) and SAFARI-92 (Southern Africa Fire/Atmosphere Research Initiative). STARE was inspired by a large-scale phenomenon: the concentrations of ozone, carbon monoxide, and other pollutants that build up annually in the troposphere over the South Atlantic (Fishman *et al.*, 1991), which had been discovered by satellite remote sensing. In contrast to the ozone hole that develops seasonally over the poles in the stratosphere, these seasonally strongly elevated ozone concentrations in the tropics occur in the troposphere. The highest values of tropospheric ozone, up to 50 Dobson units, were found in what had once been thought of as "pristine air," without industrial influence, over the South Atlantic between South America and southern Africa. With increasing knowledge about the use of fire in tropical land ecosystems the hypothesis that the "ozone high" over the South

Atlantic and the surrounding countries is a result of biomass burning in the Amazon and of savanna fires in southern Africa has gathered strength. In recent years, experiments in Brazil, the Congo, and Côte d'Ivoire have shown that in the areas of high fire activity photochemical smog is formed which contains high levels of ozone (Andreae *et al.*, 1988, 1992, 1993; Cros *et al.*, 1988, 1992; Kirchoff *et al.*, 1989).

STARE/TRACE-A/SAFARI-92 was the first intercontinental, international, and interdisciplinary fire experiment. This campaign was intercontinental because the effects of forest and savanna fires in South Africa and South America on the atmosphere over and between the two continents were studied. TRACE-A involved some 200 scientists from the following nations: Brazil, Congo, South Africa, U.S. and Great Britain, and had as its main objective the exploration of the large-scale atmospheric composition of the region extending from Brazil over the South Atlantic to southern Africa. Its research was centered around the NASA DC-8 instrumented research aircraft.

The international character of SAFARI-92 was reflected in the participation of over 150 scientists from 14 nations, its interdisciplinary character through the comprehensive representation of many scientific disciplines. Besides research groups from biogeochemistry, atmospheric chemistry and fire ecology, there were research groups representing the disciplines of forestry, meteorology, climatology, soil science, pasture science, and microbiology. Researchers from Germany, Swaziland, South Africa, Zimbabwe, Zambia, Namibia, Congo, Brazil, Belgium, France, Great Britain, Canada, and the U.S. participated.

An important focus for the measurement campaign was South Africa, where controlled burns took place within the Kruger National Park in savannas whose history and ecology has been well documented by local scientists. The climax of the experiments in September 1992 was the execution of two controlled savanna fires each in the size of about 2000 ha. This size of experimental savanna fire is a typical size for naturally occurring lightning induced fires, and for "management" fires that have been set by Africans for millennia. Such large fires made possible, besides the ecology work at ground level, the measurement of emissions on the ground and from research aircraft. Emission measurements were also made over sugar cane fields in Swaziland, where, before harvest, large fields are regularly burned creating thick smoke plumes. Emission measurements were also made in the plumes from many savanna "fires of opportunity," which were encountered by the instrumented DC-3 aircraft on its atmospheric measurement flights over southern Africa. The primary purpose of these flights was to assess the impact of biomass burning on the regional atmospheric composition over the subcontinent. These measurements were supplemented with measurement series conducted at several ground sites. In Etosha National Park, Namibia, a group of researchers from the U.S. and South Africa recorded the vertical distribution of ozone, temperature, and humidity, and the composition of dust aerosols. At a similar station in Victoria Falls in Zimbabwe, a group from Germany, Zimbabwe and South Africa measured the ozone concentration, aerosols, and the meteorology of the atmospheric boundary layer. The following sections contain a brief discussion of some initial results from STARE/TRACE-A/SAFARI-92.

THE EFFECT OF VEGETATION AND SOILS ON FIRE REGIME AND PYROGENIC EMISSIONS IN SOUTHERN AFRICA

To understand the role of fires on the atmospheric environment, a quantitative understanding of the fire regime on the African continent is essential. The term "fire regime" refers to the frequency, season and intensity of fires that prevail in a region. It is related to the basic pattern of vegetation distribution in southern Africa, which is controlled by the interaction of climate and soils. This pattern ultimately controls fire frequency, fuel load and fuel type, which in turn determine the nature, amount, timing and distribution of emissions. Natural fires are caused by lightning, but with expanding human populations, the relative importance of humans as a source of ignition is increasing. The reasons for setting fires, and the areas involved, are poorly quantified at present.

The frequency of occurrence of fire is proportional to the rate of fuel accumulation, which depends on the primary productivity of the area, which in turn is linked to rainfall. The climates of southern Africa range from very wet near the equator, to very arid in the southwestern part of the subcontinent. Rainfall seasonality is generally strong, and the prolonged dry season is the primary reason for the widespread occurrence of fire. Rainfall in the equatorial region is only weakly seasonal, and the vegetation there seldom burns naturally. A small part of the southern Cape receives winter rainfall, and supports a species-rich fire-dependent vegetation type called fynbos.

The soils of southern Africa can be broadly classified into fertile and infertile types. This is partly due to their geological origin, but is mostly a consequence of the erosional history of the subcontinent. Soils on the ancient erosion surfaces of the high interior of the continent are generally infertile, while the soils of the younger surfaces around the edges are fertile. The widespread occurrence of infertile soils is the second reason for the large pyrogenic emissions: grasses growing on these soils are indigestible for cattle and wildlife, and are therefore burned to promote new and more palatable growth.

Savannas are the most extensive vegetation type supporting fire in Africa. Frequent intense fires, on the other hand, are the primary reason for the occurrence of savannas as a vegetation type. The savannas on infertile soils, especially in the higher-rainfall areas of central Africa, accumulate large amounts of grass fuel. They burn annually to biennially, and provide most of the pyrogenic emissions. The fertile savannas are also more arid, and the grass, once mature, is mostly eaten by cattle and wildlife, so fires are less frequent. In savanna fires the principal fuel is dry grass, plus a small amount of tree leaf litter and fallen wood.

In savannas, most combustion takes place in the flaming phase, so emissions of methane and carbon monoxide are low relative to fires in forested biomes. The nitrogen content of the grass is very low, especially in infertile savannas, resulting in relatively low, but ecologically very significant nitrogen emissions. Grasslands make up the second largest burned area, and again those on infertile soils burn most frequently and intensely. Burning is part of the cropping system in infertile savannas and on the forest fringe, but the total areas under cultivation are small. Much of the interior and west of the southern part of Africa consists of arid shrub lands and desert grasslands, which seldom burn. The fynbos is confined to mostly very infertile soils in winter rainfall areas, but occurs also in high mountain regions elsewhere. It burns once every 10 to 40 years with a high fuel load of woody material.

Areas with higher rainfall can burn annually, whereas areas with low or erratic rainfall burn much less frequently, sometimes only once in 10 years. A geographical quantification of fire frequency is needed. Fires occur in the dry season, which varies geographically. For example, fires occur along the southern fringe of the Sahara in January, in central Africa in June, and in east Africa in September (Cahoon *et al.*, 1992). In southern Africa, savanna and grassland fires are concentrated in the southern winter (July–September). while fynbos fires occur in summer (December–February) (van Wilgen *et al.*, 1990).

Fire intensity is important for atmospheric dynamics, since it is directly related to the rate of energy release and thus the combustion chemistry. Fire intensity is related to fuel type (the structure of the vegetation) and moisture content, its continuity in space, and the weather conditions under which fires would burn.

Finally, the size of fires is also important. Highly settled areas, or intensively managed areas, are characterized by smaller fires. Fires in sparsely settled areas can be very extensive, covering tens of thousands of hectares per fire. The SAFARI initiative has started to shed light on these issues, but more is needed, and will follow as the program develops.

FUELS AND FIRE BEHAVIOR DYNAMICS ON LARGE-SCALE SAVANNA FIRES

The SAFARI-92 Experiment was the first major investigation of atmospheric emissions from biomass burning that included a ground component to analyze fuel characteristics and fire dynamics, thereby recognizing the important link between fire behavior/impact and emissions to the atmosphere. As a part of the SAFARI-92 ground component, fuels and fire behavior parameters were documented on two large (>2000 ha) experimental fires carried out in the arid savannas of Kruger National Park in northeastern South Africa. Pre-fire fuel loads and distribution, fuel consumption, fire spread rates, flame zone characteristics, and in-fire and perimeter wind field dynamics were measured to determine total energy release rates for each fire. This information was combined with atmospheric profile data to project convection column dynamics in support of airborne trace gas and particulate measurements.

Hand ignition of both fires followed the same pattern, with backfires being set along the leeward side of each block, followed immediately by headfire ignition along the windward edge. Grass fuel loads averaged approximately 4 tons/ha on each fire, with 92% of this being consumed during flame front passage. Grass moisture content levels for each fire were in the 8–9% range. Headfire spread rates, measured over runs 1–3 km in length, were higher on the second fire (5.8 km/h) than the first (2.3 km/h), due primarily to a slight variation in the ignition pattern sequence between the two fires. Flame lengths often exceeded 4–5 m in wind gusts, and strong in-fire horizontal and vertical wind speeds were recorded. Energy release rates, as determined through the calculation of frontal fire intensities, averaged 4000 kW/m for the first fire and 11,000 kW/m for the second. Despite upper air conditions that were favorable for convection column development, the low energy release-rates on the first fire translated into a weak and poorly-defined smoke plume, while the more vigorous second fire created a well-developed, 2500 m high column with a capping cumulus top.

The ground based studies of the savanna fires were made more difficult due to an extreme drought within the region, which was declared the most intense in a hundred years. In contrast to years with normal precipitation, very little grass had grown in most of the savannas of South Africa, Mozambique, southern Zambia, Zimbabwe, Botswana, and Namibia. Consequently, significantly less fuel was available to burn than in normal years. Under these conditions neither the experimental fires nor the otherwise normally occurring fires in southern Africa could spread as evenly or as extensively as they would in a non-drought year. In some areas, there was not enough biomass to even sustain a fire.

REMOTE SENSING

There is an increasing awareness of the role that satellite data might have in monitoring fire activity and contributing to improved regional to continental scale estimates of trace-gas emissions (Justice *et al.*, 1992; Cahoon *et al.*, 1992). Daily satellite data can provide information on the distribution and timing of fires as well as on the basis for estimating the areal extent of the burning. The SAFARI-92 Experiment was used to improve our understanding of current capabilities for satellite fire detection and to examine how satellite data can contribute to regional trace gas emission estimates. Advanced Very High Resolution Radiometer (AVHRR) 1 km data from the NOAA-11 satellite and data from the Landsat Thematic Mapper were collected by the South African ground receiving station at Hartebeeshook during the SAFARI-92 field program. Data from the Operational Linescan System of the Defense Mapping Satellite Program (DMSP), which provides night-time information on light sources including fires, were also collected during this period. Research groups at the University of Maryland, University of Virginia, NASA-Goddard Space Flight Center and NASA-Langley Research Center are currently analyzing these satellite data to develop and test algorithms for fire detection. Daily data on fire distributions for southern Africa during the SAFARI-92 Experiment are being used as an input to transport modeling studies. One of the two large controlled burns in the Kruger National Park during SAFARI-92 was detected by the AVHRR satellite and provides a useful calibration point (Figure 1); the other was obscured by cloud.

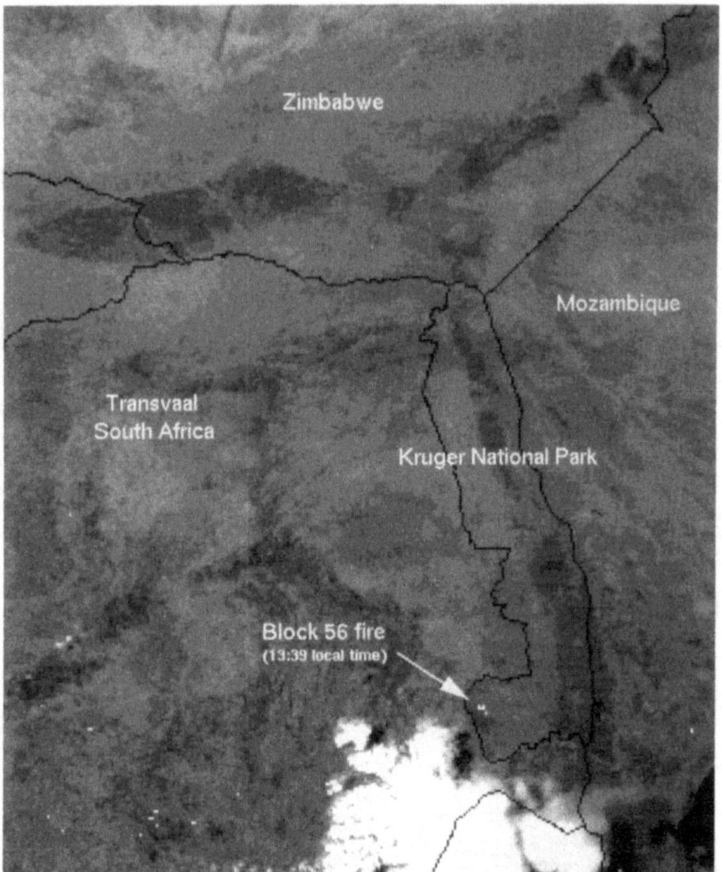

Figure 1. NOAA/AVHRR derived active fires in Transvaal province (S. Africa) and southwestern Mozambique on 18 September 1992. The high temperature active fires are identified using the middle and thermal infrared channels of the AVHRR instrument. The fires, shown in white overlay, are super-imposed on data from the near-infrared channel. (The large white area in the bottom center represents clouds).

EMISSIONS OF TRACE GASES AND AEROSOLS

Trace gas emissions from the experimental fires in Kruger National park were analyzed by ground-based sampling and measurement directly at the fire front by teams from the Universities of Toulouse and Paris, and by aircraft at different heights in the fire plume, using a helicopter (NASA), a Cessna 310, and an instrumented DC-3 (both from Max Planck Institute for Chemistry). The DC-3 was also used to measure the emission characteristics of numerous savanna and grassland fires set by the local population in South Africa, Zambia, Zimbabwe, and Angola, as well as the composition of smoke from sugar can fires in Swaziland. Measurements by the NASA DC-8 in large smoke plumes over northern Zambia complemented these data sets.

The large number of investigators and the diversity of research platforms involved in STARE/TRACE-A/SAFARI-92 made possible the most comprehensive analysis of fire emissions made in any field experiment to date. The gaseous and particulate chemical species analyzed in the fresh smoke are listed in Table 2.

The emissions showed the characteristics typical of highly efficient combustion, dominated by the flaming phase, and with relatively little smoldering. The $CO:CO_2$ emission ratio was low (~0.05) similar to previous observations from savanna regions (Ward *et al.*, 1991; Lacaux *et al.*, 1993a, b). Consequently, the emission ratios for the hydrocarbons were also as low, and the NO_x:hydrocarbon ratio relatively high, providing a basis for efficient ozone production. An extremely complex mixture of hydrocarbons and partially oxidized hydrocarbons was observed in the smoke. For the first time, the emissions of methyl bromide and methyl iodide, as well as several minor hydrocarbons from fires could be quantitated in the field. Figure 2 shows the correlation between methyl chloride and carbon monoxide observed in the smoke from the Kruger Park savanna fires.

Table 2. Gaseous and particulate species measured during STARE/TRACE-A/SAFARI-92.

Gases						Particulates	
aromatics	alkanes	NO	CO_2	O_3	H_2	total carbon	total mass
aldehydes	alkenes	NO_2	CO	CH_3Cl	H_2O_2	black carbon	soluble ions
carbonyls	alkynes	NO_y	COS	CH_3Br	H_2O	carbon isotopes	number size distribution
organic acids	alcohols	NH_3	PAN	CH_3I	HNO_3	nitrogen isotopes	volume size distribution
various halocarbons		SO_2	PPN	CH_4		trace elements	

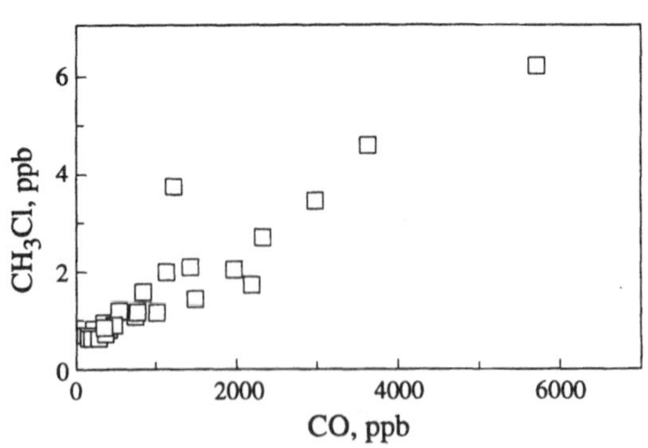

Figure 2. Methyl chloride (CH_3Cl) versus carbon monoxide (CO) in smoke and background air samples collected during the experimental burns in Kruger National Park, South Africa.

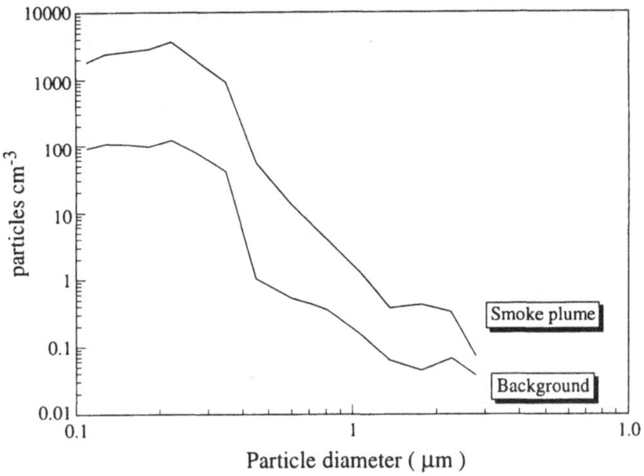

Figure 3. Aerosol particle size distributions in fresh smoke from an experimental fire in Kruger National Park, South Africa, and in background air outside of the smoke plume.

Aerosol particle number/size distributions showed a maximum at about 0.2 µm diameter in the fresh smoke, which increased slightly with aging of the plume (Figure 3). The aerosols were highly enriched in carbon, potassium and chlorine, typical characteristics of pyrogenic emissions.

Besides the emission measurements taken at the Kruger Park in South Africa, a collaborative program of field measurements was undertaken in Zambia by scientists from the U.S. Forest Service, Oregon State University, the University of Maryland and the University of Zambia. Emission factors from representative vegetation types were measured during controlled burns along with the biomass before and after the burns. These data are being used together with daily satellite data and an emissions model developed by the South African Forestek as part of a research program to develop regional estimates of trace gas emissions for southern Africa.

SOIL EMISSIONS

Biogenic emissions resulting from microbial soil processes are an important global source of the nitrogen oxides, nitric oxide and nitrous oxide. Several years ago, it was discovered that following burning, soil emissions of these two gases in midlatitude ecosystems were significantly enhanced. As part of the SAFARI-92 ground component, scientists from the NASA Langley Research Center, in collaboration with South African scientists, obtained the very first measurements of biogenic soil emissions of nitric oxide in southern African savannas. These flux chamber measurements were obtained before and after burning, and before and after artificial wetting of the flux sites. These measurements showed that wetting and burning have a very significant impact on the biogenic soil emissions of nitric oxide from the African savanna grasslands. Measurements of biogenic soil emissions of nitrous oxide were also attempted. One of the surprises of SAFARI-92 was the lack of a measurable flux of nitrous oxide, either before or after wetting and burning. The lack of a detectable nitrous oxide flux means that the flux was less than 2 nanogram of nitrogen per square meter per second (ng N $m^{-2}s^{-1}$), which is the minimum detectability of the instrumentation.

The pre-burn emissions of nitric oxide ranged from 1 to 2 ng N $m^{-2}s^{-1}$ from the dry sites to 5 to 12 ng N $m^{-2}s^{-1}$ from the wetted sites. After burning, the nitric oxide emissions from the dry sites ranged from 5 to 18 ng N $m^{-2}s^{-1}$, and from the wetted sites ranged from 35 to 75 ng N $m^{-2}s^{-1}$. The post-burn emissions of 75 ng N $m^{-2}s^{-1}$ from the wetted sites are among the highest emissions ever measured from natural sites and are comparable to emissions from agricultural lands right after fertilization.

Among the two most interesting results from the SAFARI-92 soil flux measurements are the lack of measurable emissions of nitrous oxide and the very strong post-burn emissions of nitric oxide. Both results may be related to the very severe drought experienced in South Africa preceding the SAFARI-92 Experiment. Since these measurements represent the first ever obtained in the savanna grasslands of South Africa, additional measurement during non-drought conditions are needed.

Indirect evidence of high emissions of NO from savanna soils was obtained during two of the regional atmospheric chemistry survey flights on the instrumented DC-3 aircraft (see below). High concentrations of nitrogen oxides were found at low altitudes (<100 m) over

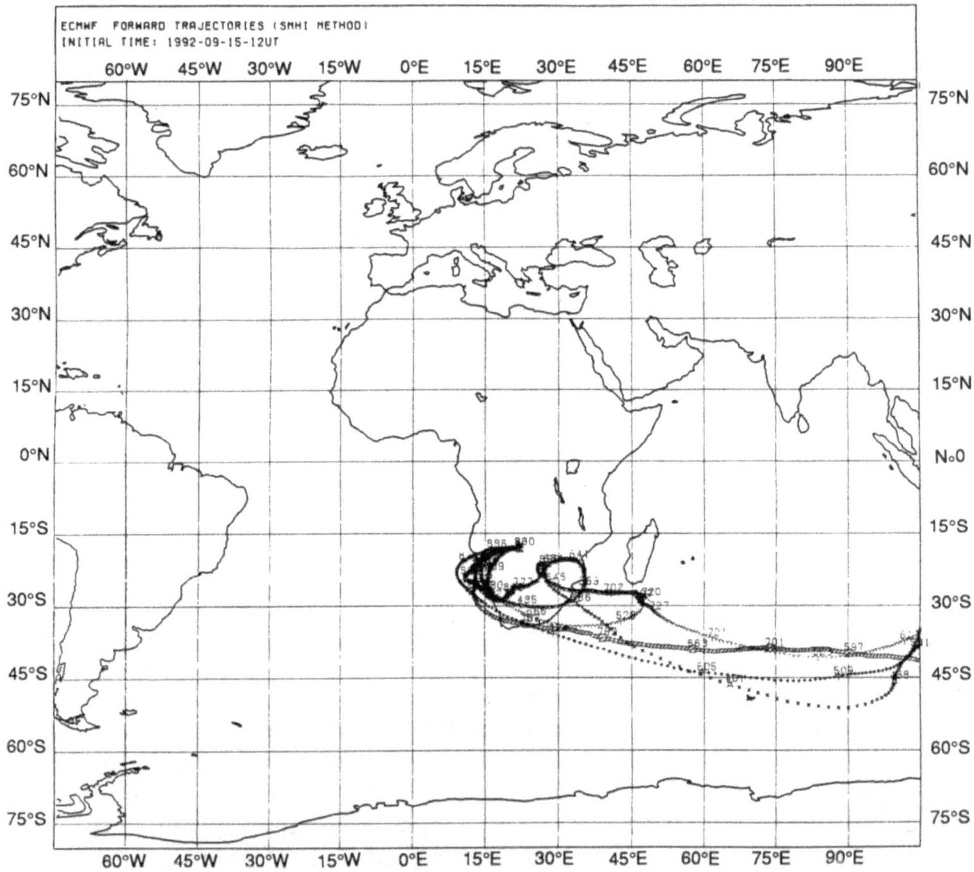

Figure 4. Airmass forward trajectories showing transport from southern Africa towards the Indian Ocean. Note that air parcels may initially leave Africa over the Atlantic coast, but then change direction and move toward the Indian Ocean. The trajectories originate at Victoria Falls (Zimbabwe) at 1200UTC on 15 September 1992. The figure shows a collection of 5 (starting at the surface, 850, 800, 700 and 500 mb) 10-day 3-D forward trajectories, which were calculated by P. Kållberg at the Swedish Meteorological and Hydrological Institute using ECMWF flow field analyses.

the savanna in northern Namibia, with a sharp decrease of NO concentrations with altitude. These high NO levels were not correlated with tracers of fossil fuel or biomass burning emissions, such as CO, excluding a pyrogenic or fossil fuel origin for the NO. Since rains had occurred in this area in the days preceding the flights, it is likely that water-stimulated soil emissions caused the high NO levels observed.

Soil emissions of CO and CO_2 were investigated by a team from the U.S. Environmental Protection Agency. They found that emissions of CO_2 from the soils were not influenced by burning, but showed a dramatic increase following a simulated rain event. CO emission, on the other hand, increased strongly after a fire, especially when water was added to the soil. This effect disappeared again after a few days, however. CO fluxes in Kruger Park were considerably larger than values that had been previously reported from savanna sites. There was not significant methanogenic or methanotrophic activity in the soils before or after fires, even when they were wetted. This is attributed to low bacterial populations due to drought. During the rainy season following the experiment, the soils consumed methane at a rate of 1 to 2 mg CH_4–C $m^{-2}d^{-1}$.

ATMOSPHERIC CIRCULATION AND TRANSPORT OVER SOUTHERN AFRICA

The meteorological environment over southern Africa was investigated using standard meteorological data obtained from the Weather Service, augmented by detailed investigations at several sites, especially at Etosha Park, Namibia, and Victoria Falls, Zimbabwe. At these sites, the thermodynamic structure of the boundary layer was studied as well as the wind regimes and transport mechanisms for pyrogenic emissions in the troposphere.

From early winter (May) to spring (October) the circulation fields over southern Africa are dominated by anticyclonic flow. During SAFARI-92 anticyclonic circulation dominated 60% of the time. By October 1992 westerly based troughs were increasing in frequency. These dominant anticyclonic circulation fields resulted in three typical transport patterns in the lower (surface to 3 km) and middle (3 to 6 km) troposphere: (1) Indian Ocean transport (Figure 4; air from the African continent is ejected over the Indian Ocean), (2) continental gyre (Figure 5; air is being recirculated over the continent), and (3) Atlantic transport (Figure 6; air is being ejected over the Atlantic Ocean).

Mixed modes of transport that consisted of more than one of the above circulation modes being present, at either different levels in the lower troposphere or at different locations on the subcontinent, occurred more than 50% of the time on any given day. On more than half of these occasions, air parcels exited at either different levels or different locations into both the Indian Ocean and the Atlantic Ocean. Air parcels entering the oceans on these trajectories therefore (1) originated at locations well removed from the points of exit and do not reflect only input at the immediate point of exit, and (2) are chemically aged and have been subjected to chemical reactions and transformations over a period of days in an environment of high solar insolation.

For about one-third of the time during SAFARI-92, air parcels, after spending 5 to 10 days in an anticyclonic gyre over the subcontinent, exited the subcontinent on the east coast into the Indian Ocean. The remaining roughly 20% of the trajectories crossed the subcontinent from west to east entering the Indian Ocean, and spending as little as 24 hours over land. Only a few trajectories during SAFARI-92 followed a direct E to W path from the continent into the Atlantic Ocean. Eastward and westward transports were at times quite protracted reaching eastward almost around the southern hemisphere in 10 days and westward into the Amazon Basin. Exit from the subcontinent, however, tended to concentrate over Angola and over central and northern Natal.

REGIONAL ATMOSPHERIC CHEMISTRY

Despite the drought in southern Africa and the resultant reduced biomass, researchers found very high concentrations of ozone, carbon monoxide, oxides of nitrogen, formaldehyde, and smoke particles. These atmospheric pollutants originated in the smoke plumes of the large and numerous savanna fires that were burning primarily in southern Africa, particularly in countries such as Zaire, Zambia, Angola, and Tanzania. The smoke aerosols were often so thick that the ground was difficult to identify from an altitude of 2 to 3 km. Ozone concentrations of up to 100 ppb were encountered up to 4 km, the maximum flight altitude of the unpressurized DC-3 (Figure 7). Later flights with the NASA DC-8

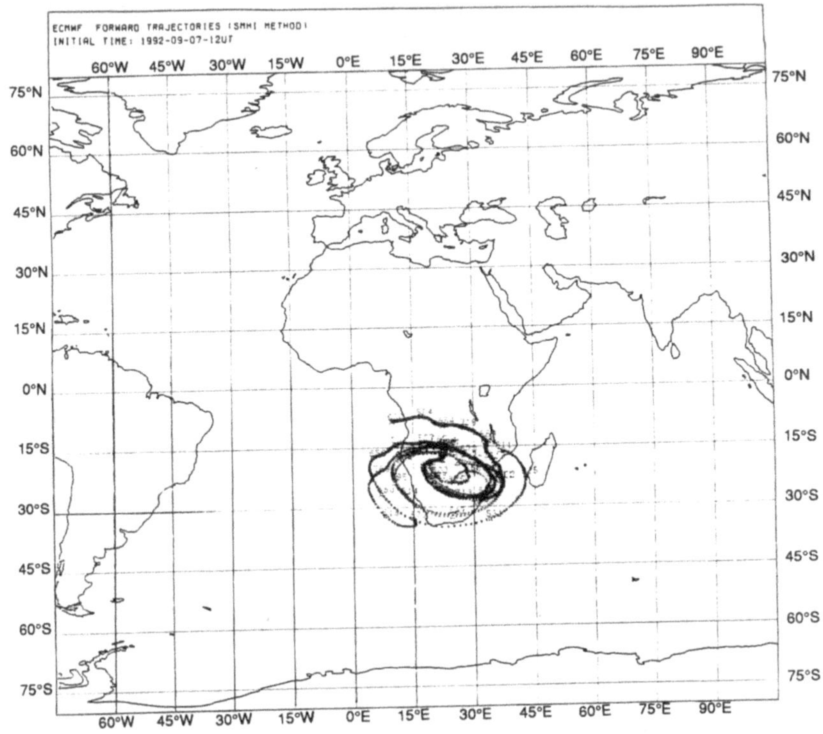

Figure 5. Airmass forward trajectories showing recirculation over southern Africa. The trajectories originate at Victoria Falls, Zimbabwe, at 1200UTC on 7 September 1992.

confirmed these results and extended the altitude to which measurements were made to 12 km. Long-term measurements at stations at Victoria Falls, Zimbabwe, and at Etosha Park, Namibia, showed that these observations were typical for the entire season investigated. In flights made with the DC-8 over the South Atlantic, similar high values were recorded. Such widely distributed atmospheric pollutant concentrations were previously thought to be limited to the anthropogenically burdened regions of Europe and North America.

The data from the field measurements in combination with data from remote sensing show that the vegetation fires in southern Africa and in South America have a considerably strong influence on the atmospheric environment and climate. The influence on climate is

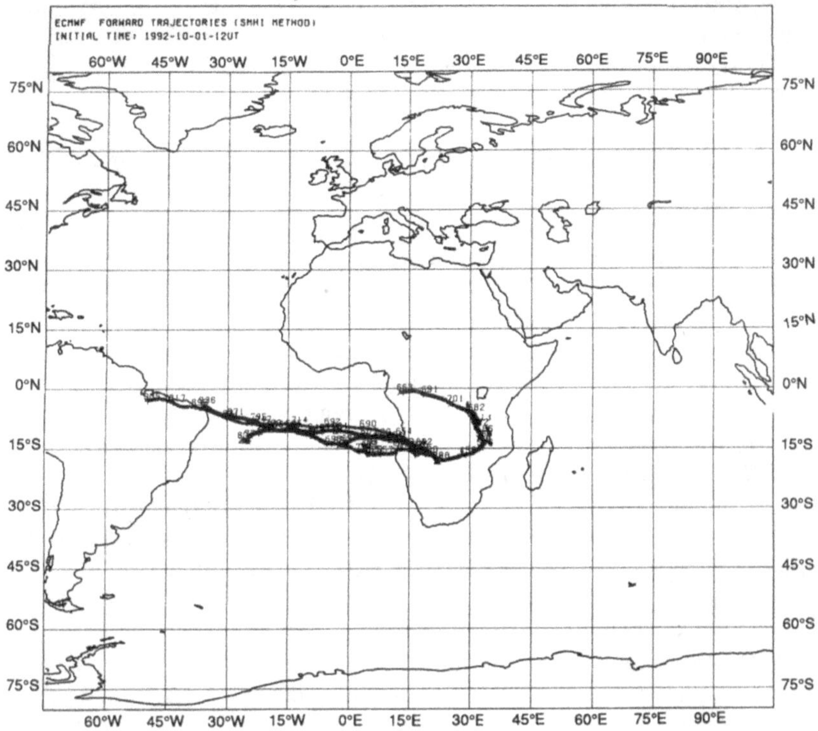

Figure 6. Airmass forward trajectories showing transport from southern Africa towards the Atlantic Ocean and eventually South America. (The trajectories originate at Victoria Falls, Zimbabwe, at 1200UTC on 1 October 1992.)

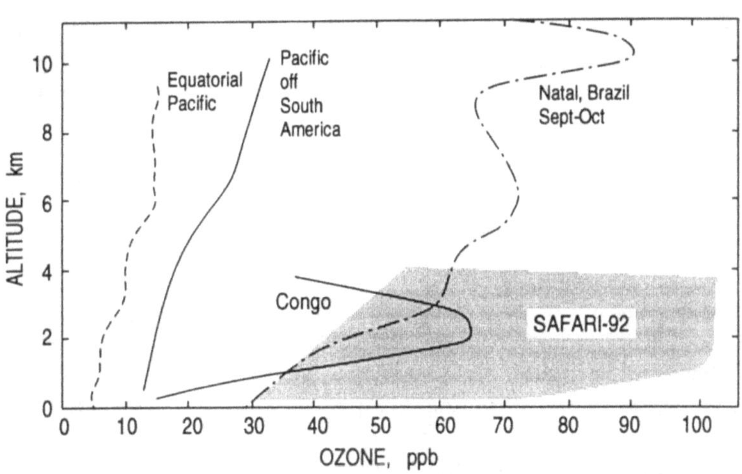

Figure 7. Ozone vertical distribution in the tropics. The profile over the equatorial Pacific shows no influence from biomass burning, while the profile over the Pacific off S. America suggests ozone enhancement due to long-range transport from the tropical continents (Fishman, 1988). The ozone profiles over the Brazilian coast at Natal and over the Congo show high ozone concentrations due to photochemical production in biomass burning plumes documented in earlier studies (Andreae *et al.*, 1988; Kirchoff *et al.*, 1991). The SAFARI-92 data are the highest regional ozone levels observed in the tropics and subtropics to date.

very complex, since it contains both warming components due to the greenhouse gases emitted in the smoke, and a cooling effect due to the aerosol particles that reflect sunlight back into space and alter the properties of clouds.

OZONE STUDIES BY REMOTE SENSING AND IN SITU MEASUREMENTS

In preparing for STARE/TRACE-A/SAFARI-92, the NASA-Goddard group improved the method for deriving TOMS (Total Ozone Mapping Spectrometer) total O_3 in regions of persistent marine stratocumulus clouds, such as the eastern tropical South Atlantic (Thompson *et al.*, 1993). At the same time, Hudson and Kim (1993) developed a method for deriving the tropospheric component of the O_3 column in these regions from TOMS. Much of the near-Africa Atlantic O_3 maximum lies over marine stratocumulus clouds, so that TRACE-A and SAFARI-92 flights in these regions presented excellent validation opportunities for the TOMS products. Using real-time TOMS processing it was possible to transmit both TOMS O_3 products to the TRACE-A Science Team in Africa to assist in flight planning. Examples of TOMS maps for the DC-8 flight that took place on 18 October 1992 (Flight #15) are shown in Figure 8a-c. Also shown is the flight track along with locations of two O_3 profiles taken from the DC-8. Integrating the O_3 profile to obtain an estimate for tropospheric O_3 gives values in fairly good agreement with the TOMS O_3 method.

The TOMS algorithm obtains a total O_3 column by adding up contributions from standard (climatological) profiles, usually taken from ozonesondes. There are very few ozonesonde observations for the tropics and southern hemisphere. Thus, the large body of remotely sensed and in situ O_3 observations collected on TRACE-A and SAFARI-92 are being used to enhance that climatology. These include ozonesondes launched at Brazzaville (Congo), Ascension Island, Pretoria (South Africa) and Etosha Park (Namibia). During TRACE-A, the DC-8 measured roughly a dozen ozone profiles off southern Africa over the persistent marine stratocumulus as well as serving as a platform for upward- and downward-looking UV-DIAL ozone. Integrated O_3 profiles from these instruments will be compared to TOMS O_3. In addition, in situ ozone measurements from aircraft and ground-based measurements from the German and South African groups taken during SAFARI-92 are being studied to obtain better profiles for the TOMS algorithm during the biomass burning season.

THE TRACE-A CAMPAIGN IN SOUTHERN AFRICA

As part of the TRACE-A Experiment, the NASA DC-8 was based in southern Africa from 4–20 October, flying six missions (46 flight hours) out of Johannesburg and Windhoek. One of these flights was from Johannesburg to northern Zambia to characterize the emissions from widespread burning taking place there on 6 October 1992. Numerous fires were observed near 10°S, 30°E, and the DC-8 penetrated several plumes down to a flight level less than 300 m above the ground. Even en route to the primary region of burning, concentrations of many trace gases associated with biomass burning were above the detection limit at the flight altitude of 10 km. For example, methyl chloride rose from near 600 to above 700 ppt from South Africa through Zimbabwe and southern Zambia.

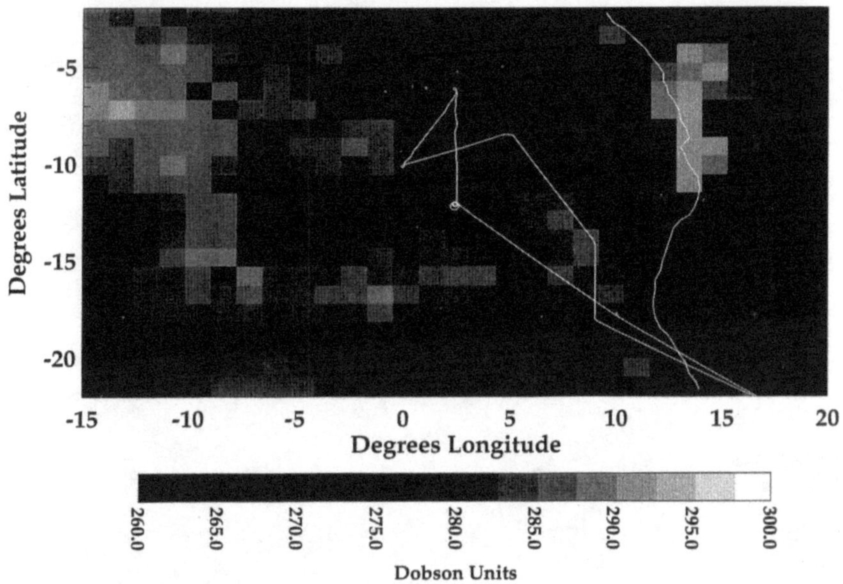

Figure 8a. Total ozone in TRACE-A study region from corrected Nimbus-7 TOMS gridded data for 18 October 1992.

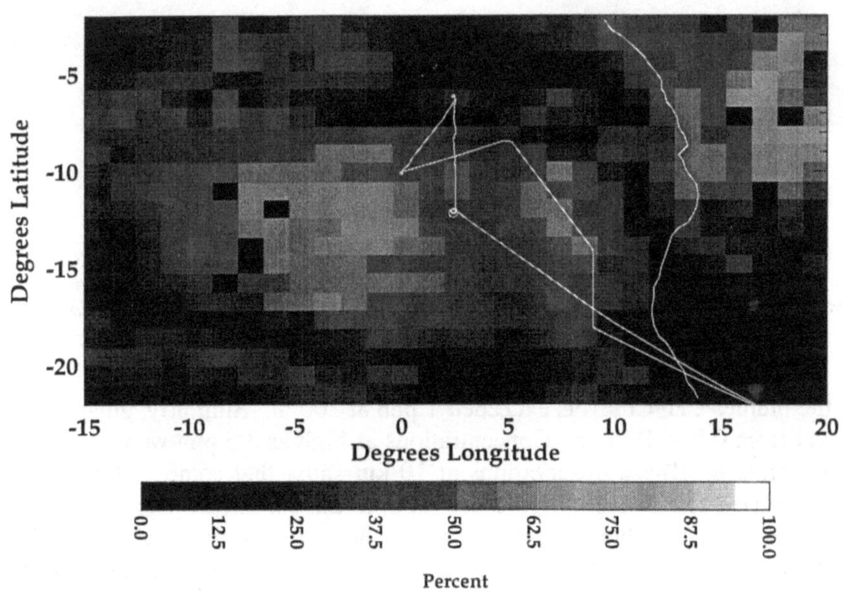

Figure 8b. Reflectivity measurements used to correct ozone data. Flight trace (Flight #15) on the same day is shown. Integrating aircraft O_3 observations from two airborne O_3 sensors taken during the flight gives tropospheric ozone estimates in very good agreement with the satellite-derived estimate of tropospheric O_3.

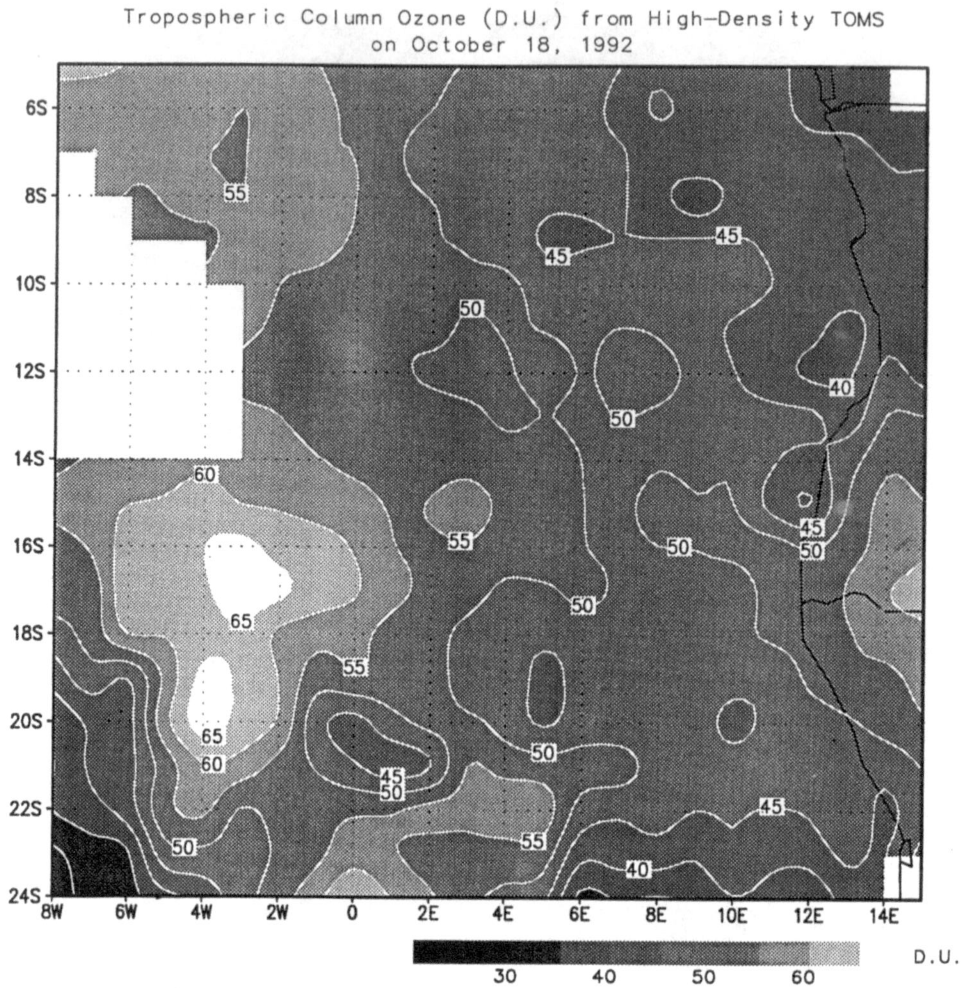

Figure 8. (c) Tropospheric O_3 over marine stratocumulus region obtained from high-density TOMS data. Integrating aircraft O_3 observations from two airborne O_3 sensors taken during the flight gives tropospheric ozone estimates in very good agreement with the satellite-derived estimate of tropospheric O_3.

Within the plumes CH_3Cl levels exceeded 1 ppb at 300 m. Similarly, ethane increased at flight level from 0.8 to 1.2 ppb. Concentrations as high as 3.5 ppb were measured during plume penetration. These observations at 10 km show that some of the emissions are efficiently carried to relatively high altitudes over the continent. Some very short-lived hydrocarbons such as ethene remained below detection limit at flight level until we descended into northern Zambia. In the intense plumes, concentrations exceeded 5 ppb.

The oxides of nitrogen showed some interesting behavior. The DC-8 had two low passes during the 6 October flight. During the first low pass, NO concentrations exceeded 1.1 ppb whereas the maximum NO_2 concentration was about 0.6 ppb. During the second low altitude pass, NO concentrations averaged about 0.2 ppb, whereas NO_2 concentrations were considerably higher at values ranging between 1.0 and 1.2 ppb. Both penetrations were within one hour of solar noon, so their ratio should not have been affected by widely varying photolysis rates. Nitric acid levels peaked at about 2.6 ppb during the first pass and somewhat higher at concentrations approaching 3.5 ppb during the second pass. The reasons for these different nitrogen ratios are yet to be found, but differences between types of vegetation and intensity of burning cannot be ruled out. The first plume also showed lower concentrations of formic acid and acetic acid compared with the second one. Representative concentrations of formic acid were 6 and 9 ppb, respectively. For acetic acid, the concentrations for the two plumes were 6 and 12 ppb, respectively.

Within the two plumes, formaldehyde concentrations rose to 3.0 and 4.5 ppb, whereas hydrogen peroxide concentrations were fairly close to 5 ppb for both plumes. PAN and PPN were also measured during TRACE-A. At the 10 km flight level, PAN increased from approximately 0.1 ppb to nearly 1 ppb and then rose sharply during both low level excursions to concentrations near 4 ppb. PPN generally stayed below detection limit at 10 km during the entire flight, but rose to 100 to 140 ppt during plume penetrations.

Out over the Atlantic Ocean, relatively high concentrations of ozone were the rule as the DC-8 flew parallel to the coasts Namibia and Angola. Throughout the troposphere, ozone concentrations of 60 to 110 ppb were commonplace. The structure of these plumes was very heterogeneous and nearly every time enhancements in ozone were observed, corresponding increases in carbon dioxide and methane were also seen. Because both CO_2 and CH_4 were measured with high response instruments (>1 Hz), detailed structure could be seen, and sharp variations in these trace gases very often paralleled similar variations in ozone. Carbon monoxide concentrations did not increase every time an increase in ozone was observed, suggesting that some air encountered was very aged. Again, a complete explanation of these findings must await further analysis.

FUTURE PLANS

The preliminary findings of this measurement campaign were presented at the 1993 Fall Meeting of the American Geophysical Union. Further studies are currently being evaluated by the participating institutions, and subsequently, they will be published in the form of a special issue of the *Journal of Geophysical Research* and a synthesis monograph. A major component of the analysis phase will consist of the application of various types of modeling approaches to the data set.

Planning for future campaigns is simultaneously underway. The measurement campaign is anticipated to continue in 1994 with a study of the atmospheric chemistry and circulation over southern Africa outside the fire season (SA'ARI-94: SAFARI without the Fire). Also during 1994-95, BIBEX will study the behavior and emissions from fires in the wooded savannas of Zambia and Zimbabwe, ecosystems with vegetation and fuel characteristics quite different from the South African savannas. A subsequent campaign (probably 1995) will investigate the large-scale distribution of fire impacted air masses over southern hemisphere Africa with emphasis on the east Africa/Indian Oceans (WINDOS: Western Indian Ocean Study). The BIBEX community is also currently preparing for the measurement campaign SEAFIRE (South East Asian Fire Experiment) planned to take place in Southeast Asia in the second half of the 1990s during which forest clearing fires in Borneo and burning of rice fields in the whole of Southeast Asia will be measured.

ACKNOWLEDGMENTS

Since the STARE/TRACE-A/SAFARI-92 Science Team is too numerous to be included in the author list, only those who have contributed text to this paper are listed on the title page. The others are: A.-L. Ajavon, C. Anderson, T.W. Andreae, H.J. Annegarn, C.B. Archer, P. Artaxo, E. Atlas, R.E. Babbitt, J. Barsby, J. Beer, R.J. Bendura, D. Bergmann, D.R. Blake, G.E. Bodeker, T. Boyle, J.D. Bradshaw, J.K. Broadbent, E.V. Browell, E.G. Brunke, R.A. Burke, H. Cachier, J. Cafmeyer, D.J. Cahoon, R. Chadyendiya, M. Chaitwa, T.-Y. Chen, G.J.R. Coetzee, W.R. Cofer III, J.E. Collins, B. Cros, P. Cunningham, G. de Beer, A. de Kock, R. Delmas, R.D. Diab, P. Dowty, B.L. Duigan, F. Echalar, M. Edwards, W. Elbert, T. Fickinger, A. Gaudichet, S.J. Godefroy, G.L. Gregory, M. Guest, G.W. Harris, G. Helas, G. Held, J.L. Hery, J.M. Hoell, R. Hudson, C. Jambert, A. Joubert, M.R. Jury, P. Kållberg, R.P. Karimanzira, J.B. Kauffman, J. Kendall, J. Kim, V.W.J.H. Kirchhoff, M.A. Kneen, R. Koppmann, T.N. Krishnamurti, F. Kruger, T. Kuhlbusch, C. Labuschagne, J.P. Lacaux, C. Liousse, E. Lynch, S.A. Macko, W. Maenhaut, C. Manickum, B. Martincigh, P. Masclet, J.A. Mason, G.K. Mather, M.A. Mazurek, D.P. McNamara, D.J. McRae, F. Meixner, W.L. Miller, E. Mpunduma, E. Mravlag, W. Munyanyiwa, A. Mwale, S. O'Beirne, U. Parchatka, D. Parsons, K. Pickering, J.J. Pienaar, S. Piketh, J.P. Pinto, W. Pollock, A. Potgieter, R.A. Preston-Whyte, M.W. Raynor, R. Rorich, J. Rudolph, G.W. Sachse, I. Salma, S.T. Sandholm, W. Schneider, M.C. Scholes, M. Schormann, G.C. Schulze, M. Scourfield, D.I. Sebacher, M.K. Seely, R. Shea, H.B. Singh, N. Snow, F. Sokolic, B. Stefanski, R. Swap, R.W. Talbot, I. Taviv, A. Tegen, M. Thompson, G.R. Tosen, L. Trollope, W.S.W. Trollope, M.M. Truter, S. Tsure, C. Turner, P. Tyson, J. van Heerden, D. Walmsley, D.E. Ward, M.G. Weber, F. Weirich, M. Welling, F.G. Wienhold, E.L. Winstead, T. Zenker, R.G. Zepp, and M. Zunckel.

REFERENCES

Andreae, M.O., 1993, The influence of tropical biomass burning on climate and the atmospheric environment, *in:* "Biogeochemistry of Global Change: Radiatively Active Trace Gases," R.S. Oremland (ed.), Chapman & Hall, New York, pgs. 113-150.

Andreae, M.O., B.E. Anderson, D.R. Blake, J.D. Bradshaw, J.E. Collins, G.L. Gregory, G.W. Sachse and M.C. Shipham, 1993, Influence of plumes from biomass burning on atmospheric chemistry over the equatorial Atlantic during CITE-3, *J. Geophys. Res.*, submitted.

Andreae, M.O., E.V. Browell, M. Garstang, G.L. Gregory, R.C. Harriss, G.F. Hill, D.J. Jacob, M.C. Pereira, G.W. Sachse, A.W. Setzer, P.L. Silva Dias, R.W. Talbot, A.L. Torres and S.C. Wofsy, 1988, Biomass-burning emissions and associated haze layers over Amazonia, *J. Geophys. Res.*, 93:1509-1527.

Andreae, M.O., A. Chapuis, B. Cros, J. Fontan, G. Helas, C. Justice, Y.J. Kaufman, A. Minga and D. Nganga, 1992, Ozone and Aitken nuclei over equatorial Africa: Airborne observations during DECAFE 88, *J. Geophys. Res.*, 97:6137-6148.

Cahoon, D.R., Jr, B.J. Stocks, J.S. Levine, W.R. Cofer, III and K.P. O'Neill, 1992, Seasonal distribution of African savanna fires, *Nature*, 359:812-815.

Cros, B., R. Delmas, D. Nganga, B. Clairac and J. Fontan, 1988, Seasonal trends of ozone in equatorial Africa: Experimental evidence of photochemical formation, *J. Geophys. Res.*, 93:8355-8366.

Cros, B., D. Nganga, A. Minga, J. Fishman and V. Brackett, 1992, Distribution of tropospheric ozone at Brazzavillle, Congo, determined from ozonesonde measurements, *J. Geophys. Res.*, 97:12,869-12,875.

Crutzen, P.J. and M.O. Andreae, 1990, Biomass burning in the tropics: Impact on atmospheric chemistry and biogeochemical cycles, *Science*, 250:1669-1678.

Crutzen, P.J. and J.G. Goldammer, 1993, "Fire in the Environment: The Ecological, Atmospheric, and Climatic Importance of Vegetation Fires," J. Wiley & Sons, Chichester, England, 400 pp.

Fishman, J., 1988, Tropospheric ozone from satellite total ozone measurements, *in:* "Tropospheric Ozone: Regional and Global Scale Interactions," I.S.A. Isaksen (ed.), Reidel, Dordrecht, pgs. 111-123.

FIRESCAN Science Team, 1993, Fire in boreal ecosystems of Eurasia: First results of the Boreal Forest Island Fire Experiment, Fire Research Campaign Asia-North (FIRESCAN), *J. World Resources Rev.*, in press.

Fishman, J., K. Fakhruzzaman, B. Cros and D. Nganga, 1991, Identification of widespread pollution in the southern hemisphere deduced from satellite analyses, *Science*, 252:1693-1696.

Hudson, R.D. and J. Kim, 1993, "Direct measurements of tropospheric ozone using TOMS data," Proceedings of the Quadrennial Ozone Symposium, Charlottesville, VA, June 1992.

Justice, C.O., J.-P. Malingreau and A.W. Setzer, 1993, Satellite remote sensing of fires: Potential and limitations, *in:* "Fire in the Environment: The Ecological, Atmospheric, and Climatic Importance of Vegetation Fires," P.J. Crutzen and J.G. Goldammer (eds.), J. Wiley & Sons, Chichester, England, pgs. 77-105.

Lacaux, J.-P., H. Cachier and R. Delmas, 1993a, Biomass burning in Africa: An overview of its impact on atmospheric chemistry, *in:* "Fire in the Environment: The Ecological, Atmospheric, and Climatic Importance of Vegetation Fires," P.J. Crutzen and J.G. Goldammer (eds.), J. Wiley & Sons, Chichester, England, pgs. 159-191.

Lacaux, J.P., J.M. Brustet, R. Delmas, J.C. Menaut, L. Abbadie, B. Bonsang, H. Cachier, J. Baudet, M.O. Andreae and G. Helas, 1993b, Biomass burning in the tropical savannas of Ivory Coast: An overview of the field experiment Fire of Savannas (FOS/DECAFE 91), *J. Atmos. Chem.*, in press.

Thompson, A.M., K.E. Pickering, D.P. McNamara and R.D. McPeters, 1993, Effect of marine stratocumulus on TOMS ozone, *J. Geophys. Res.*, in press.

van Wilgen, B.W., W.S.W. Trollope and C.S. Everson, 1990, Fire management in southern Africa: Some examples of current objectives, practices and problems, *in:* "Fire in the Tropical Biota: Ecosystem Processes and Global Challenges," J. Goldammer (ed.), Springer Verlag, Berlin, pgs. 179-215.

Ward, D.E., A.W. Setzer, Y.J. Kaufman and R.A. Rasmussen, 1991, Characteristics of smoke emissions from biomass fires of the Amazon region—BASE-A experiment, *in:* "Global Biomass Burning: Atmospheric, Climatic, and Biospheric Implications," J.S. Levine (ed.), MIT Press, Cambridge, MA.

BIOSPHERE-ATMOSPHERE EXCHANGE OF TRACE GASES IN THE TROPICS: EVALUATING THE EFFECTS OF LAND USE CHANGES

Michael Keller[1] and Pamela A. Matson[2]

[1]International Institute of Tropical Forestry
Rio Piedras, Puerto Rico
[2]NASA Ames Research Center
Moffett Field, California

ABSTRACT

Tropical soils and vegetation represent globally significant sources of a range of atmospheric gases, including CO_2, CO, N_2O, NO, CH_4, and volatile organic compounds. Conversion of tropical forests to agriculture and other uses, and intensification of agricultural practices in both forest and savanna areas are occurring very rapidly; yet the impacts of these land use practices on biogeochemical cycles, trace gas emissions, and atmospheric chemistry are not well understood. We present early results of IGAC's Biosphere-Atmosphere Trace Gas Exchange in the Tropics (BATGE) Activity that evaluate the importance of tropical land use change on fluxes of a number of gases.

Soils are important sources and sinks for trace gases, especially for oxides of nitrogen. On-going studies in Brazil, Costa Rica, Puerto Rico, and Mexico suggest that conversion of forest to pasture results in elevated soil emissions of nitrogen oxides in the first decade or less after clearing, but that older pastures, unless intensively managed, have lower fluxes than forests. Forest conversion to pasture also diminishes the soil sink strength for methane; in some cases the direction of the flux is reversed and pasture soils become methane sources. Likewise, results from a number of tropical regions suggest that agricultural sites recently converted from native vegetation have elevated fluxes of nitrogen and carbon gases, but that high emissions of nitrogen oxides and reduced uptake of methane in older agricultural systems are maintained only with intensive management and fertilizer use. Moreover, the magnitude of fluxes occurring in response to fertilization depends on soil type and management considerations such as crop type, fertilizer type, application rate, and application methods. The great spatial and temporal variability in trace gas responses to tropical agricultural practices complicates estimates of their importance at regional and global scales.

Few attempts have been made to study vegetation sources and sinks for trace gases in the tropics. It is clear that tropical vegetation can produce large quantities of volatile

Global Atmospheric-Biospheric Chemistry
Edited by R.G. Prinn, Plenum Press, New York, 1994

organic compounds. Forests may be strong sinks for oxidants and odd-nitrogen compounds. Changes in the species composition and structure of vegetation will almost certainly change these system properties; but our ability to predict the magnitude of these changes is severely hampered by a dearth of relevant data.

INTRODUCTION

Tropical terrestrial ecosystems cover approximately 42% of global land area. This region supports lush tropical rain forests with the highest levels of biological diversity on earth as well as a variety of other important terrestrial ecosystems with markedly different characteristics. Tropical forests and woodlands cover 25×10^6 km^2, 17% of earth's land surface, and account for over 40% of terrestrial net primary production (Rodin *et al.*, 1975). Tropical wetlands cover nearly 2×10^6 km^2, representing 36% of global wetlands (Matthews and Fung, 1987). Tropical savannas cover 19×10^6 km^2 (Bolin *et al.*, 1979).

Within these broad vegetation categories, there is great variation. Wet and humid tropical forests (rain forests) account for nearly 50% of all tropical forests and are among the most productive terrestrial systems with net primary production (NPP) in these systems of 13 to 28 Mg ha^{-1}yr^{-1}) (Murphy and Lugo, 1986). Seasonally dry tropical forests and woodlands receive far less attention than rain forests, but they once covered approximately 40% of the original forest area in the tropics. They are less productive systems (NPP ranging from 8 to 21 Mg ha^{-1}yr^{-1} (Murphy and Lugo, 1986) than are the humid forests, but they have been much more intensively cleared and managed (Janzen, 1988). These different tropical forest types are defined by the amount of precipitation, evapotranspiration, and the length of the dry season they experience. Within each climatic zone there are important variations in soil chemical and physical characteristics that lead to enormous differences in productivity and biogeochemical characteristics (Richter and Babbar, 1991; Vitousek and Sanford, 1986; Sanchez, 1981). Similarly, tropical wetlands and tropical savannas also differ in well-characterized ways as functions of climate and soils (Gore, 1983; Huntley and Walker, 1982). Such variations in climate and soil result in major variations in ecosystem biogeochemical cycling (Vitousek and Sanford, 1986). A growing number of studies suggest that these variations also affect the exchange of trace gases between terrestrial ecosystems and the atmosphere. Ecological variations have been used as the basis for regional and global extrapolations of trace gas fluxes (e.g., Matson and Vitousek, 1990; Sanhueza *et al.*, 1990; Bartlett *et al.*, 1988; Bartlett and Harriss, 1993).

In addition to the naturally occurring components of variation, land use changes resulting from human activity have become an increasingly important force in the tropics. Large areas of the humid tropics are being converted from forest to pasture, agriculture, or other land uses. The FAO estimates that an average of 1.7×10^5 km^2 of tropical closed and open forest were cleared annually from 1980 to 1990 (FAO, 1991). A greater proportion of the dry tropical forests has undergone conversion to human-dominated systems already (Janzen, 1988). Savanna ecosystems are widely managed for grazing (especially by regular burning) and are increasingly being converted to more intensive agricultural use. Given that the human population is likely to double by 2100 (with nearly all growth occurring in the developing world) and that a significant proportion of existing cropland is already subject to degradation due to erosion, salinization, and desertification in dryland areas (Crosson and Rosenberg, 1990), the demand for conversion of natural tropical systems to agricultural land will increase (Meyer and Turner, 1992). At the same time, intensification of tropical agriculture through the use of fertilizers, irrigation, herbicides, and other management practices is inevitable (Meyer and Turner, 1992; Naylor and Matson, 1993).

The effects of tropical land conversions on the biogeochemical cycling of carbon and in particular on carbon dioxide (CO_2) exchange between terrestrial ecosystems and the atmosphere have been the subject of intense analysis (e.g., Houghton *et al.*, 1987, 1991; Detwiler and Hall, 1988; Crutzen and Andreae, 1990). The effects of these conversions on cycling and emissions of other gases have received much less attention (Keller *et al.*, 1991). Likewise, the effects of tropical agricultural practices on trace gas fluxes have been largely unstudied (Vitousek and Matson, 1993), with the exception of rice agriculture, which has received considerable attention with respect to methane emissions (see Neue and Sass, this volume).

Because of the increasing importance of tropical land use change, including both conversion of natural to managed lands and intensification of managed land use, IGAC's Biosphere-Atmosphere Trace Gas Exchange in the Tropics (BATGE) Activity focuses on the exchanges of gases and the physical and biological factors that control them both in undisturbed ecosystems and in those undergoing change. The goals of BATGE are to:

- determine the fluxes of trace gases between tropical biomes and the atmosphere,
- determine the factors that control these fluxes,
- assess the effects of land use change on the exchange of trace gases, and
- develop the ability to predict the impact of both climate and land use changes on these fluxes.

The BATGE Activity excludes from its scope studies of wetland rice agriculture, carbon storage in tropical soils and vegetation and the direct effects of biomass burning. These problems are covered by the IGAC-RICE and BIBEX Activities and the IGBP-GCTE Core Project, but with appropriate communication with BATGE. In the following sections, we discuss the state of knowledge regarding tropical contributions to the budgets of the long-lived trace gases methane (CH_4) and nitrous oxide (N_2O) and the biological sources and sinks of short-lived trace gases such as nitric oxide (NO), ozone (O_3) and volatile organic carbon compounds (VOC). We present early data from several BATGE studies that address the importance of land use change in forested systems and describe a study illustrating the importance of 40 years of land use change in a tropical forest region.

LONG LIVED GASES

Methane

Natural tropical wetlands are a globally significant source of methane. In a recent review of methane emissions from natural wetlands, Bartlett and Harriss (1993) estimated that tropical wetlands (20°N to 30°S) emit 66 Tg yr^{-1} versus total releases from natural wetlands of 109 Tg yr^{-1} or the total source of 400 to 500 Tg yr^{-1} (1 Tg = 10^{12} g) (Cicerone and Oremland, 1988; Vaghjiani and Ravishankara, 1991; Fung *et al.*, 1991). Natural tropical wetland area is relatively limited, accounting for about 1.9×10^6 km^2 (Matthews and Fung, 1987). However, average methane fluxes from these areas are very high, 148 to 233 mg CH_4 m^{-2} d^{-1} (Bartlett and Harriss 1993). High fluxes of methane from sediments to the atmosphere are especially common in tropical areas. High temperatures and biological productivity promote the formation of methane bubbles in sediments. Bubbles may escape directly to the atmosphere bypassing oxidation at the sediment water interface (Keller and Stallard, 1993; Martens and Klump, 1980). Plants that inhabit permanently or seasonally inundated areas possess adaptations to bring oxygen to their roots. These structures and associated active and passive transport mechanisms also serve as efficient conduits for methane emissions to the atmosphere (Dacey, 1981; Chanton and Dacey, 1991).

Knowledge of methane emissions from tropical wetlands has been increasing rapidly through field studies in the basins of the worlds largest rivers, the Amazon and Congo (e.g., Bartlett *et al.*, 1988; Tathy *et al.*, 1992). Other important wetland habitats in the tropics have received little or no study. These include some of the world's most extensive wetlands such as the Pantanal of South America, the Okavango and Sudd of Africa and the peat-forming swamp-forests of Malaysia and Indonesia (Thompson and Hamilton, 1983; Breunig, 1990). The latter areas with an extent of 2×10^5 km^2 are under considerable pressure for timber extraction.

As with other tropical habitats, wetland areas are subject to pressure from development. Wetland areas are being created, destroyed and modified. In densely populated South Asia and South East Asia (mainly tropical) over 8×10^5 km^2 are planted to wetland rice. While much of this area may result from conversion of natural wetland habitats, over 35% is irrigated (Neue, 1992). The methane source from rice is discussed in an accompanying paper in this volume (Neue and Sass). Drainage of wetlands for agriculture or flooding of large areas due to hydroelectric projects and other water management schemes may provoke considerable changes in wetland areas and hence methane emissions. The trend towards increased hydroelectric development appears to be accelerating. For example, Brazil's hydroelectric capacity increased by 57% from 1980 to 1985 (Shea, 1988). In dams planned for areas of low relief such as the Amazon Basin, large areas may be flooded for small power returns. The Balbina dam alone floods over 1,500 km^2 even though it yields only 85 megawatts of power (Fearnside, 1989). Shallow waters of tropical reservoirs can emit large amounts of methane. In Gatun Lake, Panama, the reservoir that forms the Panama Canal, methane emissions from shallow waters (<6 m) average 195 to 1302 mg CH_4 m^{-2} d^{-1} (Keller and Stallard, 1993). Average emissions from deeper waters are an order of magnitude lower. An estimate of the current and future methane sources from tropical reservoirs requires information regarding their hypsography.

There is little information regarding the effects of human disturbance and water management from tropical wetlands. Some examples from the temperate and sub-tropical areas point to important considerations for future studies. Wetland methane emissions can correlate with primary productivity as demonstrated for South Florida marshlands by Whiting *et al.* (1991). The productivity effect on methane emissions may explain why methane emissions from flooded forests in the South East U.S. increase at sites receiving sewage effluents (Harriss and Sebacher, 1981). The effects of water management may be complex. Harriss *et al.* (1989) evaluated the effect of drainage and agricultural and urban development for the Everglades wetland area of South Florida. Surprisingly, they found that although large areas of the wetland were drained, overall annual methane emissions for the region increased. A major cause for this increase was the creation of high biological productivity water impoundments that remained flooded permanently.

Well-drained soils typically consume methane. In most upland environments, methane consumption is limited by rates of diffusion of atmospheric methane into the soil. The best global estimate for soil methane consumption based on diffusivities for soils classified in a digitized global soil database is 29 Tg CH_4 yr^{-1}, about 7% of the current estimates of the annual methane source (Dörr *et al.*, 1993). Methane uptake, about 1 mg CH_4 m^{-2} d^{-1}, has been found in a range of tropical forest and savanna habitats (e.g., Seiler *et al.*, 1984; Keller *et al.*, 1986; Delmas *et al.*, 1992). A number of measurements of dry soils from Venezuelan savanna sites have shown small methane emissions (Hao *et al.*, 1988) The source of these emissions remains mysterious.

The conversion of natural soils to agricultural use generally results in a diminished capacity of these soils to consume methane (Mosier *et al.*, 1991). The conversion of tropical forest to pasture and agriculture in Panama resulted in a 3-fold decrease in soil methane consumption (Keller *et al.*, 1990). In pastures, soil compaction impedes drainage.

Wetter soils have lower gas diffusivities and hence lower methane uptake. Unpublished studies in Costa Rica (Keller, private communication) show that conversion of forest to pasture may even reverse the direction of methane flux, i.e. a small sink becomes a small source. Fertilization also diminishes methane consumption in temperature environments (Steudler *et al.*, 1989; Mosier *et al.*, 1991). The mechanism for this inhibition has not been unambiguously identified for any site nor has this fertilization effect been observed at tropical field sites.

Nitrous Oxide

Early budgets for nitrous oxide (McElroy and Wofsy, 1986; Keller *et al.*, 1986) concluded that tropical forest soils (as they were prior to human disturbance) are the single most important source of nitrous oxide to the atmosphere. Since then, many more flux measurements (e.g., Goreau and de Mello, 1988; Matson and Vitousek, 1987; Keller *et al.*, 1988; Matson *et al.*, 1990) and alternative regional budgets (e.g., Matson and Vitousek, 1990; Sanhueza *et al.*, 1990) have generally supported the early conclusions, although in later budgets the magnitude of the tropical forest source has been reduced. The importance of the undisturbed tropical forest source for nitrous oxide can be explained in the context of nitrogen biogeochemistry. In most lowland tropical forests, rates of nitrogen transformations in the soil and cycling through the vegetation are much greater than in most temperate or boreal forests (Vitousek and Sanford, 1986; Vitousek and Matson, 1988). Nitrogen functions as an excess nutrient, and losses via gas flux appear to be quantitatively and proportionally larger than in higher latitude forests.

We have found that while tropical forests as a whole may be the most important source of nitrous oxide, fluxes vary significantly among tropical forest types. Matson and Vitousek (1987) and Matson *et al.* (1990) reported a strong positive relationship between soil nitrogen cycling rates and nitrous oxide fluxes across a range of tropical soil types; they used the areal extent and distribution of those soil types to develop a tropical budget for nitrous oxide (Matson and Vitousek, 1990). Seasonally dry tropical forests have significantly lower annual fluxes than humid forests (Garcia-Mendez *et al.*, 1991; Matson and Vitousek, 1993), and exhibit variation within the climatic type as well (Vitousek *et al.*, 1989).

The early McElroy and Wofsy (1986) budget was essentially a balanced budget; in it, N_2O from fossil fuel combustion accounted for the annual atmospheric increment of 3.5 Tg. It is now clear that combustion accounts for less than 10% of the annual increase in tropospheric nitrous oxide (Watson *et al.*, 1990). A number of other possible sources for the increase have been identified, including biomass burning, degassing of irrigation water, agricultural activity, emissions from natural systems receiving large inputs of anthropogenic nitrogen deposition, and specialized industrial processes (cf., Vitousek and Matson, 1993). Even assuming the upper limits on the ranges of each of these sources, it is difficult to account for the global sink and accumulation (Watson, 1990). There are a number of reasons why tropical land use change and intensification of tropical agriculture may be significant contributors to the missing source.

While most soils lose carbon and nitrogen after conversion from natural vegetation to agriculture, loss rates in tropical soils are particularly rapid. Matson *et al.* (1987) estimated that 20 to 25 Tg of nitrogen are lost annually by tropical land-clearing, not including the amount lost in biomass burning. There are relatively few studies on the pathways by which this nitrogen is lost, but some have reported increased nitrous oxide emissions following disturbance (Keller *et al.*, 1986; Matson *et al.*, 1990). Luizão *et al.* (1989) reported elevated nitrous oxide fluxes from tropical pastures ranging in age from 3 to 10 years following forest clearance and estimated a 3-fold greater annual flux from pastures than forests. Assuming that all tropical pastures of all ages had similar increases over forest fluxes, Matson and Vitousek (1990) estimated that tropical forest conversion could account for one-fourth to one-half of the annual global increase in N_2O.

In a more recent study of forest lands converted to pasture in Costa Rica, Keller *et al.* (1993) found that young pastures (under 10 years old) had significantly elevated fluxes over forests, but that fluxes in older pastures were lower than those in forests. Luizão *et al.* and Keller *et al.*'s studies were both done in humid tropical areas on relatively fertile soils. Garcia-Mendez *et al.* (1991) measured N_2O fluxes in pastures converted from seasonally dry tropical forest, and found that upland pastures did not have elevated fluxes. More research spanning the range of climate and soils in the tropics will be necessary to accurately estimate the global importance of forest conversion to pasture.

The effects of agricultural practices (especially nitrogen fertilization) on nitrous oxide emissions has received considerable attention, but only in the temperate zone (Eichner, 1991). The concentration of studies in the temperate zone is not surprising, considering that less than 20 years ago nearly all fertilizer N was applied in temperate zone agriculture. Now, however, nearly half of the approximately 80 Tg N applied each year is used in the developing world, including most of the tropics. This trend of increasing use in the developing world and stable use in the developed world is projected to continue (FAO, 1985; EPA, 1990). As Vitousek and Matson (1993) point out, even the current amount of fertilizer N nearly doubles the amount added to terrestrial ecosystems via natural processes. Such an alteration of the global budget of nitrogen is likely to affect all the fluxes of the nitrogen cycle, including fluxes out of terrestrial systems to the atmosphere.

Current estimates based on temperate zone data suggest that direct losses of N_2O related to fertilization may account for 0.5 to 1.0 Tg N_2O–N yr^{-1}. If tropical agricultural systems function as tropical forests do, then a greater proportion of the total quantity of N cycled between soil and plants (and applied in fertilizer) is likely to be emitted as N_2O in tropical systems than is measured in temperate systems (Vitousek and Matson, 1993). There have been very few studies of nitrogen gas flux from tropical agricultural systems. Garcia-Mendez *et al.* (1991) measured N_2O fluxes from maize and sorghum and from a managed pasture; they found fluxes elevated over forest both in maize (in the first season after forest clearing) and in fertilized pastures, but lower fluxes in sorghum that had not received fertilizer or organic matter inputs for many years. Matson (private communication) measured large pulses of nitrous oxide immediately after fertilization events in tropical sugar cane systems; the proportion of the fertilizer N lost as N_2O was very small (<1%) in fields using sub-surface irrigation/fertilization, but much higher in fields receiving broadcast applications of nitrogen. Keller (private communication) measured N_2O losses of as great as 20% of the applied N in palm plantations in Costa Rica where ammonium nitrate fertilizer was broadcast. However, N_2O losses varied greatly among fields. It appears that management practices (and not just quantity or type of N applied) are critical to the control of nitrous oxide flux. Clearly, there is need for more research on trace gas fluxes in tropical agriculture before conclusions can be drawn concerning the importance of the tropical agricultural systems as sources of N_2O.

REACTIVE GASES

Nitric oxide

Nitric oxide (NO) is a short-lived, reactive gas; thus, its global budget is not well constrained by atmospheric measurements, in contrast to the budgets for nitrous oxide, methane or carbon dioxide. Moreover, estimates of NO flux at the soil-air interface do not necessarily equate to inputs to the atmosphere because NO can react and be deposited in plant canopies. Internal deposition within the forest was the fate of about 80% of the soil NO emission observed in forests during the wet season in forest near Manaus, Brazil (Bakwin *et al.*, 1990a, b). Despite the uncertainties in the global budget of NO, it does appear that undisturbed tropical forests and savannas are stronger sources than are temperate systems (Johansson *et al.*, 1988; Johansson and Sanhueza, 1988; Kaplan *et al.*, 1988; Sanhueza *et al.*, 1990; Williams *et al.*, 1992; Müller, 1992).

In temperate systems, NO flux is positively correlated with inorganic nitrogen concentrations in the soil (Williams and Fehsenfeld, 1991), and agricultural systems tend to have higher fluxes than do undisturbed ecosystems (Williams *et al.*, 1992). Data on flux of NO from managed tropical soils, whether in pasture or agriculture, are extremely limited. Davidson *et al.* (1991) reported elevated NO fluxes from a fertilized pasture and maize field in Mexico where Garcia-Mendez *et al.* (1991) also found elevated N_2O fluxes. Likewise, Matson (private communication) found concomitant increases in NO and N_2O after fertilization of sugar cane fields, although the proportional increase in NO was less than that of N_2O. Because the plant canopy is at times absent or limited in agricultural systems, it is likely that a larger fraction of emitted NO reaches the atmosphere in agricultural systems than in forests. Thus, it is probable that the increase of nitrogen fertilization in the tropics will increase nitric oxide fluxes. Likewise, the biomass burning associated with tropical land clearing represents a very significant source of NO, and represents one of the important ways tropical land use change effects a whole range of reactive gases (Crutzen and Andreae, 1990; Lobert *et al.*, 1990).

Vegetation Emissions of Reactive Gases

Terrestrial vegetation emits more volatile organic carbon compounds (VOC) to the atmosphere than industrial processes (Müller, 1992). These compounds are important to regulating atmospheric oxidants at global and regional levels, balancing the global carbon cycle, and producing organic acids that contribute to acidic deposition in remote areas (Fehsenfeld *et al.*, 1992). Isoprene and the monoterpene compounds have been investigated most extensively and a variety of emission algorithms have been developed to relate fluxes to environmental conditions, the most critical being light intensity and temperature (Guenther *et al.*, 1991). Both temperature and light intensity are higher in the tropics versus the temperate and boreal zones. Accordingly, tropical emissions probably account for more than half of global emissions of isoprene (Rasmussen and Khalil, 1988). Oxygenated hydrocarbons are also found in abundance in rural air; their sources, which may be biogenic, are poorly understood (Fehsenfeld *et al.*, 1992).

Investigation of the vegetation emissions of VOC to the atmosphere in the tropics is exceedingly limited. Data collected as part of the Brazilian-American Amazon Boundary Layer Expedition (ABLE-2A) in forest outside Manaus, Brazil may be the only measurements of VOC emissions that permit area flux estimation. Based on data collected at the ABLE experimental sites, isoprene emissions are estimated in the range of 25 to 38 mg m^{-2} d^{-1} and monoterpenes at about 6 mg m^{-2} d^{-1} (Zimmerman *et al.*, 1988; Jacob and Wofsy, 1988). We do not know of published estimates for vegetation VOC emissions from tropical dry forests or tropical savannas.

We cannot estimate the effects of land use change in the tropics on vegetation emissions of VOC. Replacement of natural vegetation with pastures and crops certainly leads to lower leaf areas. Taken alone, the reduction of leaf area would lead to lower fluxes were VOC emissions independent of species. However, VOC emissions are highly species dependent and may vary within species depending upon growing conditions such as nutrient availability or herbivory by insects (Lerdau, 1991).

Vegetation also emits carbon monoxide through light dependent and light-independent pathways (Lüttge and Fischer, 1980; Siegel and Siegel, 1987). The importance of this emission from tropical vegetation is unknown. Measurements of boundary layer CO concentrations in tropical forest as part of ABLE-2B indicate the presence of a strong source, 4×10^{11} molecules $cm^{-2}\,s^{-1}$, for this gas (Kirchhoff and Marinho, 1990). This CO source cannot be explained by hydrocarbon oxidation in the model of Jacob and Wofsy (1990). The effects of land use change on the vegetation source of CO is unknown.

Deposition of Reactive Gases

Although there are few measurements, it appears that tropical forests may be strong sinks for ozone. A summary of measurements of canopy-level ozone deposition for tropical forests is given in Table 1. The measured deposition velocities are at the high end of the range of deposition velocities measured for O_3 over vegetation surfaces (Finlayson-Pitts and Pitts, 1986). Given the expression for deposition velocity recommended by Baldocchi *et al.* (1987),

$$V_d = 1/(r_a + r_b + r_c) \tag{1}$$

we recognize three "resistances" to trace gas deposition: the aerodynamic resistance (r_a), the boundary layer resistance (r_b), and the net-canopy resistance (r_c). Two of these terms, r_a and r_c, are likely to change both from day to night and from wet season to dry season. Fan *et al.* (1990) showed that changes in stomatal resistance (part of the term r_c) could explain the day to night differences in O_3 deposition observed at the Ducke tower site outside of Manaus Brazil. In comparison to the wet season results, dry season nighttime O_3 deposition velocities measured at a nearby site by Kaplan *et al.* (1988) are quite large. Andreae *et al.* (1992) found a similar deposition velocity for nighttime measurements over forest in the Congo Basin. These results cannot be explained by changes in stomatal resistance; stomata close at night increasing stomatal resistance. Greater nighttime deposition velocities for O_3 in the dry season probably results a from smaller aerodynamic resistance (r_a), a function of wind speed, atmospheric stability, and surface roughness.

Deforestation would change the resistance to O_3 deposition through changes to r_a and r_c. Leaf area would decrease as a result of deforestation leading to a greater canopy resistance term. Using the boundary-layer model of Jacob and Wofsy (1990), Fan *et al.* (1990) calculated that halving the leaf area index leads to a 16% decrease in O_3 flux to the canopy. Changes in aerodynamic resistance had a proportionally greater effect. Doubling r_a resulted in a 30% decrease in O_3 flux while halving r_a caused a 40% increase in O_3 flux. The effect of deforestation on r_a is an open question.

Table 1. A summary of mean values for canopy level ozone deposition velocities (V_d) and fluxes (F) at a given reference height (z_r) at tropical forest sites.

Location	Time (Day/Night)	Season (Wet/Dry)	V_d cm s^{-1}	z_r m	F molec. $cm^{-2}\,s^{-1}$	Source
Brazil	Night	Dry	1.8	6	5.6×10^{11}	Kaplan *et al.* (1988)
Brazil	Day	Wet	1.8	39	2.3×10^{11}	Fan *et al.* (1990)
Brazil	Night	Wet	0.3	39	2.2×10^{10}	Fan *et al.* (1990)
Congo	Night	Dry	1.2	n.s.	1.5×10^{11}	Andreae *et al.* (1992)

n.s. = Reference height is not specified.

One study indicates that tropical forests may also be a strong sink for reactive nitrogen compounds. Bakwin *et al.* (1990b) observed that the forest outside of Manaus, Brazil was a net sink for NO_y (8×10^9 molecules $cm^{-2} s^{-1}$). Deposition was greater during the daytime than at night. It is unclear whether stomatal or aerodynamic resistance has the greater effect on this diel pattern.

TROPICAL LAND USE CHANGE AT A REGIONAL SCALE: A CASE STUDY

For long-lived trace gases such as methane and nitrous oxide, soil-atmosphere fluxes are often not sufficiently large to cause measurable perturbations in local and regional atmospheric concentrations. Therefore it is difficult to make areal or regional estimates for fluxes of these gases without resorting to some extrapolation scheme (Stewart *et al.*, 1989). Generally, small scale chamber measurements (<1 m^2) are used to quantify soil-atmosphere flux (Mosier, 1989). The simplest approach may be to sample an area randomly and to estimate the areal flux based upon the random sample. Because trace gas fluxes can be highly variable in space and time, a large numbers of samples may be required to achieve an estimate with an acceptable level of error (Folorunso and Rolston, 1984). A modification of this approach requires stratifying the study area according to parameters that control trace gas fluxes (Matson *et al.*, 1989). The resulting sub-units may then be sampled randomly and greatest effort may be placed in sub-units of greatest flux or greatest variability in order to reduce the absolute error of the areal flux estimate. Areal flux (F) is then estimated as

$$F = \sum A_i f_i \tag{2}$$

where A_i represent the areas of sub-units and f_i are the flux estimates for those areas. Land cover may be an appropriate criterion for choosing areal sub-units since controlling variables such as soil moisture or soil fertility will be associated with a land cover class. Remote sensing provides a technique for classifying land cover (Matson *et al.*, 1990).

We apply the approach described above to the Sarapiqui Canton of Costa Rica, a rain forest area subject to rapid development. We use data from the Costa Rican agricultural census (Censo Agropecuario, 1983) for classification of the Sarapiqui Canton according to land use. Available data span three decades, from 1963 through 1983. Sarapiqui was almost entirely forested prior to colonization in the 1950s (Hall, 1984). We have extrapolated land use trends to the present decade. The agricultural census divides land use according to three categories: forest, pasture, and crops. We sub-divided the pasture category into young and old pastures. Young pastures are those that have been converted from forest use less than 10 years prior to a given date. Old pastures were converted more than 10 years ago. To simplify our analysis further, we assume that the use of land passes from forest directly to pasture and that old pastures are converted to crop land (Figure 1). Except in limited cases such as the banana expansion of the 1960s (cf., Veldkamp *et al.*, 1992), capital requirements limited the pace of agricultural development. In the Sarapiqui Canton, as in other parts of Latin America, cattle raising is a preferred economic activity as a result of a complex of factors including cultural heritage, government incentives, land tenure requirements, and low capital and labor requirements (Hecht, 1992). Land use distributions for five decades based upon census data and our assumptions are presented in Table 2.

Fluxes for given land uses were determined using chamber techniques. For forest and pasture we conducted monthly surveys of flux over at least one year at four forest sites and 10 pastures covering a range of ages from 2 to 35 years. All sites were on similar inceptisols, which cover most of the Sarapiqui Canton. Details of sampling and analysis are described by Keller *et al.* (1993). We use unpublished data based on 8 crop sites (bananas, palms, manioc) on inceptisols and andisols to estimate flux from the crop category. These sites are heavily fertilized; these crops generally receive more than 200 kg-N annually in the form of ammonium nitrate and urea. Farmers apply fertilizer 4 to 6 times per year. Average annual fluxes for methane and nitrous oxide according to land use type are compiled in Table 3.

Figure 2 presents the results of this flux × area extrapolation. In the first decade of colonization, nitrous oxide emissions for the Canton jump to 3 times forest levels. By 1993, emissions of nitrous oxide are still double pre-development levels owing to the substantial areas of crops and a small area of new pastures. In the case of nitrous oxide, fluxes from young pastures dominate the balance. For methane, a weak soil sink is converted to a nearly negligible level as pasture areas become predominant.

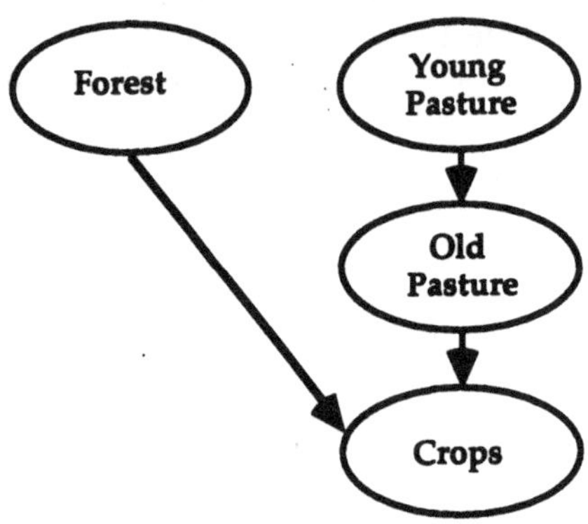

Figure 1. The pattern of land use change observed in the Sarapiqui Canton of Costa Rica from 1953 to 1993.

Table 2. Land use in Sarapiqui Canton of Costa Rica from 1953 to 1993. Percentage of area is shown for each land use category.

Year	Forest	Young Pasture	Old Pasture	Crops
1953	100	0	0	0
1963	70	24	0	6
1973	50	20	22	8
1983	32	18	38	12
1993	24	8	52	16

Table 3. Estimates of methane and nitrous oxide fluxes from soil to the atmosphere from four land use categories in Sarapiqui Canton of Costa Rica. Positive fluxes indicate net transfer of a gas from the soil to the atmosphere (emission from soil). Negative fluxes indicate net transfer of a gas from the atmosphere to the soil (consumption by soil).

Land use	CH_4 Flux (mg m^{-2} d^{-1})	N_2O Flux (ng-N m^{-2} h^{-1})
Forest	−1.22	6.78
Young Pasture	+0.65	48.95
Old Pasture	+0.65	2.20
Crops	−1.22	29.00

This preliminary analysis of land use change is necessarily simplistic. A more comprehensive analysis would use process-based models to estimate fluxes from a database of soil and management characteristics. Soil-atmosphere exchange of methane and nitrous oxide also represents a simple case because these long-lived gases are not lost rapidly by atmospheric reaction or uptake on vegetation surfaces. However, even from this preliminary regional analysis we learn that we must take a dynamic view of the role of land use change in atmospheric processes. In the near-term, and even in the long-term, tropical land use change will contribute significantly to changing the composition of the atmosphere.

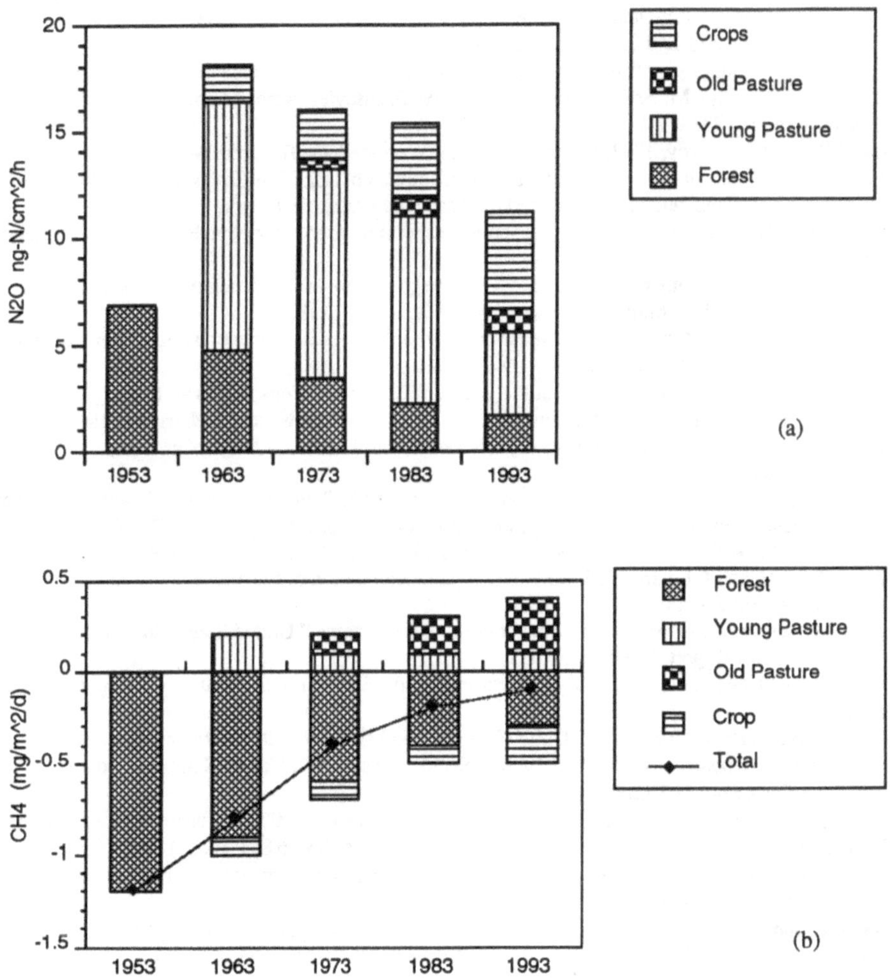

Figure 2. Fluxes of **(a, upper panel)** N_2O (ng–N cm^{-2} h^{-1}) and **(b, lower panel)** CH_4 (mg m^{-2} d^{-1}) for the Sarapiqui Canton of Costa Rica from 1953 through 1993. Net transfers of a gas from soil to the atmosphere are indicated as positive fluxes. Net transfers of a gas from the atmosphere to the soil are indicated as negative fluxes.

REFERENCES

Andreae, M.O., A. Chapuis, B. Cros, J. Fontan, G. Helas, C. Justice, Y.J. Kaufman, A. Minga, and D. Nganga, 1992, Ozone and aitken nuclei over equatorial Africa: Airborne observations during DECAFE 88, *J. Geophys. Res.*, 97:6137-6148.

Bakwin, P.S., S.C. Wofsy, S.-M. Fan, M. Keller, S.E. Trumbore and J.M. da Costa, 1990a, Emission of nitric oxide from tropical forest soils and exchange of NO between the forest canopy and atmospheric boundary layers, *J. Geophys. Res.*, 95:16,755-16,764.

Bakwin, P.S., S.C. Wofsy and S.-M. Fan, 1990b, Measurements of reactive nitrogen oxides (NO_y) within and above a tropical forest canopy in the wet season, *J. Geophys. Res.*, 95:16,765-16,772.

Baldocchi, D.D., B.B. Hicks, and P. Camara, 1987, A canopy stomatal resistance model for gaseous depostion to vegetated surfaces, *Atmos. Environ.*, 22:869-884.

Bartlett, K.B., P.M. Crill, D.I. Sebacher, R.C. Harriss, J.O. Wilson, and J.M. Melack, 1988, Methane flux from the Central Amazonian floodplain, *J. Geophys. Res.*, 93:1571-1581.

Bartlett, K.B. and R.C. Harriss, 1993, Review and assessment of methane emissions from wetlands, *Chemosphere*, 26:261-320.

Bolin, B., E.T. Degens, P. Duvignaud, and S. Kempe, 1979, *in* "The Global Carbon Cycle," B. Bolin, E.T. Degens, S. Kempe, and S. Kertner (eds.),Wiley, Chichester.

Bruenig, E.F., 1990, Oligotrphic forested wetlands in Borneo, *in:* "Ecosystems of the World 15: Forested Wetlands," A.E. Lugo, M. Brinson, and S. Brown (eds.), Elsevier, Amsterdam.

Censo Agropecuario, 1983.

Chanton, J.P. and J.W. Dacey, 1991, Effects of vegetation on methane flux, reservoirs, and carbon isotopic composition, *in:* "Environmental and Metabolic Controls on Trace Gas Emissions from Plants," T. Sharkey, E. Holland, and H. Mooney (eds.), Academic Press, New York.

Cicerone, R. and R. Oremland, 1988, Biogeochemical aspects of atmospheric methane, *Global Biogeochem. Cycles*, 2:299-327.

Crosson, P.R. and N.J. Rosenberg, 1990, Strategies for Agriculture, *in:* "Managing Planet Earth. Readings from Scientific American Magazine," Freeman and Co., New York.

Crutzen, P.J. and M.O. Andreae, 1990, Biomass burning in the tropics: Impact on atmospheric chemistry and biogeochemical cycles, *Science*, 250:1669-1678.

Dacey, J.W., 1981, Pressurized ventilation in the yellow waterlilly, *Ecology*, 62:1137-1147.

Davidson, E.A., P.M. Vitousek, P.A. Matson, R. Riley, G. Garcia-Mendez and J.M. Maass, 1991, Soil emissions of nitric oxide in a seasonally dry tropical forest of Mexico, *J. Geophys. Res.*, 96:15,439-15,445.

Delmas, R.A., J. Servant, J.P. Tathy, B. Cros, and M. Labat, 1992, Sources and sinks of methane and carbon dioxide exchanges in the mountain forest in Equatorial Africa, *J. Geophys. Res.*, 97:6169-6179.

Detwiler, R.P. and C.A.S. Hall, 1988, Tropical forests and the global carbon cycle, *Science*, 239:42-47.

Dörr, H., L. Katruff, and I. Levin, 1993, Soil texture parameterization of the methane uptake in aerated soils, *Chemosphere*, 26:697-713.

EPA, 1990, "Greenhouse gas emissions from agricultural ecosystems," United States Environmental Protection Agency Report, Washington, D.C.

Eichner, M.J., 1990, Nitrous oxide emissions from fertilized soils: Summary of the available data, *J. Env. Qual.*, 19:272-280.

FAO, 1985, "FAO Fertilizer Yearbook," United Nations Food and Agriculture Organization, Rome.

FAO, 1991, "Forest Resources Assessment 1990 Project," Forestry N.7, United Nations Food and Agriculture Organization, Rome.

Fan, S.M., S.C. Wofsy, P.S. Bakwin, D.J. Jacob, and D.R. Fitzjarrald, 1990, Atmosphere-biosphere exchange of CO_2 and O_3 in the Central Amazon forest, *J. Geophys. Res.*, 95:16,851-16,864.

Fearnside, P.M., 1989, Brazil's Balbina Dam: Environment versus the legacy of the pharoahs in Amazonia, *Environm. Mgt.*, 13:401-423.

Fehsenfeld, F. and 10 others, 1992, Emissions of volatile organic compounds from vegetation and the implications for atmospheric chemistry, *Global Biogeochem. Cycles*, 6:351-388.

Finlayson-Pitts, B.J. and J.N. Pitts, 1986, "Atmospheric Chemistry: Fundamentals and Experimental Techniques," John Wiley and Sons, New York.

Folorunso, O.A. and D.E. Rolston, 1984, Spatial variablity of field-measured denitrification gas fluxes, *Soil Sci. Soc. Am. J.*, 48:1214-1219.

Fung, I., J. John, J. Lerner, E. Matthews, M. Prather, L.P. Steele, and P.J. Fraser, 1991, Three-dimensional model synthesis of the global methane cycle, *J. Geophys. Res.*, 96:13,033-13,065.

Garcia-Mendez, G., J.M. Maass, P.A. Matson and P.M. Vitousek, 1991, Nitrogen transformations and nitrous oxide flux in a tropical deciduous forest in Mexico, *Oecologia*, 88:362-366.

Gore, A.J.P. (ed.), 1983, "Mires: Swamp, Bog, Fen, and Moo," Elsevier, Amsterdam.

Goreau, T.J. and W.Z. de Mello, 1988, Tropical deforestation: Some effect on atmospheric chemistry, *Ambio*, 17:275-281.

Guenther, A.B., R.K. Monson, and R. Fall, 1991, Isoprene and monoterpene emission rate variability: Observations with eucalyptus and emission rate algorithm development, *J. Geophys. Res.*, 96:10,799-10,808.

Hao, W.M., D. Scharffe, and P.J. Crutzen, 1988, Production of N_2O, CH_4, and CO_2 from soils in the tropical savanna during the dry season, *J. Atm. Chem.*, 7:93-105.

Harriss, R.C. and D.I. Sebacher, 1981, Methane flux in forested swamps of the southeastern United States, *Geophys. Res. Lett.*, 8:1002-1004.

Harriss, R.C., D.I. Sebacher, K.B. Bartlett, D.S. Bartlett, and P.M. Crill, 1989, Sources of atmospheric methane in the South Florida environment, *Global Biogeochem. Cycles*, 2:231-243.

Hecht, S.B., 1992, Logics of livestock and deforestation: The case of Amazonia, *in:* "Development or Destruction: The Conversion of Tropical Forest to Pasture in Latin America," T.E. Downing, S.B. Hecht, H.A. Pearson, and Carmen Garcia-Downing, (eds.), Westview Press, Boulder, Colorado.

Houghton, R.A., R.D. Boone, J.R. Fruci, J.E. Hobbie, J.M. Melillo, C.A. Palm, B.J. Peterson, G.R. Shaver, G.M. Woodwell, B. Moore, D. Skole and N. Meyers, 1987, The flux of carbon from terrestrial ecosystems to the atmosphere in 1980 due to changes in land use: geographic distribution of the global flux, *Tellus*, 39B:122-139.

Houghton, R.A., D.L. Sklole and D.S. Lefkowitz, 1991, Changes in the landscape of Latin America between 1950 and 1985. II. Net release of CO_2 to the atmosphere, *Forest Ecol. and Mgt.*, 38:173-199.

Janzen, D.H., 1988, Tropical dry forests: The most endangered major tropical ecosystem, *in:* "Biodiversity," E.O. Wilson (ed.), National Academy Press, Washington, D.C.

Jacob, D.J. and S.C. Wofsy, 1988, Photochemistry of biogenic emissions over the Amzon forest, *J. Geophys. Res.*, 93:1477-1486.

Jacob, D. J. and S.C. Wofsy, 1990, Budgets of reactive nitrogen, hydrocarbons, and ozone over the Amazon forest during the wet season, *J. Geophys. Res.*, 95:16,737-16,754.

Johansson, C., H. Rodhe and E. Sanhueza, 1988, Emission of NO in a tropical savanna and cloud forest during the dry season, *J. Geophys. Res.*, 93:7180-7192.

Johansson, C. and E. Sanhueza, 1988, Emission of NO from savanna soils during the rainy season, *J. Geophys. Res.*, 93:14,193-14,198.

Kaplan, W.A., S.C. Wofsy, M. Keller, and J.M. da Costa, 1988, Emission of NO and depostion of O_3 in a tropical forest system, *J. Geophys. Res.*, 93:1389-1395.

Keller, M., W.A. Kaplan, and S.C. Wofsy, 1986, Emissions of N_2O, CH_4, and CO_2 from tropical forest soils, *J. Geophys. Res.*, 91:11,791-11,802.

Keller, M., W. Kaplan, S.C. Wofsy and J.M. da Costa, 1988, Emissions of N_2O from tropical forest soils: Response to fertilization with NH_4^+, NO_3^- and PO_4^{-3}, *J. Geophys. Res.*, 93:1600-1604.

Keller, M. and R.F. Stallard, 1993, Methane emission by bubbling from Gatun Lake, Panama, *J. Geophys. Res.*, in press.

Keller, M. D.J. Jacob, S.C. Wofsy and R.C. Harriss, 1991, Effects of tropical deforestation on global and regional atmospheric chemistry, *Climatic Change*, 19:139-158.

Keller, M., E. Veldkamp, A.M. Weitz, and W.A. Reiners, 1993, Pasture age effects on soil-atmosphere trace gas exchange in a deforested area of Costa Rica, submitted.

Kirchhoff, V.W.J.H. and E.V.A. Marinho, 1990, Surface carbon monoxide measurements in Amazonia, *J. Geophys. Res.*, 95:16,933-16,943.

Lerdau, M., 1991, Plant function and biogenic terpene emissions, *in:* "Environmental and Metabolic Controls on Trace Gas Emissions from Plants," T. Sharkey, E. Holland, and H. Mooney (eds.), Academic Press, New York.

Lobert, J.M., D.H. Scharffe, W.M. Hao and P.J. Crutzen, 1990, Importance of biomass burning in the atmospheric budget of nitrogen-containing trace gases, *Nature*, 346:552-554.

Luizão, F., P. Matson, G. Livingston, R. Luizão and P.M. Vitousek, 1989, Nitrous oxide flux following tropical land clearing, *Global Biogeochem. Cycles*, 3:281-285.

Lüttge, U. and K. Fischer, 1980, Light-dependent net CO-evolution by C_3 and C_4 plants, *Planta*, 149:59-63.

Martens, C.S. and J.V. Klump, 1980, Biogeochemical cycling in an organic-rich coastal marine basin. 1. Methane sediment-water exchange processes, *Geochim. et Cosmoch. Acta*, 44:471-490.

Matson, P.A. and P.M. Vitousek, 1987, Cross-system comparisons of soil nitrogen transformations and nitrous oxide flux in tropical forest ecosystems, *Global Biogoechem. Cycles*, 1:163-170.

Matson, P.A., P.M. Vitousek, J.J. Ewel, M.J. Mazzarino and G.P. Robertson, 1987, Nitrogen transformations following tropical forest felling and burning on a volcanic soil, *Ecology*, 68:491-502.

Matson, P.A., P.M. Vitousek, and D.S. Schimel, 1989, Regional extrapolation of trace gas flux based on soils and ecosystems, *in:* "Exchange of Trace Gases Between Terrestrial Ecosystems and the Atmosphere," M.O Andreae and D.S. Schimel (eds.), John Wiley and Sons, New York.

Matson, P.A., P.M. Vitousek, G.P. Livingston, and N.A. Swanberg, 1990, Sources of variation in nitrous oxide flux from Amazonian ecosystems, *J. Geophys. Res.*, 95:16,789-16,798.

Matson, P.A. and P.M. Vitousek, 1990, Ecosystem approach to a global nitrous oxide budget, *Bioscience*, 40(9):667-672.

Matson, P.A. and P.M. Vitousek, 1993, Biosphere-atmosphere interactions in a tropical deciduous forest ecosystem, *in:* "The Tropical Deciduous Forest Ecosystem," H.A. Mooney, E. Medina and S.H. Bullock (eds.), Springer-Verlag, New York, in press.

Matthews, E. and I. Fung, 1987, Methane emissions from natural wetlands: Global distribution, area, and environmental characteristics of sources, *Global Biogeocheem. Cycles*, 1:61-86.

McElroy, M.B. and S.C. Wofsy, 1986, Tropical forests: Interaction with the atmosphere, *in:* "Tropical Rain Forests and the World Atmosphere," G.T. Prance (ed.), Westview Press, Boulder, Colorado.

Meyer, W.B. and B.L. Turner, II., 1992, Human population growth and global land use/cover change, *Annual Rev. of Ecol. and Systemat.*, 23:39-61.

Mosier, A., 1989, Chamber and isotope techniques, *in:* "Exchange of Trace Gases between Terrestrial Ecosystems and the Atmosphere," M.O Andreae and D.S. Schimel (eds.), John Wiley and Sons, New York.

Mosier, A., D. Schimel, D. Valentine, K. Bronson, and W. Parton, 1991, Methane and nitrous oxide fluxes in native, fertilized, and cultivated grasslands, *Nature*, 350:330-332.

Müller, J.-F., 1992, Geographical distribution and seasonal variation of surface emissions and deposition velocities of atmospheric trace gases, *J. Geophys. Res.*, 97:3787-3804.

Murphy, P.G. and A.E. Lugo, 1986, Ecology of tropical dry forest, *Annual Rev. of Ecol. and Systemat.*, 17:67-88.

Naylor, R. and P.A. Matson, 1993, Environmental Impact of Population Growth, *EOS*, 74:178-179.

Neue, H.-U., 1992, Agronomic practices affecting methane fluxes from rice cultivation, *Ecolog. Bull. (Copenhagen)*, 42:174-182.

Rasmussen, R.A. and M.A.K. Khalil, 1988, Isoprene over the Amazon Basin, *J. Geophys. Res.*, 93:1417-1421.

Richter, D.D. and I. Babbar, 1991, Soil diversity in the tropics, *Adv. in Ecol. Res.*, 21:315-389.

Rodin, L.E., N.I. Bazilevich and N.N. Rogov, 1975, The productivity of the world's main ecosystem, *in:* "Productivity of World Ecosystem," W. National Academy Press, Washington, D.C.

Sanchez, P.A., 1981, Soils of the humid tropics, *in:* "Blowing in the Wind: Deforestation and Long-Range Implications," Department of Anthropology (ed.), College of William and Mary, Williamsburg, VA.

Sanhueza, E., W.M. Hao, D. Scharffe, L. Donoso and P.J. Crutzen, 1990, N_2O and NO emissions from soils of the northern part of the Guayana Shield, Venezuela, *J. Geophys. Res.*, 95:22,481-22,488.

Seiler, W., R. Conrad, and D. Scharffe, 1984, Field studies of methane emission from termite nests into the atmosphere and measurements of methane uptake by tropical soils, *J. Atm. Chem.*, 1:171-186.

Shea, C.P., 1988, Shifting to renewable energy, *in:* "The State of the World 1988," World Watch Institute, W.W. Norton & Company, New York.

Siegel, S.M. and B.Z. Siegel, 1987, Biogenesis of carbon monoxide: Production by fungi and seed plants in the dark, *Phytochem.*, 26:3117-3119.

Steudler, P., R.D. Bowden, J.M. Mellilo, and J.D. Aber, 1989, Influence of nitrogen fertilization on methane uptake in temperate forest soils, *Nature*, 341:314-316.

Stewart, J.W.B., and 11 others, 1989, Extrapolation of flux measurements to regional and global scales, *in:* "Exchange of Trace Gases between Terrestrial Ecosystems and the Atmosphere," M.O Andreae and D.S. Schimel (eds.), John Wiley and Sons, New York.

Tathy, J.P., B.Cros, R.A. Delmas, A. Marenco, J. Servant, and M. Labat, 1992, Methane emission from flooded forest in Central Africa, *J. Geophys. Res.*, 97:6159-6168.

Thompson, K. and A.C. Hamilton, 1983, Peatlands and swamps of the African continent, *in:* "Mires: Swamp, Bog, Fen, and Moor," Gore, A.J.P. (ed.), Elsevier, Amsterdam.

Vaghjiani, G.L. and A.R. Ravishankara, 1991, New measurement of the rate coefficient for the reaction of OH with methane, *Nature*, 350:406-409.

Veldkamp, E., A.M. Weitz, I.G. Staritsky, and E.J. Huising, 1992, Deforestation trends in the Atlantic zone of Costa Rica: A case study, *Land Degrad. and Rehab.*, 3:71-84.

Vitousek, P.M. and R.L. Sanford, Jr., 1986, Nutrient cycling in moist tropical forest, *Annual Rev. of Ecol. and Systemat.*, 17:137-167.

Vitousek, P.M. and P.A. Matson, 1988, Nitrogen transformations in tropical forest soils, *Soil Biol. Biochem.*, 20:361-367.

Vitousek, P.M., P.A. Matson, C. Volkmann, J.M. Maass and G. Garcia-Mendez, 1989, Nitrous oxide flux from seasonally-dry tropical forests: A survey, *Global Biogeochem. Cycles*, 3:375-382.

Vitousek, P.M. and P.A. Matson, 1993, Agriculture, the global nitrogen cycle, and trace gas flux, "Proceedings of the 10th International Symposium on Environmental Biogeochemistry," August 19-23, 1991, in press.

Watson, R.T., H. Rodhe, H. Oeschger and U. Siegenthaler, 1990, Greenhouse gases and aerosols, *in:* "Climate Change: The IPCC Scientific Assessment," J.T. Houghton, G.J. Jenkins and J.J. Ephrams (eds.), Cambridge University Press, Cambridge.

Williams, E.J. and F.C. Fehsenfeld, 1991, Measurement of soil nitrogen oxide emissions in three north American ecosystems, *J. Geophys. Res.*, 96:1033-1042.

Williams, E.J., G.L. Hutchinson and F.C. Fehsenfeld, 1992, NO_x and N_2O emissions from soil, *Global Biogeochem. Cycles*, 6:351-388.

Whiting, G.J., J.P. Chanton, D.S. Bartlett, J.D. Happell, 1991, Relationships between CH_4 emission, biomass, and CO_2 exchange in a subtropical grassland, *J. Geophys. Res.*, 96:13,067-13,071.

Zimmerman, P.R., J.P. Greenberg, and C.E. Westberg, 1988, Measurments of atmospheric hydrocarbons and biogenic emission fluxes in the Amazon boundary layer, *J. Geophys. Res.*, 93:1407-1416.

TRACE GAS EMISSIONS FROM RICE FIELDS

Heinz-Ulrich Neue[1] and Ronald L. Sass[2]

[1]Soil and Water Sciences Division
International Rice Research Institute
Los Baños, Laguna, Philippines
[2]Department of Ecology and Evolutionary Biology
Rice University
Houston, TX 77251, USA

ABSTRACT

Wetland rice cultivation is considered to be one of the larger sources of atmospheric methane, a gas which is an important potential driver of global warming. The atmospheric methane concentration is increasing at about 1% per year and it is an unanswered question as to how much of this increase is due to increased emissions from wetland ricefields. Objectives in current research are to reduce uncertainties concerning how much methane and other climatically active trace gases are annually emitted from irrigated, rainfed, and flood prone rice ecosystems at present, to predict future emissions for given management scenarios, and to develop feasible rice technologies that will reduce emissions and yet will meet the required increase in rice production.

Recent global estimates of methane emission from ricefields range from 20 to 100 Tg/yr corresponding to 6 to 30% of total anthropogenic methane emission. A part of the methane emitted from naturally flooded ricefields may not be considered anthropogenic. Because of the limited number and locations of comprehensive seasonal flux measurements, global extrapolations of emission rates from ricefields are still highly uncertain and tentative. They do not account for varying floodwater regimes, soil properties, organic amendments, cultural practices, and rice cultivars. Irrigated ricefields seem to be the major potential source for increased methane emission. Methane emissions are lower and highly variable in rainfed rice because of periodic droughts during the growing season. Flood prone rice may also emit less methane because of deep flooding or tidal influence. Upland rice is not a source of methane because it is grown like wheat on aerobic soils.

Flooding a ricefield cuts off the oxygen supply from the atmosphere causing an anaerobic fermentation of organic matter in the soil. Methane is a major end product of this process. Zero to over 90% of the methane produced may be oxidized in the soil depending

Global Atmospheric-Biospheric Chemistry
Edited by R.G. Prinn, Plenum Press, New York, 1994

on flood condition and time of growing season. Methane is released to the atmosphere by diffusion, ebullition, and through rice plants. A well developed vascular system, common to wetland plants, provides an effective vent to supply atmospheric oxygen to the rice roots for respiration and to release methane from the soil. Methane fluxes are influenced by: temperature; water regime; low molecular carbon supply from decomposing soil organic residues and root exudates; soil physical, chemical and biological properties; plant physiology; rice cultivars; and cultural practices. Methane emissions from ricefields show distinct diurnal and seasonal variations. Diurnal variation strongly correlates with soil temperature while seasonal variation seems to be more influenced by plant development.

The world's annual rice production must increase by 65% in the next 30 years to feed the expected population. With present agronomic practices, such increased production will lead to further increases in methane emission. Promising mitigation candidates that are in accord with increased production are: shortening of flooding periods through direct seeding and multiple-drainage aeration, minimizing application of easily decomposable organic matter, use of sulfate-containing fertilizer, application of chemicals that inhibit nitrification and methane formation at the same time, breeding of rice cultivars with a lower methane emission potential, and cultural practices that cause less soil disturbance.

INTRODUCTION

Recent atmospheric measurements indicate that concentrations of greenhouse gases are increasing. The present atmospheric methane (CH_4) concentration of 1.72 ppmv is more than double its pre-industrial value of about 0.8 ppmv (Khalil and Rasmussen, 1987). Data from polar ice cores indicate an approximately exponential increase in tropospheric methane concentrations over the past 300 years (Khalil and Rasmussen, 1989). The resulting effect on global temperature is highly significant because the warming efficiency of methane is up to 30 times that of carbon dioxide (Dickinson and Cicerone, 1986). Carbon dioxide contributes currently about 50% and methane 20% to increased radiative forcing of global warming (US-EPA, 1990). The total annual global emission of methane is estimated to be 420 to 620 Tg/yr (Khalil and Rasmussen, 1990). Isotope measurements of atmospheric methane show that 70 to 80% is of biogenic origin (Wahlen *et al.*, 1989; Manning *et al.*, 1990; Quay *et al.*, 1991). Flooding a soil causes anaerobic decomposition of organic matter that leads to the formation and emission of various trace gases not found in well aerated upland soils. These gases include methane, ammonia, hydrogen sulfide, mercaptans, and dimethylsulfide (Neue and Scharpenseel, 1984; Kanda *et al.*, 1992). Wetland rice soils that are alternately flooded and drained emit significant amounts of N_2O (Reddy and Patrick, 1986). With CH_4 emissions up to 100 Tg/yr flooded rice fields are probably one of the largest agricultural sources of atmospheric methane, aside from ruminant enteric digestion (up to 100 Tg/yr), biomass burning (up to 80 Tg/yr), and animal wastes (up to 30 Tg/yr) (Bouwman, 1990; IPCC, 1992). Projected global population levels indicate that the demand for rice will increase by 65% over the next 30 years, from 460 Tg/yr today to 760 Tg/yr in the year 2020 (IRRI, 1989). The growing demand has most likely to be met through intensifying rice production on the existing 144 million ha of harvested riceland, especially in irrigated and rainfed rice ecologies of Asia. This will likely increase methane fluxes to the atmosphere from wetland rice agriculture if current technologies are continued.

Recent estimates provide a fairly good overall balance between sources, sinks and atmospheric accumulation of methane. But global emission rates from many individual sources, including rice fields, are still quite uncertain. The rate of growth of the atmospheric concentration of methane has slowed during recent years from 20 ppbv per year (about 1.3% per year) in the late 1970s to a rate of about 13 ppbv (about 0.75% per year) in 1989 (IPCC, 1992). Whether the declining growth rate is due to decreasing emissions or to increasing atmospheric oxidation of methane is not known. More information and mechanistic understanding of fluxes of methane from individual sources, and of its fate in the atmosphere are needed to reduce uncertainties of current and future emissions and to provide a reliable basis for mitigation. This paper discusses trace gas emissions from rice fields with special emphasis on methane. International research in this area is coordinated by the IGAC Rice Cultivation and Trace Gas Exchange (RICE) Activity.

RICE ENVIRONMENTS

Rice is cultivated under a wider variety of climatic, soil, and hydrological conditions than any other crop. Rice is grown from the equator to as far as 50°N and 40°S, and from sea level to altitudes of more than 2500 m. The temperature may be as low as 4°C during the seedling stage and as high as 40°C at flowering. Rice is irrigated in arid areas and is grown in rainfed areas with only 500 mm rain/yr. Rice is cultivated like wheat on upland soils and in soils that are submerged more than 1 m. Rice is the major crop grown on wetland soils.

Neue (1989) subdivided ricelands into wetlands and uplands. Applying floodwater source hierarchically and floodwater depth as diagnostic criteria he discriminated three major rice ecologies (irrigated rice, rainfed rice and upland rice) with a total of seven subecologies (Table 1). Further differentiation was done by modifiers related to climate, landform, floodwater regime, soil, and cropping system. Wetland rice soils are submerged for at least the major part of one rice growing season. Free water may occur naturally or may be retained by field bunds, puddled plow layers, or traffic pans from rainfall, run-off, or irrigation sources. Wetland rice fields may be without surface water, moist, or dry in other seasons and may therefore alternately support wetland rice and upland crops. The boundary between wetland and upland is often gradual and may fluctuate from year to year depending on variations in precipitation. If water (drainage and irrigation) can be fully controlled it is within farmers' discretion to establish wetlands or uplands. But in most cultivated wetlands of the tropics drainage capacities are insufficient to prevent prolonged soil submergence during rainy seasons.

Table 1. Classification of rice ecologies.

Floodwater source	Irrigation		Pluvial, phreatic, surface flow or tidal				
Floodwater depth (cm)	1–5	5–25	0–25	25–50	50–100	>100	<0
Rice ecologies	Irrigated rice			Rainfed rice			Upland rice
Subecologies	Shallow	Medium	Shallow	Medium	Deep	Very Deep	Upland rice
Land ecosystem	Wetland			Wetland			Upland

The development of high-yielding rice cultivars, large investments in irrigation schemes, and improved soil, water, and crop management dramatically increased rice production since the 1960s. Access to irrigation water and shorter growth duration of modern cultivars enabled the growth of 2 to 3 rice crops per year. But expansion of residential and industrial areas as well as diversification of rice cropping systems resulted only in a slight increase of the harvested rice area (Neue and Roger, 1993). Though many factors will determine the relative contribution of each rice ecology to rice supplies in the future, irrigated areas will continue to dominate rice production. At present, about 50% of the harvested area is devoted to irrigated rice but it contributes about 70% to total production. The distribution of harvested ricelands by rice ecologies is given in Table 2.

Table 2. Distribution of harvested ricelands (million ha) by rice ecologies (FAO, 1988).

Region	Irrigated	Rainfed	Deep-water	Upland	Total area	Yield (t/ha)	Rough rice production (Tg)
East Asia[a]	34.0	2.8	–	–	36.8	5.4	200.0
Southeast Asia[b]	13.9	13.7	3.75	4.65	36.0	2.9	102.5
South Asia[c]	19.4	20.0	7.3	6.7	53.4	2.0	105.5
Near East[d]	1.25	–	–	–	1.25	3.3	4.1
South/Central American Carribean and U.S.	2.5	0.5	0.4	5.65	9.05	2.9	26.5
Africa	0.9	1.95	–	2.70	5.5	1.8	9.9
USSR	0.66	–	–	–	0.66	4.1	2.7
Europe	0.42	–	–	–	0.42	5.4	2.3
Oceania	0.12	–	–	–	0.12	6.6	0.79
Australia	0.11	–	–	–	0.11	7.1	0.76
World Total	73.26	38.95	11.45	19.70	143.36	3.2	455.05

[a]China, Taiwan, Korea, DPR, Korea, RP, Japan
[b]Burma, Cambodia, Indonesia, Laos, Malaysia, Philippines, Thailand, Vietnam
[c]Bangladesh, Bhutan, India, Nepal, Pakistan, Sri Lanka
[d]Afghanistan, Iran, Iraq

Irrigated rice has the highest potential to produce CH_4 because flooding and, consequently, anoxic conditions are assured and controlled. The potential for CH_4 production in rainfed rice should widely vary in time and space since floodwater regimes are primarily controlled by rainfall within the watershed. Periods of severe droughts and/or floods during the growing season are characteristic for rainfed rice. The potential of upland rice for methane production is not significant since upland rice is never flooded for a significant period of time. Emission rates and harvested area of each rice ecology determine the global methane emission.

WETLAND RICE SOILS

Flooding an air-dried cultivated soil drastically changes the hydrosphere, atmosphere, and biosphere of that soil. Flooding highly limits diffusion of air into the soil. The O_2 supply cannot meet the demand of aerobic organisms and facultative and anaerobic organisms proliferate using oxidized soil substrates as electron acceptors in their respiration. Consequently, the redox potential falls sharply according to a sequence predicted by thermodynamics, and CO_2 and HCO_3^- concentrations increase to very high levels. As a result, the soil pH of acid soils increases while that of sodic and calcareous soils decreases, stabilizing between 6.5 and 7.2. Flooding and puddling render most soils an ideal growth medium for rice by supplying abundant water, buffering soil pH near neutral, enhancing N_2 fixation, and increasing diffusion rates, mass flow, and availability of most nutrients. In less favorable soils, flooding may result in toxicities of Fe, H_2S, or organic acids, or deficiencies of Zn or S. The chemistry and biology of wetland rice soils have been reviewed frequently (Ponnamperuma, 1972, 1981, 1984a, 1985; Patrick and Reddy, 1978; De Datta, 1981; Patrick *et al.*, 1985; Watanabe and Roger, 1985; Yu, 1985; Roger *et al.*, 1987; Neue, 1991).

The major gaseous end products of anaerobic fermentation in submerged soils produces an array of organic substances, many of them transitory and not found in aerobic soils. The major gaseous end products are CO_2, H_2S, and CH_4. Methane formation mainly takes place in the reduced topsoil where easily degradable organic substrates are enriched. If the subsoil is saturated with water (aquic moisture regime) the B horizon may also become a source of methane. But in general, carbon contents of B horizons are low and their organic matter is less degradable. If only the topsoil is saturated with water (epiaquic moisture regime) methane oxidation may predominate in the B horizon. Methane oxidation also takes place in the floodwater-soil interface (Bont *et al.*, 1978) and in the rice rhizosphere (Sass *et al.*, 1991a). Methane may also be oxidized in shallow floodwater since it is often oversaturated with O_2 due to assimilation of the aquatic flora. In deepwater rice fields, the deeper layers of the floodwater may also become anoxic during the crop cycle (Whitton and Rother, 1988). Methanogenic bacteria are strict anaerobes and intolerant of O_2 exposure for growth, although they may not be killed by some exposure (Knowles, 1993). Twenty genera of methane-producing bacteria have been described but only a few, including *Methanobacterium* and *Methanosarcina* have been isolated from rice soils (Rajagopal *et al.*, 1988). *Methanospirillum*, and *Methanocorpusculum*, which were isolated from freshwater sediments, as well as methanogens found as endosymbionts in sapropelic amoebae should also be present in wetland rice fields (Neue and Roger, 1993). Most described species are hydrogenotrophs and use formate or acetate while methylotrophs use 1–C compounds such as methanol or methylated amines (Garcia, 1990). Most can fix CO_2 and need H_2 as their electron donor (energy source) even if they use acetate as a C source (Knowles, 1993). In wetland rice soils, the major substrates for CH_4 production are acetate and H_2–CO_2. The microbiology, biochemistry, taxonomy, and ecology of methanogens have been reviewed frequently (Zeikus, 1977; Daniels *et al.*, 1984; Winfrey, 1984; Oremland, 1988; Conrad, 1989; Garcia, 1990; Boone, 1993; Neue and Roger, 1993).

Aerobic methanotrophs oxidize CH_4 in wetland rice soils at the interface of aerobic and anaerobic environments (floodwater-soil interface and rice rhizosphere) where both of the required methane and oxygen are abundant. Several reviews have been published on methane-oxidizing bacteria and aerobic methane oxidation (Whittenbury *et al.*, 1970a, b; Anthony, 1982; Higgins *et al.*, 1981; Crawford and Hanson, 1984). All aerobic methanotrophs sequentially oxidize CH_4 to CO_2 via methanol, formaldehyde, and formate. Oxygen is essential for the growth of methane-oxidizing bacteria but the required partial pressure may be low (Cicerone and Oremland, 1988). There is evidence that certain, mostly sulfate-containing habitats exist in which anaerobic oxidation of methane occurs (Hanson, 1980; Alperin and Reeburgh, 1984; Iversen *et al.*, 1987). The process has also been reported to occur in freshwater systems (Panganiban *et al.*, 1979), but the organisms responsible have not been isolated yet.

Soils can be classified into four groups according to methane produced during anaerobic incubation for some weeks (Neue *et al.*, 1993; Wang *et al.*, 1993b). Each group has a distinct pattern of methane formation and differs significantly in the total amount of CH_4 produced (Figure 1). Differences are probably more related to substrate (hydrogen, acetate) limitation rather than to soil conditions. No general simple correlation between any property, except electrical conductivity (EC), of air-dried soils and methane production was found. The rate of reduction and the Eh and pH of reduced soils are significantly

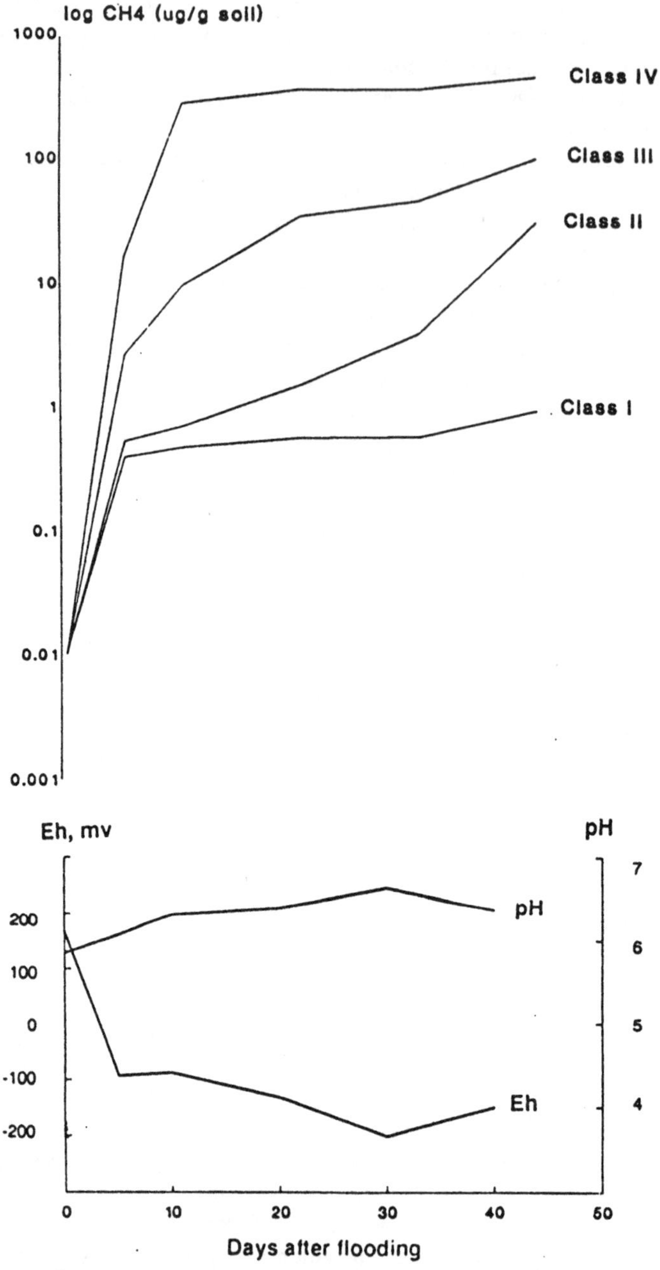

Figure 1. Effect of soil properties, Eh and pH on methane production (IRRI, 1992).

correlated to methane production (Neue and Roger 1993, Neue *et al.*, 1993; Wang *et al.*, 1993b). Total organic carbon and, especially, water soluble carbon are correlated with methane production if other limiting factors are mitigated (Wang *et al.*, 1993b). Since methane formation requires a redox potential lower than −150 mV, NO_3^-, Mn^{4+}, Fe^{3+} and SO_4^{2-} in the soil must be reduced (Patrick and Reddy, 1978). As long as a soil contains significant amounts of active oxidants methane formation is restricted and more decomposable carbon may be mineralized to CO_2. Methane production is limited in sandy soils if water percolation and the resultant redox potential are high. Reduced sandy soils high in organic carbon produce more methane than clay soils with similar carbon contents (Neue and Roger, 1993; Neue *et al.*, 1993). The negative impact of clay-like texture on methane production may be due to formation of organo-mineral complexes. Sandy soils show lower entrapped methane (Wang *et al.*, 1993b) because their pore size distribution enhances ebullition and diffusion (Neue *et al.*, 1993). Methane fluxes in clay-like soils may also be reduced because entrapped methane may be oxidized before it can escape to the atmosphere. Increasing the soil bulk density of flooded soils retards organic matter decomposition, increases the concentration and residence time of organic acids, and reduces the speed of Eh and pH changes as well as methane formation. Neue *et al.* (1990) amalgamated conditions for high methane production in wetland rice soils into six crucial parameters: water regime, Eh/pH buffer, carbon supply, temperature, texture and mineralogy, and salinity. They suggested that Oxisols, most of the Ultisols, and some of the Aridisols, Entisols, and Inceptisols are less favorable to methane production when flooded. Rice soils that are prone to methane production mainly belong to the orders of Entisols, Inceptisols, Alfisols, Vertisols, and Mollisols.

TEMPERATURE

Rice is grown under widely differing temperature regimes. The temperature of flooded rice soils at planting may range from 15°C in northern latitudes to 40°C in equatorial wetlands. The soil temperature of flooded soils varies in response to the meteorological regime acting upon the atmosphere-floodwater and floodwater-soil interfaces. Changing properties of soil and floodwater (i.e. temporal changes in reflectivity, heat capacity, thermal conductivity, temperature of rain and irrigation water, and water flow) as well as vegetation, interact with these external influences. Hackman (1979) reported that floodwater temperatures are above the minimum air temperature but below the maximum air temperature if diurnal amplitudes of air temperature are high, like in subtropical regions or at high altitudes in the tropics. Floodwater temperatures are above maximum air temperatures if diurnal fluctuations are low as in tropical lowlands. The temperature of the puddled topsoil layer (0 to 15 cm) closely follows the temperature of the floodwater and decreases with depth. The high thermal conductivity of flooded and puddled topsoils, in which the soil bulk density may be reduced to only 0.2 to 0.5 g cm^{-3}, enhances the downward conduction. Diurnal amplitudes reached up to 8°C and decreased to less than 1°C in and below the dense plow pan (Neue, 1991). In tropical rice soils maximum soil temperatures occur mostly early in the afternoon and minimum soil temperatures are found early in the morning. Schntz *et al.* (1990) found that the diurnal soil temperature pattern of flooded soils in Italy changed with depth and maximum temperatures showed an increasing phase shift to later daytime probably because of time-dependent heat transport. Rice soils in the U.S. also reveal maximum temperatures in the later afternoon Sass *et al.* (1991a).

Variations in CH_4 production and emission from natural and agricultural wetlands have been correlated to soil temperature by a number of researchers (Crill *et al.*, 1988; Conrad, 1989; Bouwman, 1990; Schntz *et al.*, 1990; Sass *et al.*, 1991a). At soil temperatures found in flooded tropical soils, CO_2 and CH_4 formation occur sooner and in larger amounts than in cooler climates (Tsutsuki and Ponnamperuma, 1987). Holzapfel-Pschorn and Seiler (1986) reported a marked influence of soil temperature on the CH_4 flux with doubling of emission rates when temperature increased from 20 to 25°C. Sass *et al.* (1991a) observed maximum CH_4 production in rice soils at about 37°C irrespective of differences in CH_4 production potentials (Figure 2). Most isolates of methanogenic bacteria are mesophilic with temperature optima of 30 to 40°C (Vogels *et al.*, 1988).

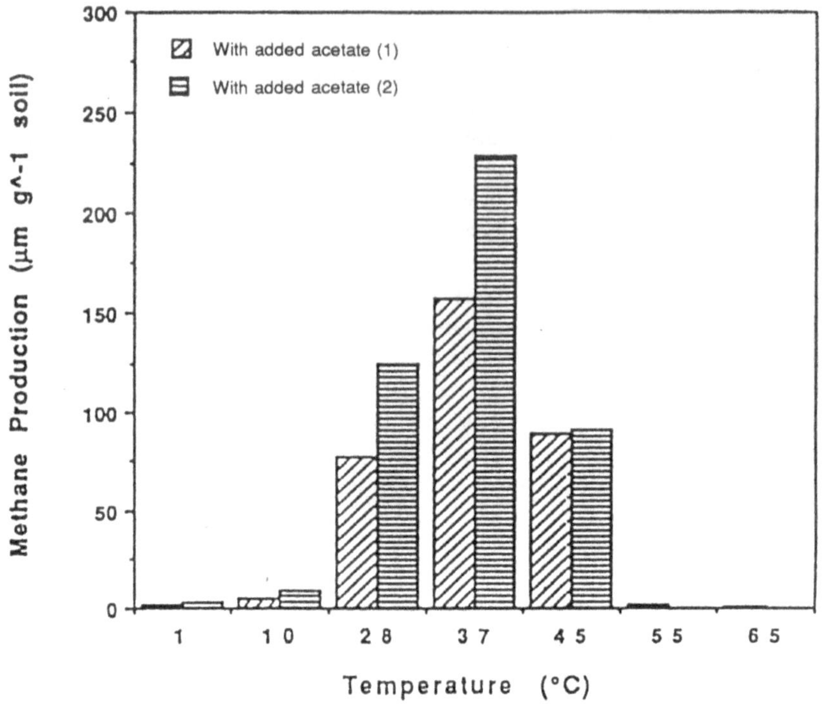

Figure 2. Effect of temperature on methane production (Sass *et al.*, 1991).

Psychrophilic acetate-utilizing methanogens with a temperature optimum below 20°C seem to occur in acidic peat (Svensson, 1984).

Diurnal variation of CH_4 fluxes (Figure 3) is highly correlated with temperature fluctuation. Temperature affects the solubility of methane in the soil solution and floodwater (Figure 4). Less CH_4 is solubilized with rising temperature during daytime. Additional factors like oxygen and substrate supply modify the relationship, especially at later growth stages of the rice crop (Schntz *et al.*, 1989). Thus a simple comparison of field measurements over a whole season may not show a clear temperature correlation. Apparent activation energy values for CH_4 production obtained from incubations of rice soils and diurnal CH_4 emissions in field experiments are in the range of 60 to 130 kJ mol^{-1}. Koyama (1963) found 90 kJ mol^{-1} in Japanese rice soils, Conrad *et al.* (1989) 68 to 90 kJ mol^{-1} in Italian rice soils, Schntz *et al.* (1990) 50 to 130 kJ mol^{-1} in Italian rice soils, and Sass *et al.* (1991a) 90 to 100 kJ mol^{-1} in rice soils of the U.S. Activation energies increased with soil depth (Schntz *et al.*, 1990), possibly indicating differences in the substrate decay process. The same may be assumed for the higher values that have been reported for natural wetlands in temperate zones (Crill *et al.*, 1988) and for upland soils (Sextone and Mains, 1990). Since up to 90% of the produced CH_4 may be oxidized within the soil-floodwater system (Holzapfel-Pschorn *et al.*, 1986, Schntz *et al.*, 1989; Sass *et al.*, 1991a) temperature effects on methanotrophs may have substantial impacts on the activation energy of CH_4 emission as the net flux. Similar activation energies for CH_4 production and emission suggest that both processes may be limited by the same production step. This would suggest that emission is a faster process than production, and a large buildup of CH_4 in a flooded soil grown to rice should not be expected (Sass *et al.*, 1991a).

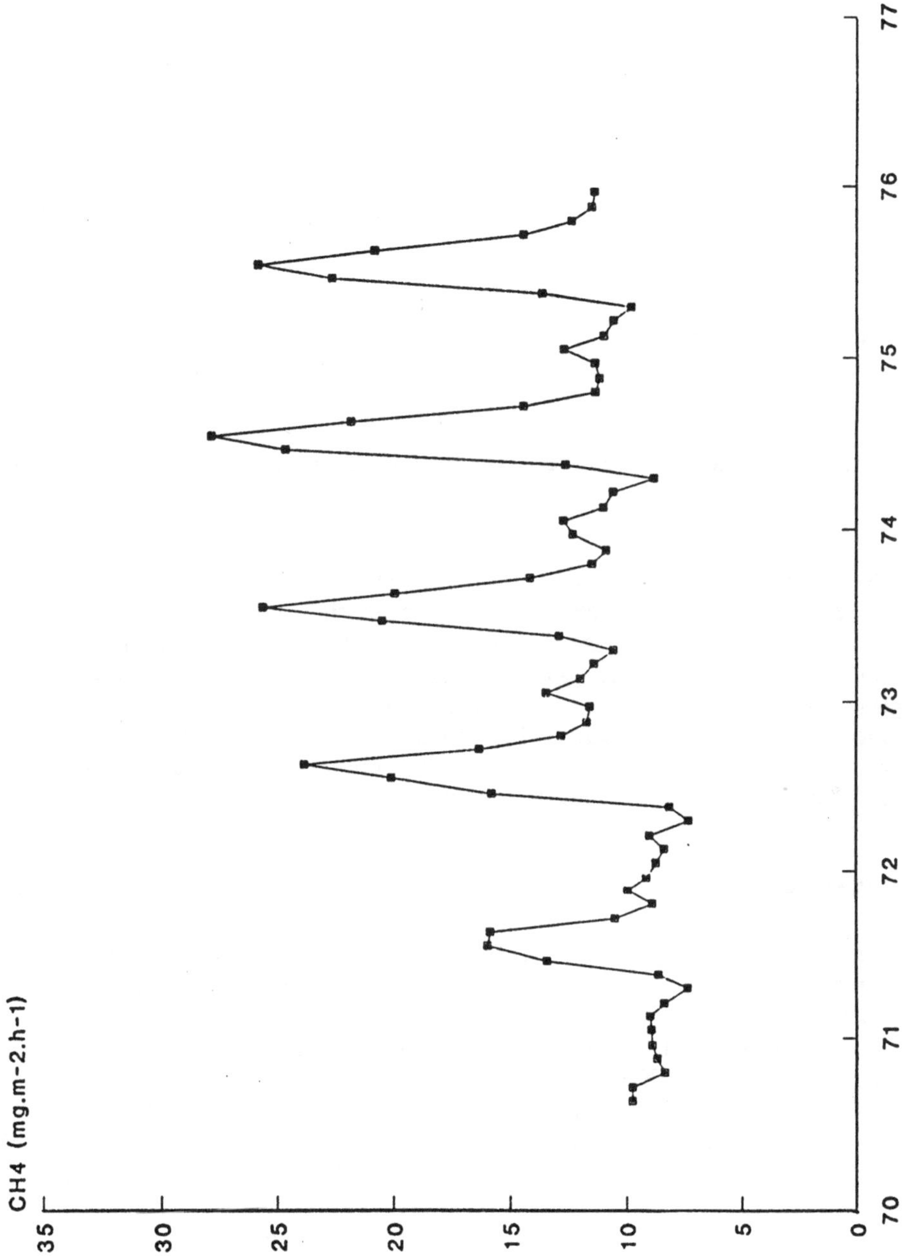

Figure 3. Diurnal variation of methane emission (from IRRI, 1992 DS).

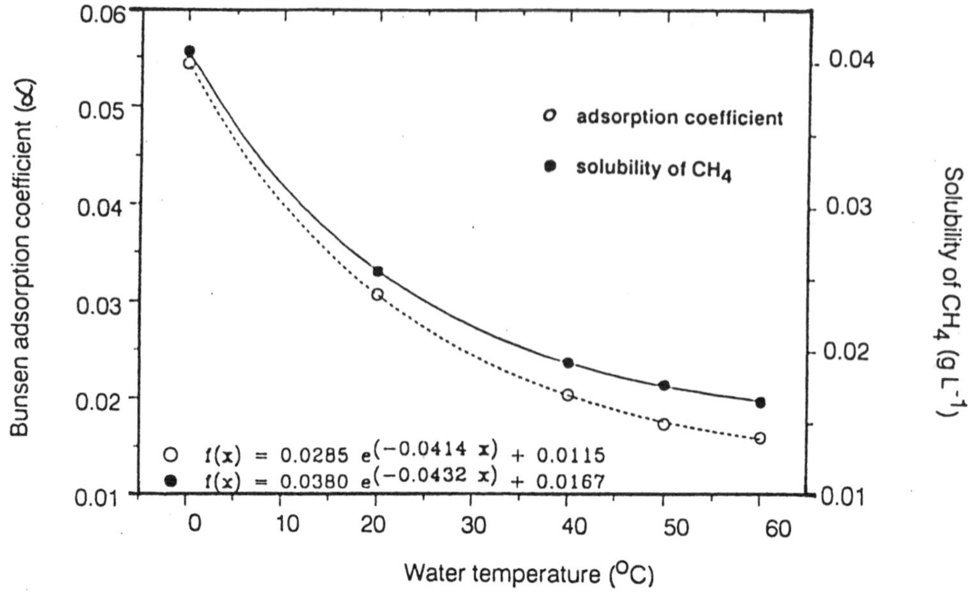

Figure 4. Effect of temperature on solubility of methane in water at 1 atm partial pressure (Butterbach-Bahl, 1993).

The temperature optimum of the production of methanogenic substrates by fermenting bacteria may not concur with the response methanogenesis. In subtropical regions or at high altitudes, the accumulation of intermediate metabolites may reach toxic levels, especially early in the rice-growing season, because of low temperatures. Specific drainage techniques with increased percolation rates and/or intermittent aeration periods are practiced to remedy such accumulations. In tropical lowlands, high temperature throughout the growing seasons stimulates organic matter degradation which favors CH_4 production and limits accumulation of intermediate metabolites.

REDOX POTENTIAL

The supply of biodegradable carbon and the activity of the edaphon are the key to most of the characteristic biochemical and chemical processes in flooded soils (Neue, 1991). These processes include soil reduction and associated electrochemical changes; N immobilization and fixation; production of an array of organic compounds, especially organic acids; and release of NH_4^+, CO_2, H_2S, and CH_4. Since CH_4 is only produced by strictly anaerobic bacteria (methanogens), a sufficiently low redox potential (Eh) is required. Wang *et al.* (1993a) showed that the critical soil Eh for initiation of CH_4 production is approximately −150 to −160 mV (Figure 5). Between −150 mV and −230 mV the relationship of CH_4 production and soil Eh becomes negatively exponential.

The magnitude of soil reduction is determined by the amount of easily degradable organic substrates, their rate of decomposition, and the amounts and kinds of reducible nitrates, iron and manganese oxides, sulfates, and organic compounds. A rapid initial decrease of Eh after flooding in most soils is caused by high decomposition rates of organic substrates and a low buffer of nitrates and Mn oxides. The most important redox buffer systems in rice soils are Fe(III) oxyhydroxides/Fe(II) and organic compounds. Soils low in active iron with high organic matter may attain Eh values of −200 to −300 mV within two weeks after submergence (Ponnamperuma, 1972). In soils high both in iron and organic matter, the Eh may rapidly fall to −50 mV and then decline slowly over weeks and level off. Soils where the redox potential is controlled by a ferritic, ferruginous, or oxidic mineralogy and/or the soil reaction is strong acidic or allic are less prone to CH_4 formation (Neue *et al.*, 1990).

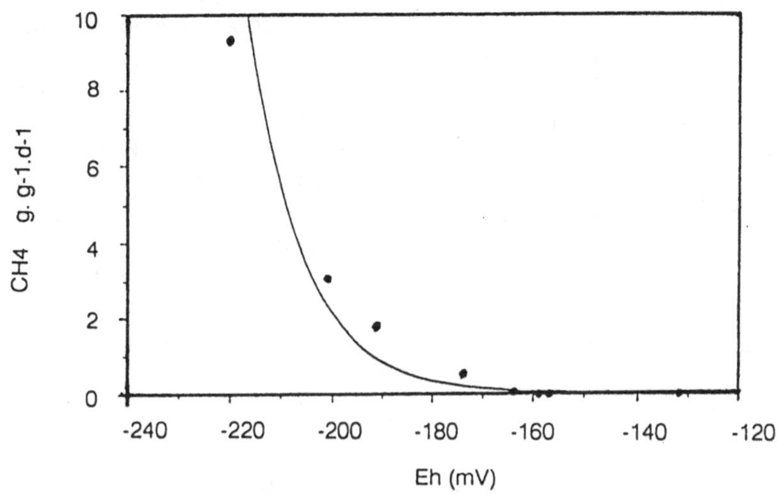

Figure 5. Relationship between soil redox potential and methane production (Wang *et al.*, 1993a).

PARTIAL PRESSURE OF CO_2 AND pH

Soil reduction and accumulation of CO_2 coupled with the formation of HCO_3^- buffers the pH near neutral in flooded soils. The pH increase of acid soils is initially brought about by soil reduction of Fe-oxyhydroxides. The pH decrease of sodic and calcareous soil and the final regulation of the pH rise in acid soils are the result of CO_2 accumulation. Accordingly, the pH values of all flooded soils are highly sensitive to the partial pressure of CO_2 at steady state. Up to 2.6 t CO_2/ha is produced in the puddled layer during the first few weeks of flooding (IRRI, 1964). After addition of organic substrates, the partial pressure of CO_2 in a flooded soil may reach almost 100 kPa (Neue and Bloom, 1989; Ponnamperuma, 1985). Typical values in flooded soils range from 5 to 20 kPa (Kundu, 1987; Patra, 1987). Carbon dioxide concentrations greater than 15 kPa retard root development and reduce nutrient uptake leading to wilting (Dent, 1986). Carbon dioxide profoundly influences the chemical equilibria of almost all divalent cations (Ca^{2+}, Mg^{2+}, Fe^{2+}, Mn^{2+}, Zn^{2+}) in flooded soils as well as CH_4 formation.

The formation of CH_4 after soil submergence is preceded by the production of volatile acids. Short-term H_2 evolution immediately follows the disappearance of O_2. Thereafter, CO_2 production increases, and finally, with decreasing CO_2, CH_4 formation increases (Takai *et al.*, 1956; Neue and Scharpenseel, 1984). The amount of CH_4 found in the soil solution and in gas bubbles of flooded soils may be up to three times higher that of CO_2 after the initial stage of flooding (Martin *et al.*, 1983). The change in favor of CH_4 is mainly caused by assimilation of CO_2 and precipitation of carbonates rather than reduction of CO_2 to CH_4. According to Takai (1970), the bulk of CH_4 in rice fields is formed through decarboxylation of acetic acid, which would result in a 1:1 ratio of CO_2 and CH_4 formation.

Most methanogens are neutrophilic with a relatively narrow pH range of 6 to 8. A few alkaliphilic isolates with optimum growth at pH 8 to 9 have been reported in the genera *Methanosarcina*, *Methanobacterium*, (Blotevogel *et al.*, 1985; Worakit *et al.*, 1986) and *Methanohalophilus* (Mathrani *et al.*, 1988). No acidophilic strains have been reported. Wang *et al.* (1993a) found highest CH_4 production rates at a pH of 6.9 to 7.1 in an acid rice soil (Figure 6). Small changes in pH sharply lowered CH_4 production. Below pH 5.75 and above 8.75 CH_4 production was inhibited completely. Acharya (1935) reported that during decomposition of organic substrates the preliminary stage of acid formation is more tolerant to pH reactions, but CH_4 formation is greatly impeded outside the range of pH 7.5 to 8.0. Parashar *et al.* (1990) found highest CH_4 emission rates at a pH of 8.2 in some Indian rice soils. The actual optimum pH for CH_4 production may slightly differ between soils because of differences in substrate supply and the intimate linkage between pH and Eh. In calcareous and alkaline soils, CH_4 production may occur within hours after flooding an air-dried soil while in acid soils it may take weeks before CH_4 is formed. In very acid soils, CH_4 may not be formed at any time (Neue and Roger, 1993).

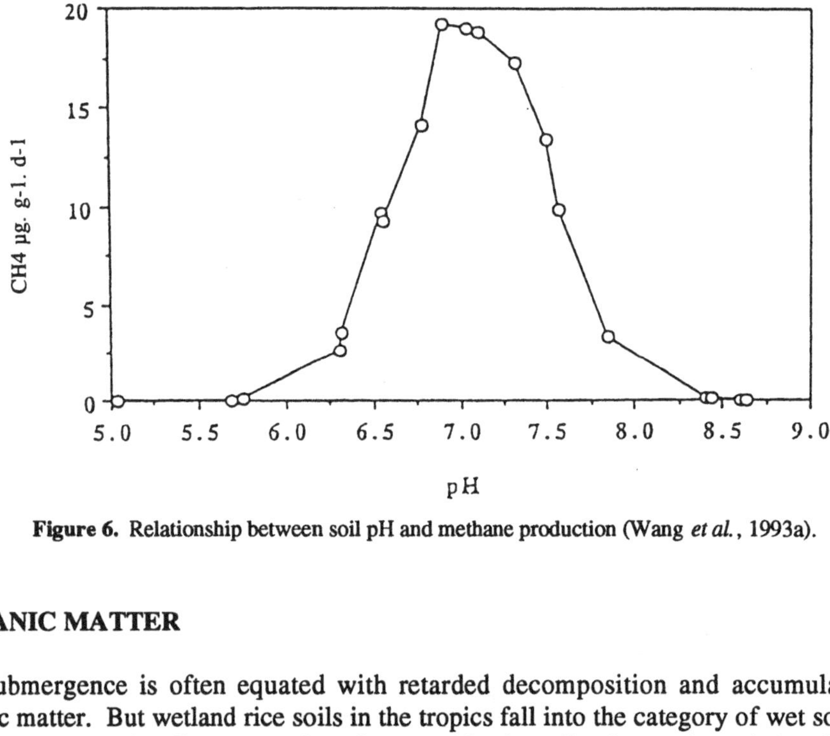

Figure 6. Relationship between soil pH and methane production (Wang *et al.*, 1993a).

ORGANIC MATTER

Submergence is often equated with retarded decomposition and accumulation of organic matter. But wetland rice soils in the tropics fall into the category of wet soils with high temperature in all seasons that show rapid mineralization and weak humification (Bonneau, 1982; Neue and Scharpenseel, 1987; Neue, 1991), both of which favor CH_4 formation. Submergence of soils retards initial decomposition of rice straw in the field only slightly compared with upland soils (Neue and Scharpenseel, 1987; Neue, 1991). The rate of decomposition decreases with soil depth (Neue, 1985). Decomposition of the remaining, more resistant metabolites and residues is also very rapid in all but very acid tropical soils irrespective of water regimes. Half-lives of about two years indicate that total fermentation in a cultivated rice soil, even when continuously flooded, is by no means exclusively anoxic. If the biological activity in flooded rice soils is restricted to bacteria, as in laboratory experiments, the decomposition of rice straw in flooded soils is highly retarded (Capistrano, 1988). Anaerobic fermentation produces an array of organic substances, many of them transitory and not found in well aerated soils. Ponnamperuma (1984a) listed various gases, hydrocarbons, alcohols, carbonyls, volatile fatty acids, nonvolatile fatty acids, phenolic acids, and volatile S compounds. Methanogens constitute the last step in the electron transfer chain generated by the anaerobic degradation of organic matter. Readily mineralizable soil organic matter is the main source for the fermentation products that finally drive the CH_4 formation in wetland rice soils. Generally, there is a strong correlation between water soluble carbon and CH_4 production (Vermoesen *et al.*, 1991).

Organic amendments of flooded soils increase CH_4 production and emission (Schntz *et al.*, 1989; Yagi and Minami, 1990; Sass *et al.*, 1991b; Cicerone *et al.*, 1992; Neue *et al.*, 1993) by enhancing the reduction of soils and providing carbon sources. Changes are more pronounced when organic substrates are added to soils low in organic matter. Based on the content of readily mineralizable carbon, rice straw or green manures produce more CH_4 per unit carbon and humified substrates like compost produce less. Wang *et al.* (1992) showed that CH_4 production increased in proportion to the application rate of rice straw (Figure 7) indicating that CH_4 production was carbon limited in that soil. Application of rice straw remarkably increased CH_4 emission irrespective of soil type while addition of compost enhanced CH_4 emission only slightly (Yagi and Minami, 1990; Minami and Neue, 1993).

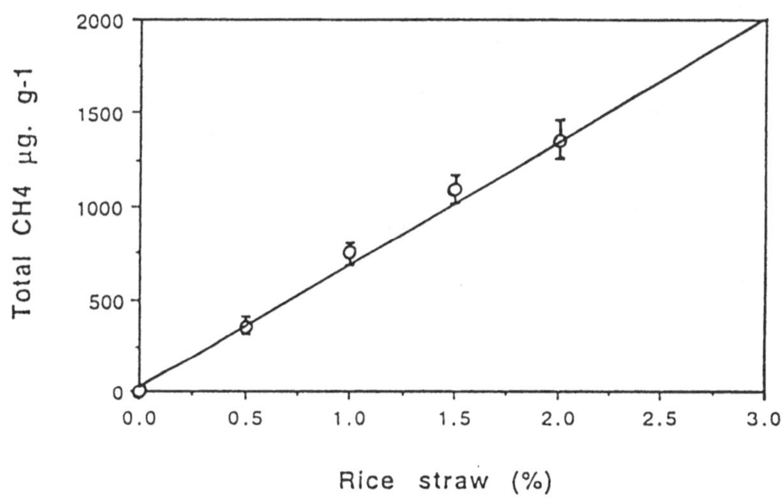

Figure 7. Effect of straw addition on methane production.

RICE CULTIVARS

Rice plants play an important role in the flux of CH_4. Up to 90% of the CH_4 released from rice fields to the atmosphere is emitted through the rice plant (Seiler, 1984; Holzapfel-Pschorn *et al.*, 1986). Well-developed intracellular air spaces (aerenchyma) in leaf blades, leaf sheath, culm and roots provide an efficient gas exchange between the atmosphere and the anaerobic soil. Atmospheric O_2 is supplied via the aerenchyma to the roots for respiration and gases enriched in the soil, like CH_4, diffuse from the reduced soil through the aerenchyma to the atmosphere (Raimbault *et al.*, 1977; Wagatsuma *et al.*, 1990). The root aerenchyma develops as lysogenous intercellular space in the cortex. In older roots the aerenchyma comprises about 50% of its cross-section. The thickness of the surrounding exodermis is about 20 to 30 μm, and the aerenchyma channels are subdivided into separate chambers by porous (2 to 15 μm pore diameter) star-parenchyma (Butterbach-Bahl, 1992). The main gas exchange between root aerenchyma and soil seems to occur via fissures in the exodermis caused by outgrowing side roots. The aerenchyma of roots is not directly connected with the aerenchyma of the shoot. The aerenchyma volume is reduced in the transition, and cell layers of 40 to 50 μm separate root from shoot aerenchyma (Butterbach-Bahl, 1992). The diffusion coefficient of the transition between root and shoot aerenchyma seems to control the gas transport through the rice plant. Cutting shoots above the floodwater or roots does not significantly affect CH_4 diffusion through rice plants within one or two days. Short term variations in humidity, light intensity, and CO_2 supply to the shoot do not change CH_4 emissions. Mariko *et al.* (1991) showed that the CH_4 concentration in the growth medium and the number of tillers per plant is positively correlated with the CH_4 emission rate (Figure 8). Older tillers within a single plant hill show higher CH_4 fluxes (Kimura, 1992). Sass *et al.* (1990) found significant positive correlations between above ground biomass and CH_4 emission and between root biomass and CH_4 production (Figures 9 and 10). Temporal and spatial variability of CH_4 production over the season coincided with distribution of roots. Root exudates and decaying roots seem to become an important carbon source for CH_4 production with progressing plant development. Neue and Roger (1993) found a positive correlation between rice grain yield and CH_4 emission. In general, rice grain yield is a function of plant biomass production. The harvest index (ratio of grain yield to total dry weight) is about 0.3 for traditional tall rice cultivars and 0.5 for improved, short cultivars (Yoshida, 1981).

Oxygen diffusion from rice roots constitutes an important part of the root-oxidizing power aside from enzymatic oxidation due to hydrogen peroxide production. Because of the abundance of CH_4-oxidizing bacteria present in the rhizosphere, its potential for CH_4 oxidation is very high. At tillering, Bont *et al.* (1978) counted 10 times more CH_4-oxidizing bacteria in the rhizosphere than in the bulk anaerobic soil and one third more than in the oxidized soil-water interface. Up to 90% of the CH_4 produced is apparently oxidized in the rhizosphere and in the oxidized soil floodwater interface (Holzapfel-Pschorn *et al.*, 1985; Schntz *et al.*, 1989a; Frenzel *et al.*, 1992). Large cultivar differences in root oxidation power (Neue and Roger, 1993) and in emission rates (Parashar *et al.*, 1990) have been reported.

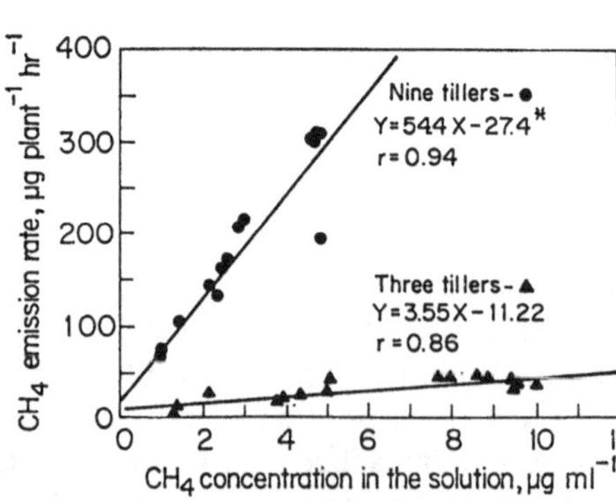

Figure 8. Relationship between methane emission rate from rice plants and methane concentration in the culture solution (Mariko *et al.*, 1991).

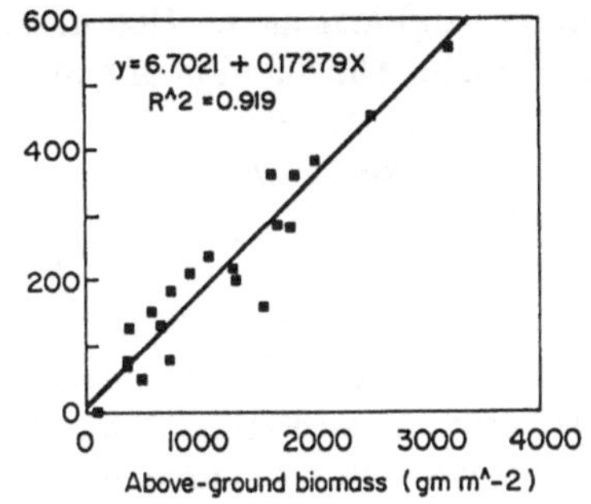

Figure 9. Correlation of measured aboveground biomass with methane emission in the Lake Charles Field (Sass *et al.*, 1990).

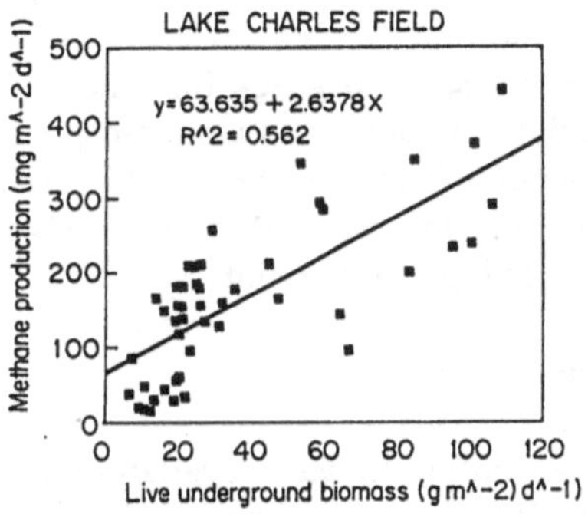

Figure 10. Correlation of measured underground biomass with methane production in the Lake Charles Field (Sass *et al.*, 1990).

AGRONOMIC PRACTICES

Various cultural and agronomic practices have been developed to suit the physical, biological, and socioeconomic conditions of different regions and environments. Effects of agronomic practices on growth and yield of rice are fairly well documented but relationships to CH_4 emission are not well established (Neue, 1992).

Water Control

Water control (irrigation and drainage) is one of the most important factors in rice production. In many rice fields, crops suffer from either too much or too little water because of rainfall pattern and topography. During the monsoonal rainy season in tropical Asia, ricelands are flooded naturally. In field operation, a total of 1240 mm is an average water requirement for an irrigated transplanted rice crop in Asia (Yoshida, 1981) while upland crops may need less than half this amount (Maesschalck *et al.*, 1985). Short aeration periods at the end of the tillering stage and just before heading may improve wetland rice yields (Wang and Zhaoqian, 1986). In the U.S. the shortest flooding period of dry seeded rice consistent with high grain yields is from 75 to 105 days after seeding (Sass *et al.*, 1992). Moisture stress of 50 kPa (slightly above field capacity) may reduce grain yield to 20 to 25% of the yield of continually flooded treatments (De Datta, 1981). The rice plant is most sensitive to water deficit during the reproductive stage causing a high percentage of sterility (Yoshida, 1981). Water deficit during the vegetative stage may reduce plant height, tiller number, and leaf area. This may also highly reduce yields if plants do not recover before flowering. The duration of a moisture stress is more important than the growth stage at which the stress occurs.

Methane emission rates vary markedly with water regimes. A single mid-season drainage may reduce seasonal emission rates by about 50% (Sass *et al.*, 1992; Kimura, 1992). Multiple aeration for 2 to 3 days at 3, 6, and 9 weeks after initial flooding reduced CH_4 emission by 88% and did not reduce rice yields compared to the normal irrigation in the southern U.S. where dry-seeded rice is flooded from 4 to 5 weeks after seedling emergence until 10 to 14 days before harvest (Sass *et al.*, 1992). However, the multiple aeration did require 2.7 times more water than required by the normal floodwater treatment. Fields flooded four weeks later than normal showed the same initial time course observed in other fields but at a later phase in plant development. Daily emissions increased to much higher values thereafter than with other water regimes, leading to the highest total emission.

Increasing percolating rates of water may supply sufficient oxygen to the soil to raise the Eh, decrease CH_4 production and increase CH_4 oxidation. Percolating water also transports organic solutes and dissolved gases into the subsoil or groundwater where leached CH_4 may be oxidized or be released to the atmosphere elsewhere (Kimura, 1992). In soils with high CH_4 production Inubushi *et al.* (1992) found that a percolation rate of about 4 mm per day reduced CH_4 emission up to 58%. They concluded that leaching of dissolved CO_2 mainly caused the high suppression.

Land Preparation and Crop Establishment

Tillage operations vary according to water availability, soil texture, topography, rice culture, and resources available. Kawaguchi and Kyuma (1977) found that 40% of the tropical rice soils they studied had at least 45% clay. Soils with such high clay content have a poor structure and are hard when dry. Since hand and animal powered tillage is still common in most Asian countries and the principal form of mechanization is only the 10 to 15 hp tiller (hand tractor), wet tillage is preferred. Advantages of wet tillage are improved weed control, ease of transplanting, less draft requirement, reduced water percolation, establishment of reduced soil condition which improves soil fertility, and higher fertilizer efficiency, especially for N fertilizer (De Datta, 1981). Wet field preparation requires at least two weeks and comprises land soaking until the soil is saturated, plowing, puddling, leveling and harrowing. One third of the total water required for a rice crop is needed for the wet field preparation. After final harrowing rice seedlings are transplanted. Transplanting rice is still the major practice of rice culture in most of tropical Asia. Direct seeding of pregerminated seeds onto puddled fields without standing water becomes popular in areas with good water control and if manpower is lacking or becomes expensive. The field is reflooded after crop establishment.

When initial crop growth at early monsoonal rainfall becomes essential because of subsequent floods, like in deepwater rice ecosystems, tillage and seeding is done in dry soils. Dryland preparation and seeding may also be practiced in rainfed rice ecosystems when rainfall for the planting month becomes less than 200 mm. Labor constraints and limited water supply may also force farmers in irrigated rice areas to dryland tillage and seeding to ensure timely crop establishment. In United States, Australia, most of Latin America, and Europe where large power units can be employed, dryland tillage and dryland seeding of wetland rice is practiced commonly. Fields are flooded after crop establishment. The crop duration (vegetative phase) is shorter in direct seeded rice, avoiding delay due to seedbed preparation, transplanting, and reduced initial growth because of the transplanting shock. The yield potential of direct seeded rice is similar to that of transplanted rice (De Datta, 1981).

Information on the effect of land preparation and crop establishment on CH_4 fluxes is lacking. Dryland tillage and direct seeding shorten the anaerobic phase and may slow down the decrease of the redox potential after flooding resulting in delayed and probably lower CH_4 emission. But results from late flooding experiments (Sass *et al.*, 1992) indicate that delayed flooding may even result in higher CH_4 emission rates because emission potentials at later plant growth stages are higher. Soil disturbances due to cultural practices during the anaerobic phase will release soil entrapped CH_4 to the atmosphere (Figure 11). Reducing the pressure head of floodwater and aeraeting soil macropores also results in the escape of soil-entrapped CH_4 (shown in Figure 15). Highest emission rates were observed during initial soil drying when soil water was still above field capacity. Thereafter, production of CH_4 declined rapidly and oxidation of CH_4 was highly favored (Neue *et al.*, 1993). Denier *et al.* (1992) estimated that cultural soil disturbances, pre-flooding, and post harvest drying contribute about 15% to the seasonal CH_4 emission.

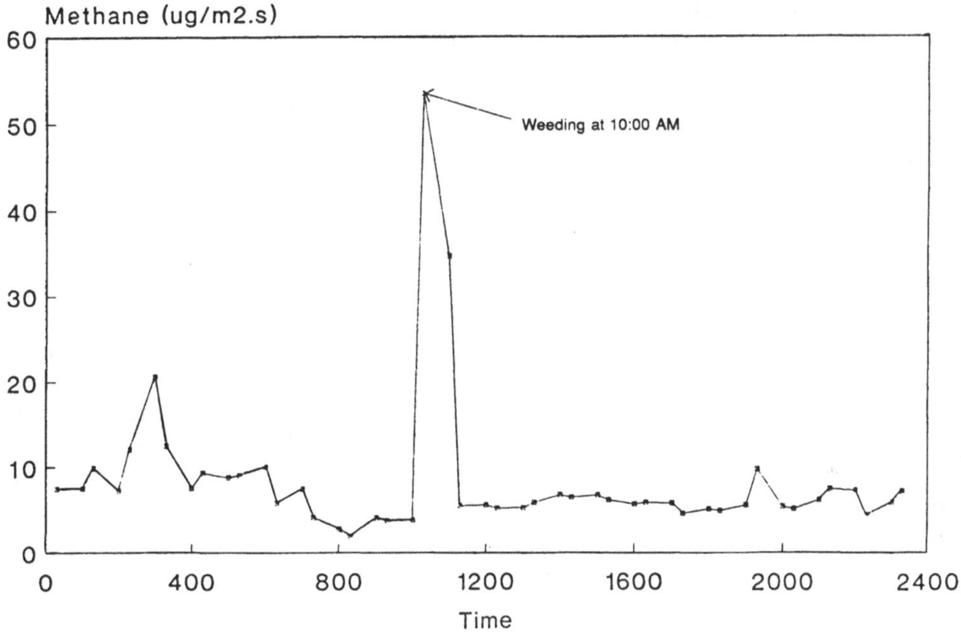

Figure 11. Effect of weeding on methane emission from a rice paddy (IRRI, 1992).

Fertilization

The most deficient nutrient for high wetland rice yields is nitrogen, followed by P, K and Zn. The choice of nitrogen source depends on the method and time of application. Most farmers apply nitrogen fertilizer in two or three splits. The first split is applied during final land preparation or shortly after transplanting. The remainder is topdressed at later growth stages, usually at the early panicle stage. The most common source of N-fertilizer in wetland rice is urea followed by ammonium-containing fertilizers like ammonium sulfate. The source of nitrogen used as topdressing at later growth stages is less critical because of rapid uptake. In general, K and P are applied basally during the final land preparation. Potassium chloride is the principal fertilizer source of K and superphosphate is the primary source of P fertilizer. On acid rice soils, phosphate rock may be applied. Zn may be added by seed treatments, dipping seedling roots in ZnO solution, or broadcasting Zn salts at the time Zn deficiency symptoms occur.

Studies on fertilizer use and rice crop management to minimize nitrogen losses and to increase the efficiency of fertilizer have recently been reviewed (De Datta, 1981; De Datta and Patrick, 1986; De Datta, 1987). Up to 60% of basally applied nitrogen may be lost by volatilization of NH_3 and due to nitrification and denitrification. For basal application, ammonium-containing or ammonium producing (urea) N fertilizer are recommended (De Datta, 1981) to minimize denitrification losses. Soil incorporation or deep placement of N fertilizer without standing water at final harrowing reduces volatilization losses. Broadcasting basal N-fertilizer into floodwater at the early growth stage of rice (low sink strength) results in extensive N losses (as ammonia) to the atmosphere due to high pH values as a result of algal assimilation or alkaline irrigation water (Fillery and Vlek, 1986).

Reports on the influence of the mineral fertilizer application (source, mode, and rate) on CH_4 production and emission are inconsistent. Improving plant growth and enhancing root growth in CH_4 enriched soil layers through fertilization will obviously increase CH_4 emission. Schntz *et al.* (1989) concluded that the type, and method of application strongly influenced CH_4 emission rates. Lindau *et al.* (1991) reported increasing CH_4 fluxes with increasing rates of urea application. Addition of ammonium to the floodwater inhibited CH_4 oxidation in the soil-floodwater interface and reversibly increased the CH_4 flux (Conrad and Rothfuss, 1991). Cicerone and Shetter (1981) reported fivefold increases in emission rates after fertilization with ammonium sulfate while others (Schntz *et al.*, 1989; Yagi and Minami, 1990; Sass *et al.*, 1990) reported a decrease. Schntz *et al.* (1989) reported a decrease of CH_4 emission by 6% when ammonium sulfate was applied to the surface and up to 62% when incorporated into the soil. Lindau *et al.* (1993) found sodium sulfate more effective than ammonium sulfate in reducing CH_4 emissions. Effects of different N-fertilizer on CH_4 production from soil incubation studies have been reported by Wang *et al.* (1992). Addition of urea did not increase total production of CH_4 but stimulated CH_4 production in the acid soil, possibly due to short term increase in pH following urea hydrolysis and resulting decrease of the Eh. A significant decrease in both the production rate and the total amount of CH_4 produced by application of nitrate-containing fertilizer was associated with short term increases in soil Eh. Sulfate-containing fertilizer decreased CH_4 production, especially when applied in larger amounts. Addition of sulfate containing N-fertilizer hardly changes soil Eh and pH. Competition of sulfate-reducing bacteria for hydrogen and hydrogen sulfide toxicity are likely mechanisms.

Inhibitors of CH_4 Formation

Mineral terminal electron acceptors like nitrate or sulfate inhibit methanogenesis in sediments by channeling the electron flow to thermodynamically more efficient bacteria like denitrifiers or sulfate reducers (Balderston and Payne, 1976; Ward and Winfrey, 1985). Manganese and iron oxides should have the same effect. Sodium chloride inhibits pure cultures of methanogens, but high concentrations (about 0.2 M) are required for several strains (Patel and Roth, 1977). Brackish water inhibits methanogenesis (Garcia *et al.*, 1974; De Laune *et al.*, 1983). Methanogens, sulfate-reducers and homoacetogenic bacteria compete for H_2 produced by fermentative bacteria. Hydrogenotrophic homoacetogens do not compete significantly with methanogens for H_2 in sediments (Lovley and Klug, 1983). Since H_2 concentration is usually very low in such environments (Strayer and Tiedje, 1978), sulfate reducers are able to out-compete hydrogenotrophic methanogens in the presence of sulfate because of their higher affinity for H_2 and faster growth (Winfrey and Zeikus, 1977; Abram and Nedwell, 1978; Kristjansson *et al.*, 1982). Freney *et al.* (1982) has also suggested that the reoxidation of S^{2-} to $SO_4{}^{2-}$ in the rice plant rhizosphere may inhibit methanogens over longer periods. However, methanogenesis and sulfate reduction are not mutually exclusive when CH_4 is produced from methanol or methylated amines for which sulfate reducers show little affinity (Oremland *et al.*, 1982; Oremland and Polcin, 1982; Kiene and Visscher, 1987).

Chemical substances inhibiting methanogenesis have been reviewed by Oremland and Capone (1988). The 2-bromoethane-sulfonic acid (BES), an analog to Coenzyme M, is a specific inhibitor of methanogenesis. Several chlorinated CH_4 analogs like chloroform and methyl chloride have been identified to inhibit methanogenesis. Chloroform completely suppressed CH_4 production in a paddy soil but did not hamper the turnover of glucose (Krumböck and Conrad, 1991) although evidence for glucose-utilizing H_2-syntrophic methanogenic bacterial associations has been found in glucose amended paddy soil (Conrad *et al.*, 1989). Substances that inhibit methanogenesis and CH_4 oxidation include DDT (McBride and Wolfe, 1971) and inhibitors of nitrification like acetylene (Raimbault, 1975), nitrapyrin (Salvas and Taylor, 1980), and dicyanidiamide (Lindau *et al.*, 1993). Slow release of acetylene from calcium carbide, encapsulated in fertilizer granules, significantly reduced CH_4 emission (Bronson and Mosier, 1991).

SEASONAL VARIATION IN CH_4 EMISSION

A very clear dependence on temperature is observed in the diurnal variation of CH_4 emission irrespective of emission rates. The temperature dependence of seasonal emission patterns is not as obvious. When the ricefield is flooded during the entire growing season CH_4 emission increases over the growing season and generally shows three distinct seasonal maxima (Schntz *et al.*, 1989; Yagi and Minami, 1990; Wang *et al.*, 1990; Neue *et al.*, 1993). Seasonal patterns of emission from rice fields in the Philippines are shown in Figure 12. The first maximum develops shortly after flooding, the second during the vegetative stage of the rice plant, and the third during the grain filling and maturity stage. If the ricefield is flooded only after full crop establishment, the first two maxima may coincide, as shown in reports of Sass *et al.* (1990) and Lindau *et al.* (1991). The first maximum of CH_4 emission is apparently caused by fermentation of easy degradable soil organic matter accumulated during the fallow period, incorporated crop residues, and organic amendments.

If the amount of easily degradable carbon is low at the beginning of the season no initial peak of CH_4 emission develops (Figure 12). The difference in CH_4 emission rates between plots treated with urea only and amended additionally with rice straw decreases over time and becomes non-significant at the end of the growing season. This clearly indicates the increasing impact of the rice plant over the season not only in mediating CH_4 emission but also in providing carbon for CH_4 production. Patterns of CH_4 ebullition are also affected by plants (Figure 13). After an initial peak, CH_4 ebullition declines because of increasing plant mediated emission. Most methane generated from root exudates and decaying roots seems to be emitted through the rice plant during this period of vigorous vegetative growth. After flowering ebullition increases again until the end of maturity. Although the general pattern of CH_4 fluxes was similar in the wet and the dry season CH_4 emission rates were much higher in the dry season except during the maturity stage (Figure 14). Better plant growth in the dry season seems to cause the difference. Emission rates and diurnal patterns remained similar when only stubbles remained in the field after harvest and the floodwater started to recede (Figure 15). When the soil became only saturated without standing water, emission rates increased sharply without showing diurnal variations anymore. Methane emission became zero when the soil was fully aerated (about field capacity).

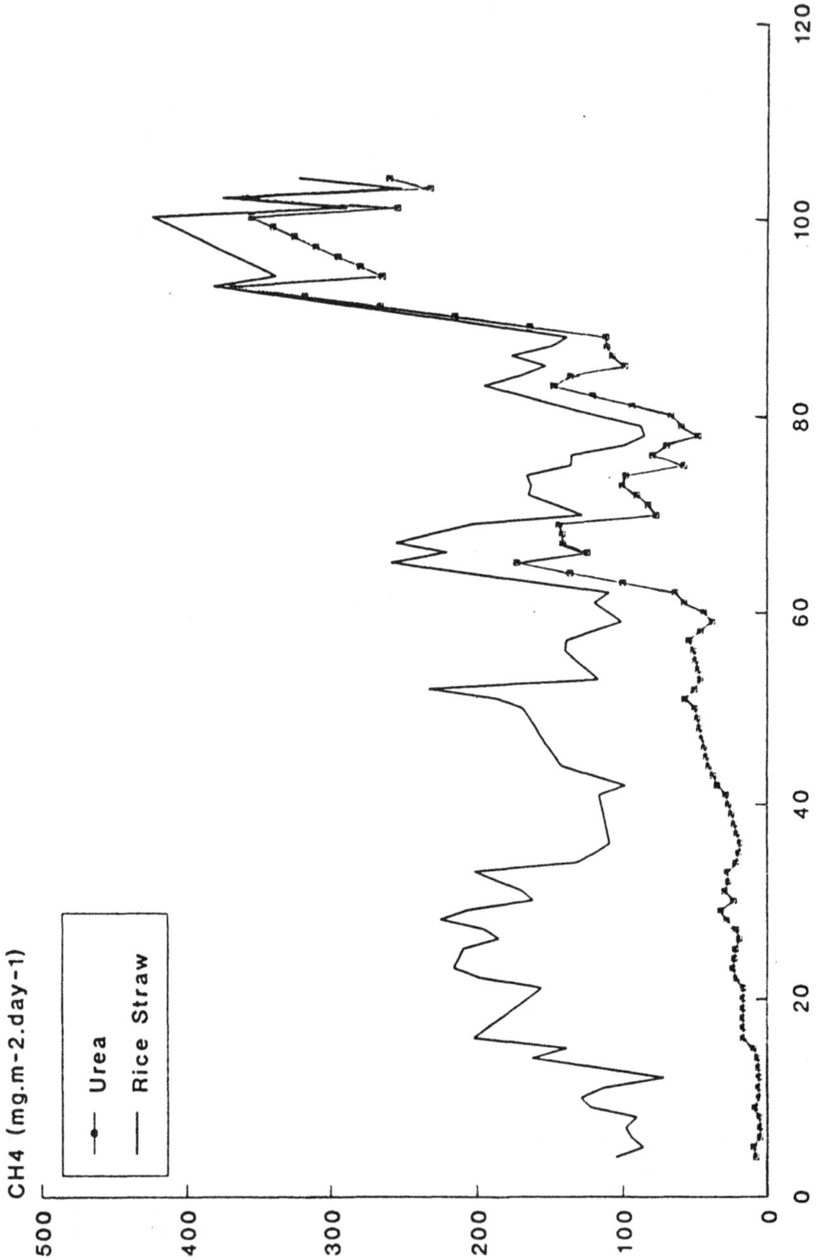

Figure 12. Methane emission under two N sources (from IRRI, 1992 WS).

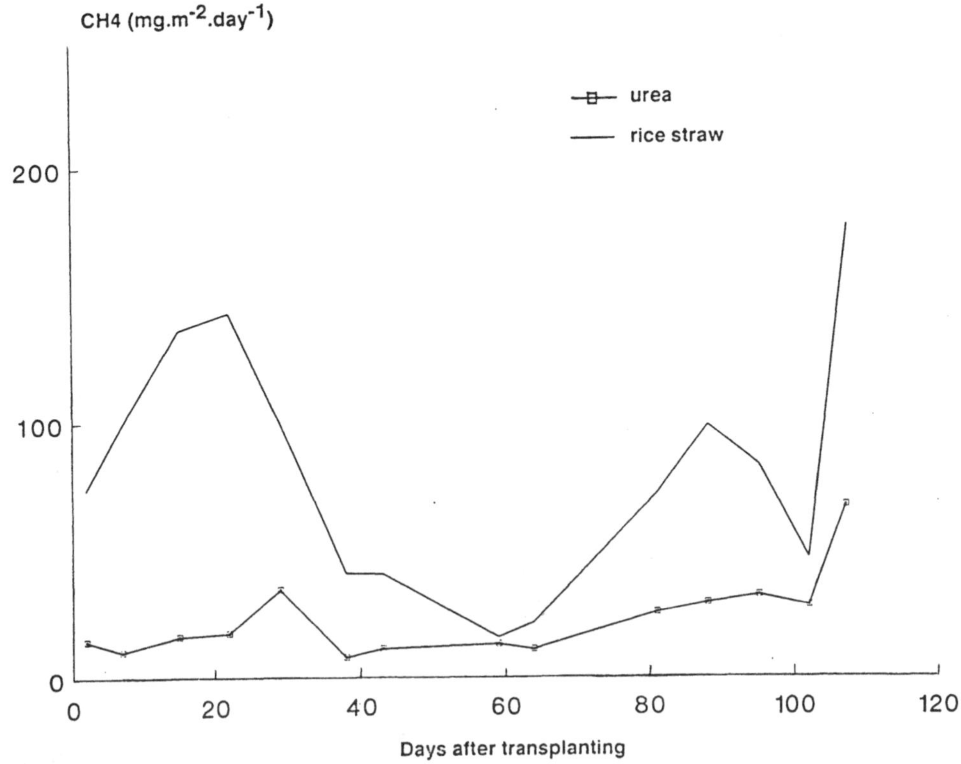

Figure 13. Methane ebullition under two N sources (IRRI, 1992).

CONCLUSIONS

Recent estimates of global CH_4 emission from rice agriculture range from 20 to 100 Tg/yr with a best estimate of 60 Tg/yr (IPCC, 1992). Global and regional estimates vary greatly with the assumptions made on the importance of different factors affecting methane emission, and with the information (mechanistic, distributional) on factors currently available. Predicting future CH_4 emission trends from rice agriculture and developing mitigation technologies that are in accord with the needed increase in rice production requires more information on processes affecting CH_4 fluxes and the geographic distribution of factors controlling these processes. Only three studies have attempted to calculate detailed regional CH_4 emissions. Aselmann and Crutzen (1990) computed monthly distributions of global CH_4 emissions from linearly-temperature-dependent CH_4 fluxes of wetland rice based on 2.5° × 5° grids. Matthews *et al.* (1991) combined land use data on a 1° grid and a rice crop calendar to calculate CH_4 fluxes for each rice growing country, assuming an annual global emission of 100 Tg. Bachelet and Neue (1993) superimposed on Matthews *et al.*'s rice database, information about carbon cycling and turnover and soil type-based coefficients modifying methane emission potentials.

All current global and regional extrapolations of CH_4 emission rates are still highly uncertain and tentative. Accounting for variations of emission rates due to climate, soil properties, duration and pattern of flooding, and rice growth reveals that most published extrapolations seem too high. But reported estimates and measurements of CH_4 fluxes in rice fields do not reflect rice cultural practices, organic amendments, and fertilization that

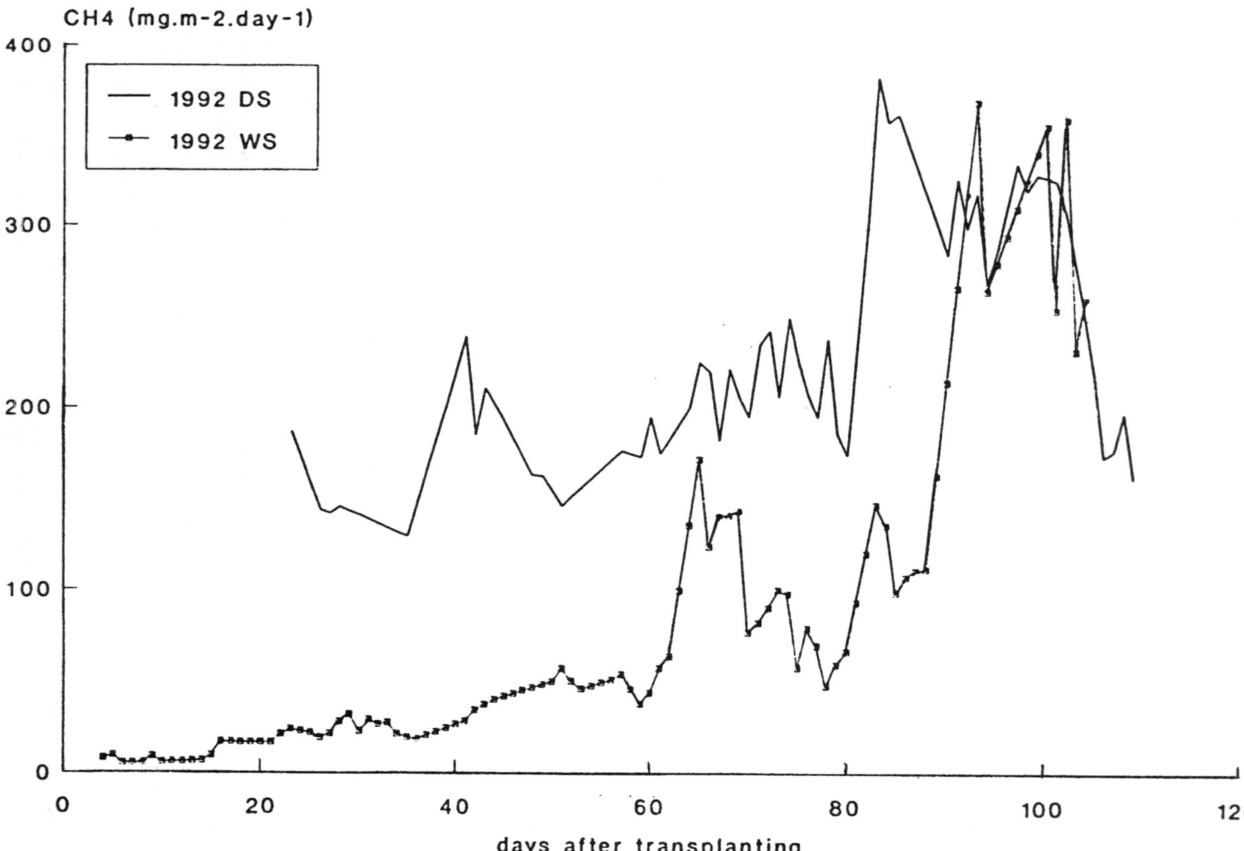

Figure 14. Methane emission during two seasons under urea fertilization (from IRRI, 1992 DS, WS).

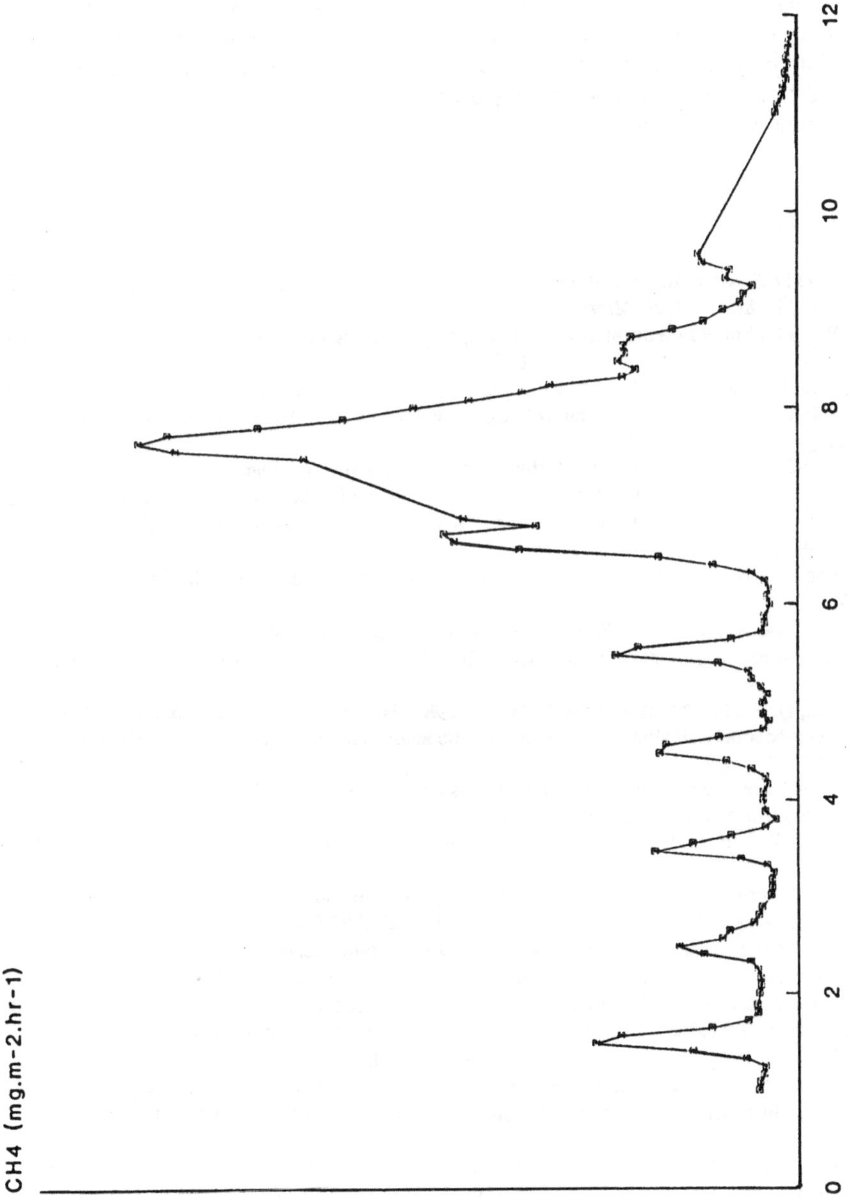

Figure 15. Methane emission after harvest (from IRRI, 1992).

increase CH_4 fluxes. Only further integration of mechanistic modeling of CH_4 fluxes with geographic information systems of factors controlling these processes will improve estimates and predictions. This will be an on-going task for the IGAC-RICE Activity in the future.

In recent years tremendous progress has been made through international and interdisciplinary research collaborations to understand processes of methane fluxes, its temporal and spatial variation, and its relationship to rice agriculture. The achieved results are an excellent base to reduce uncertainties about the CH_4 source strength of different rice ecologies and rice systems and to develop environmentally and socioeconomically sound mitigation technologies where needed.

REFERENCES

Abram, J.W. and D.B. Nedwell, 1978, Inhibition of methanogenesis by sulfate reducing bacteria competing for transferred hydrogen, *Arch. Microbiol.*, 117:89-92.

Acharya, C.N., 1935, Studies on the anaerobic decomposition of plant materials. II. Some factors influencing the anaerobic decomposition, *Biochem. J.*, 29:953-960.

Alperin, M.J. and W.S. Reeburgh, 1984, Geochemical observations supporting anaerobic methane oxidation, *in:* "Microbial growth on C-1 compounds," R.L. Crawford and R.S. Hanson (eds.), American Society of Microbiology, Washington, D.C.

Anthony, C., 1982, "The Biochemistry of Methylotrophs," Academic, San Diego California.

Aselmann, I. and P.J. Crutzen, 1990, Global inventory of wetland distribution and seasonality of net primary production and estimated methane emission, *in:* "Soils and the Greenhouse Effect," A.F. Bouwman (ed.), John Wiley & Sons, Chichester England.

Bachelet, D. and H.U. Neue, 1993, Methane emission from wetland rice areas of Asia, *Chemosphere*, 26:219-238.

Balderston, W.L. and W.J. Payne, 1976, Inhibition of methanogenesis in salt marsh sediments and whole-cell suspensions of methanogenic bacteria by nitrogen oxides, *Appl. Environ. Microbiol.*, 32:264-269.

Blotevogel, K.H., U. Fischer, M. Mocha and S. Jansen, 1985, *Methanobacterium thermoalcapiphilum* sp. nov. A new moderately alkaliphilic and thermophilic autotrophic methanogen, *Arch. Microbiol.*, 142:211-217.

Bonneau, M., 1982, Soil temperature, *in:* "Constituents and properties of soils," M. Bonneau and B. Souchier (eds.), Academic Press, London, pgs. 366-371.

Bont, J.A.M. de, K.K. Lee and D.F. Bouldin, 1978, Bacterial oxidation of methane in rice paddy, *Ecol. Bull.*, 26:91-96.

Boone, D.R., 1993, Formation and consumption of atmospheric methane, *in:* "Global Atmospheric Methane," M.A.K. Khalil and M. Shearer (eds.), NATO ASI/ARW series, in press.

Bouwman, A.F., 1990, Exchange of greenhouse gases between terrestrial ecosystems and the atmosphere, *in:* "Soils and the Greenhouse Effect," A.F. Bouwman (ed.), John Wiley, New York, pgs. 61-127.

Bronson, K.F. and A.R. Mosier, 1991, Effect of encapsulated calcium carbide on dinitrogen, nitrous oxide, methane and carbon dioxide emissions in flooded rice, *Biology and Fertility of Soils*, 3:116-120.

Butterbach-Bahl, K., 1992, "Mechanismen der Produktion und Emission von Methan in Reisfeldern: Abhängigkeit von Felddüngung und angebauter Varietät," Diss. Techn. Univ. München. Schriftenreihe des Fraunhofer Instituts für Atmospherische Umweltforschung Bd. 14. Wiss. Verl. Mauraun, Frankfurt/M.

Capistrano, R.F., 1988, "Decomposition of ^{14}C-labeled rice straw in three submerged soils under controlled laboratory conditions," MS thesis, University of the Philippines at Los Baños Laguna, Philippines.

Cicerone, R.J. and J.D. Shetter, 1981, Sources of chemospheric methane: Measurements in rice paddies and a discussion, *J. Geophys. Res.*, 86:7203-7209.

Cicerone, R.J. and R.S. Oremland, 1988, Biogeochemical aspects of atmospheric methane, *Global Biogeochem. Cycles*, 2:299-327.

Cicerone, R.J., C.C. Delwiche, S.C. Tyler and P.R. Zimmermann, 1992, Methane emissions from California rice paddies with varied treatments, *Global Biogeochem Cycles*, 6:233-248.

Conrad, R., 1989, Control of methane production in terrestrial ecosystems, *in:* "Exchange of trace gases between terrestrial ecosystems and the atmosphere," M.O. Andrae and D.S. Schimel (eds.), S.Bernhard Dahlem Konferenzen,Wiley, New York, pgs. 39-58.

Conrad, R., H.P. Mayer and M. Wnst, 1989, Temporal change of gas metabolism by hydrogen-syntrophic methanogenic bacterial association in anoxic paddy soil, *FEMS Microbiol. Ecol.*, 62:265-274.

Conrad, R. and F. Rothfuss, 1991, Methane oxidation in the soil surface layer of a flooded rice field and the effect of ammonium, *Biol. Fert. Soils*, 12:28-32.

Crawford, R.L. and R.S. Hanson (eds.), 1984, "Microbial growth on C1 compounds," Proceedings of the 4th International Symposium American Society for Microbiology, Washington, D.C.

Crill, P.M., K.B. Bartlett, R.C. Harriss, E. Gorham, E.S. Verry, D.I. Sebacher, L. Madzar and W. Sanner, 1988, Methane flux from Minnesota peatlands, *Global Biogechem. Cycles*, 2:371-384.

Daniels, L., R. Sparling, and G.D. Sprott, 1984, The bioenergetics of methanogenesis, *Biochim. Biophys. Acta.*, 768:113-163.

De Datta, S.K., 1981, "Principles and Practices of Rice Production," John Wiley and Sons, New York.

De Datta, S.K., 1987, Advances in soil fertility research and nitrogen fertilizer management for lowland rice, *in:* "Efficiency of Nitrogen Fertilizer for Rice," International Rice Research Institute, P.O. Box 933 Manila, Philippines, pgs. 27-41.

De Datta, S.K. and W.H. Patrick (eds.), 1986, "Nitrogen Economy of Flooded Rice Soils: Development in Plant and Soil Sciences," Martin Nijhoff Publication, Dodrecht, The Netherlands.

De Laune, R.D., E.J. Smith and W.H. Patrick, 1983, Methane release from Gulf Coast wetlands, *Tellus*, 35B:8-15.

Denier van der Gon, H.A.C., H.U. Neue, R.S. Lantin, R. Wassmann, M.C.R. Alberto, J.B. Aduna and M.J.P. Tan, 1992, Controlling factors of methane emission from rice fields, *in:* "World Inventory of Soil Emission Potentials," N.H. Batjes and E.M. Bridges (eds.), WISE Report 2, ISRIC, Wageningen, The Netherlands, pgs. 81-92.

Dent, D., 1986, "Acid sulfate soils: a baseline for research and development," ILRI Publication 39, Wageningen, The Netherlands.

Dickinson, R.E. and R.J. Cicerone, 1986, Future global warming from atmospheric trace gases, *Nature*, 319:109-115.

FAO, 1988, United Nations Food and Agriculture Organization Report, Rome.

Fillery, I.R.P. and P.L.G. Vlek ,1986, Reappraisal of the significance of ammonia volatilization as a N loss mechanism in flooded rice fields, *in:* "Development in Plant and Soil Sciences," S.K. De Datta and W.H. Patrick (eds.), Martin Nijhoff Publ., Dodrecht, The Netherlands, pgs. 79-98.

Freney, J.R., V.A. Jacq and J.F. Baldensperger, 1982, The significance of the biological sulfur cycle in rice production, *Dev. Plant Soil Sci.*, 5:271-317.

Frenzel, P., F. Rothfuss and R. Conrad, 1992, Oxygen profiles and methane turnover in a flooded microcosm, *Biol. Fertil. Soils*, 14:84-89

Garcia, J.L., 1990, Taxonomy and ecology of methanogens, *FEMS Microbiol. Rev.*, 87:297-308.

Garcia, J.L., M. Raimbault, V. Jacq, G. Rinaudo and P. Roger, 1974, Activities microbiennes dans les sols de rizieres du senegal: relations avec les proprietes physicochimiques et influence de la rhizosphere, *Rev. Ecol. Biol.*, 11:169-185.

Hackman, C.W., 1979, Rice field ecology in Northeastern Thailand: The effect of wet and dry season on a cultivated aquatic ecosystem, *Monogr. Biol.*, 34, J. Illies (ed.), W. Junk Publisher.

Hanson, R.S., 1980, Ecology and diversity of methylotrophic organisms, *Adv. Appl. Microbiology*, 26:3-39.

Higgins, I.J., D.J. Best, R.C. Hammond and D.C. Scott, 1981, Methane-oxidizing microorganisms, *Microbiol. Rev.*, 45:556-590.

Holzapfel-Pschorn, A., R. Conrad and W. Seiler, 1986, Effects of vegetation on the emission of methane from submerged paddy soil, *Plant and Soil*, 92:223-233.

Holzapfel-Pschorn, A., R. Conrad and W. Seiler, 1985, Production oxidation and emission of methane in rice paddies, *FEMS Microbiol. Ecol.*, 31:343-351.

Holzapfel-Pschorn, A. and W. Seiler, 1986, Methane emission during a cultivation period from an Italian rice paddy, *J. Geophys. Res.*, 91:11,803-11,814.

Inubushi, K., Y. Muramatsu and M. Umebayashi, 1992, Influence of percolation on methane emission from flooded paddy soil, *Jpn. J. Soil Sci. Plant Nutr.*, 63:184-189

IPCC (Intergovernmental Panel on Climate Change), 1992, "Climate Change: The supplementary report to the IPCC scientific assessment," J.T. Houghton, B.A. Callender and S.K.Varney (eds.), Cambridge University Press, UK.

IRRI: International Rice Research Institute, 1964, "Annual report for 1963," P.O. Box 933, Manila, Philippines.

IRRI: International Rice Research Institute, 1989, "IRRI Toward 2000 and Beyond," P. O. Box 933, Manila, Philippines.

IRRI: International Rice Research Institute, 1991, "World Rice Statistics 1990," P. O. Box 933, Manila, Philippines.

IRRI: International Rice Research Institute, 1992, "Wet Season & Dry Season Reports," P. O. Box 933, Manila, Philippines.

Iversen, N., R.S. Oremland and M.J. Klug, 1987, Big Soda Lake (Nevada) 3 pelagic methanogenesis and anaerobic METHANE oxidation, *Limnol.Oceanogr.*, 32:804-814.

Kanda, K., H. Tsuruta and K. Minami, 1992, Emission of dimethyl sulfide, carbonyl sulfide, and carbon disulfide from paddy fields, *Plant Nutr.*, 38(4):709-716.

Kawaguchi, K. and K. Kyuma, 1977, "Paddy Soils in Tropical Asia: Their Material Nature and Fertility," The University Press of Hawaii Honolulu Hawaii, U.S.

Khalil, M.A.K. and R.A. Rasmussen, 1987, Atmospheric methane: trends over the last 10000 years, *Atmos. Environ.*, 21: 2445-2452.

Khalil, M.A.K. and R.A. Rasmussen, 1989, Climate induced feedback for the global cycles of methane and nitrous oxide, *Tellus*, 41B:554-559.

Khalil M.A.K. and R.A. Rasmussen, 1990, Constraints on the global sources of methane and an analyses of recent budgets, *Tellus*, 42B:229-236.

Kiene, R.P. and P.T. Visscher, 1987, Production and fate of methylated sulfur compounds from methionine and dimethylsulfoniopropionate in anoxic marine sediments, *Appl. Environ. Microbiol.*, 53:2426-2434.

Kimura, M., 1992, Methane emission from paddy soils in Japan and Thailand, *in:* "World inventory of soil emission potentials," N.H. Batjes and E.M. Bridges (eds.), WISE Report 2, ISRIC, Wageningen, pgs. 43-79.

Knowles, R., 1993, Methane: Processes of production and consumption, *in:* "Agricultural Ecosystem Effects on Trace Gases and Global Climate Change," ASA Special Publication No. 55, pgs. 145-156.

Kristjansson, J.K., P. Schonheit and R.K. Thauer, 1982, Different Ks values for hydrogen and methanogenic and sulfate-reducing bacteria: An explanation for the apparent inhibition of methanogenesis by sulfate, *Arch. Microbiol.*, 131:278-282.

Koyama, T., 1963, Gaseous metabolism in lake sediments and paddy soils and the production of atmospheric methane and hydrogen, *J. Geophys. Res.*, 68:3971-3973.

Krumböck, M. and R. Conrad, 1991, Metabolism of position-labeled glucose in anoxic methanogenic paddy soil and lake sediment, *FEMS Microbiol. Ecol.*, 85:247-256.

Kundu, D.K., 1987, "Chemical kinetics of aerobic soils and rice growth," PhD thesis, Indian Agricultural Research Institute, New Delhi, India.

Lindau, C.W., P.K. Bollich, R.D. DeLaune, W.H. Patrick Jr. and V.J. Law, 1991, Effect of urea fertilizer and environmental factors on methane emissions from a Louisiana rice field, *Plant and Soil*, 136:195-203.

Lindau, C.W., P.K. Bollich, R.D. Delaune, A.R. Moisier and K.F. Bronson, 1993, Methane mitigation in flooded Louisiana rice fields, *Biol. Fert. Soils*, 15:174-178.

Lovley, D.R. and M.J. Klug, 1983, Methanogenesis from methanol and from hydrogen and carbon dioxide in the sediments of a eutrophic lake, *Appl. Environ. Microbiol.*, 45:1310-1315.

Maesschalck, G., H. Verplancke and M. De Boodt, 1985, Water use and wateruse efficiency under different management systems for upland crops, *in:* "Soil Physics and Rice," International Rice Research Institute, P.O. Box 933, Manila, Philippines, pgs. 397-408.

Manning, M.R., D.C. Lowe, W. Melhuish, R. Spaarks, G. Wallace, C.A.M. Brenninkmeijer and R.C. McGill, 1990, The use of radiocarbons measurements in atmospheric studies, *Radiocarbons*, 32:37-58.

Mariko, S., Y. Harazono, N. Owa and I. Nouchi, 1991, Methane in flooded soil, water and the emission through rice plants to the atmosphere, *Environm. Experim. Botany*, 31:343-350.

Martin, U., H.U. Neue, H.W. Scharpenseel and P.M. Becker, 1983, Anaerobe Zersetzung von Reisstroh in einem gefluteten Reisboden auf den Philippinen, *Mitt. Dtsch. Bodenkdl. Gesellsch.*, 38:245-250.

Mathrani, I.M., D.R. Boone, R.A. Mah, G.E. Fox and P.P. Lau, 1988, Methanohalophilus zhilinae sp.nov. an alkaliphilic halophilic methylotrophic methanogen, *Inst. J. Sys. Bacteriol.*, 38:139-142.

McBride, B.C.and R.S. Wolfe, 1971, Inhibition of methanogenesis by DDT, *Nature*, 234:551–552.

Minami, K. and H.U. Neue, 1993, Rice paddies as methane source, *Cimate Change*, in press.

Neue, H.U., 1985, Organic matter dynamics in wetland soils, *in:* "Wetland Soils: Characterization Classification and Utilization," International Rice Research Institute, P.O. Box 933 Manila Philippines, pgs. 109-122.

Neue, H.U., 1989, Rice growing soils: constraints utilization and research needs, *in:* "Classification and Management of Rice Growing Soils," FFFTC Book Series No. 39, Food and Fertilizer Technology Center for the ASPAC Region Taiwan R.O.C., pgs. 1-14

Neue, H.U., 1991, Holistic view of chemistry of flooded soil, *in:* "Soil Management for Sustainable Rice Production in the Tropics," International Board for Soil Research and Management, IBSRAM Monograph No. 2, Bangkok, pgs. 5-32.

Neue, H.U., 1992, Agronomic practices affecting methane fluxes from rice cultivation, *Ecol. Bull.*, 42:174-182.

Neue, H.U. and H.W. Scharpenseel, 1984, Gaseous products of the decomposition of organic matter in submerged soils, *in:* "Organic Matter and Rice," International Rice Research Institute, P.O. Box 933, Manila, Philippines, pgs. 311-328.

Neue, H.U. and H.W. Scharpenseel, 1987, Decomposition pattern of 14C-labeled rice straw in aerobic and submerged rice soils of the Philippines, *Science Total Environ.*, 62:431-434.

Neue, H.U. and P.R. Bloom, 1989, Nutrient kinetics and availability in flooded soils, *in:* "Progress in Irrigated Rice Research," International Rice Research Institute, P.O. Box 933, Manila, Philippines, pgs. 173-190.

Neue, H.U., P. Becker-Heidmann and H.W. Scharpenseel, 1990, Organic matter dynamics soil properties and cultural practices in ricelands and their relationship to methane production, *in:* "Soils and the Greenhouse Effect," A.F. Bouwman (ed.), John Wiley & Sons, Chichester, England, pgs. 457-466.

Neue, H.U. and P.A. Roger, 1993, Rice agriculture: factors controlling emission, *in:* "Global Atmospheric Methane," M.A.K. Khalil and M. Shearer (eds.), NATO ASI/ARW series, in press.

Neue, H.U., R.S. Lantin, R. Wassmann, J.B. Aduna, M.C.R. Alberto and M.J.F. Andales, 1993, Methane emission from rice soils of the Philippines, *in:* "Methane and Nitrous Oxide Emission from Natural and Anthropogenic Sources," NIAES, Japan, in press.

Oremland, R.S., 1988, The biogeochemistry of methanogenic bacteria, *in:* "Biology of anaerobic microorganism," A.J.B. Zehnder (ed.), J. Wiley, New York, pgs. 641-702.

Oremland, R.S. and D.G. Capone, 1988, Use of specific inhibitors in biogeochemistry and microbial ecology, *Adv. Microbiol. Ecol.*, 10:285-383.

Oremland, R.S., L.M. Marsh and S. Polcin, 1982, Methane production and simultaneous sulfate reduction in anoxic salt marsh sediments, *Nature*, 296:143-145.

Oremland, R.S. and S. Polcin, 1982, Methanogenesis and sulfate-reduction: competitive and noncompetitive substrate in estuarine sediments, *Appl. Environ. Microbiol.*, 44:1270-1276.

Panganiban, A.T., T.E. Patt, W. Hart and R.S. Hanson, 1979, Oxidation of methane in the absence of oxygen in lake water samples, *Appl. Environ. Microbiol.*, 37:303-309.

Parashar, D., C.J. Rai, P.K. Gupta and N. Singh, 1990, Parameters affecting methane emission from paddy fields, *Indian J. Radio Space Physics*, 20:12-17.

Patel, G.B. and L.A. Roth, 1977, Effect of sodium chloride on growth and methane production of methanogens, *Can. J. Microbiol.*, 6:893.

Patra, P.K., 1987, "Influence of water regime on the chemical kinetics of soils and rice growth," PhD thesis, Indian Agricultural Research Institute, New Delhi, India.

Patrick, W.H., Jr , D.S. Mikkelsen and B.R. Wells, 1985, Plant nutrient behavior in flooded soil, *in:* "Fertilizer Technology and Use," Soil Science Society of America, Madison, Wisconsin, 3rd edition.

Patrick, W.H., Jr. and C.N. Reddy, 1978, Chemical changes in rice soils, *in:* "Soils and Rice," International Rice Research Institute, P.O. Box 933, Manila, Philippines, pgs. 361-380.

Ponnamperuma, F.N., 1972, The chemistry of submerged soils, *Adv. Agron.*, 24:29-96.

Ponnamperuma, F.N., 1981, Some aspects of the physical chemistry of paddy soils, *in:* "Proceedings of the Symposium on Paddy Soils," Science Press, Beijing People's Republic of China, pgs. 59-94.

Ponnamperuma, F.N., 1984a, Effects of flooding on soils, *in:* "Flooding and Plant Growth," T.T. Kozlowski (ed.), Academic Press, New York, pgs. 9-45.

Ponnamperuma, F.N., 1984b, Straw as a source of nutrients for wetland rice, *in:* "Organic Matter and Rice," International Rice Research Institute, P.O. Box 933, Manila, Philippines, pgs. 117-136.

Ponnamperuma, F.N., 1985, Chemical kinetics of wetland rice soils relative to soil fertility, *in:* "Wetland Soils: Characterization Classification and Utilization," International Rice Research Institute, P.O. Box 933, Manila, Philippines, pgs. 71-89.

Quay, P.D., S.L. King, J. Stutsman, D.O. Wilbur, L.P. Steele, I. Fung, R.H. Gammon, T.A. Brown, G.W. Farewell, P.M. Grootes and F.H. Schmidt, 1991, Carbon isotopic composition of atmospheric methane: Fossil and biomass burning strength, *Global Biogeochem. Cycles*, 5:25-47.

Raimbault, M., 1975, Étude de l'influence inhibitrice de l'acétylene sur la formation biologique du méthane dans un sol de riziére, *Ann. Microbiol.* (Inst. Pasteur), 126a:217-258.

Raimbault, M., G. Rinaudo, J.L. Garcia and M. Boureau, 1977, A device to study metabolic gases in the rice rhizosphere, *Biol. Biochem.*, 9:193-196.

Rajagopal, B.S., N. Belay and L. Daniels, 1988, Isolation and characterization of methanogenic bacteria from rice paddies, *FEMS Microbiol. Ecol.*, 53:153-158.

Reddy, K.R. and W.H. Patrick Jr., 1986, Denitrification losses in flooded rice fields, *in:* "Nitrogen Economy of Flooded Rice Soils: Development in Plant and Soil Sciences," S.K. DeDatta and W.H. Patrick Jr. (eds.), M. Nijhoff Publ., pgs. 99-116.

Roger, P.A., I.F. Grant, P.N. Reddy and I. Watanabe, 1987, The photosynthetic aquatic biomass in wetland rice fields and its effect on nitrogen dynamics, *in:* "Efficiency of N Fertilizers for Rice, International Rice Research Institute, P.O. Box 933, Manila, Philippines, pgs. 43-68.

Salvas, P.L. and B.F. Taylor, 1980, Blockage of methanogenesis in marine sediments by the nitrification inhibitor 2-chloro-6-(trichloromethyl) pyridine (Nitrapin or N-serve), *Curr. Microbiol.*, 4:305-306.

Sass, R.L., F.M. Fisher, P.A. Harcombe and F.T. Turner, 1990, Methane production and emission in a Texas rice field, *Global Biogeochem. Cycles*, 4:47-68.

Sass, R.L., F.M. Fisher, F.T. Turner and M.F. Jund, 1991a, Methane emission from rice fields as influenced by solar radiation, temperature, and straw incorporation, *Global Biogeochem. Cycles*, 5:335-350.

Sass, R.L., F.M. Fisher, P.A. Harcombe and F.T. Turner, 1991b, Mitigation of methane emissions from rice fields: possible adverse effects of incorporated rice straw, *Global Biogeochem. Cycles*, 5:275-287.

Sass, R.L., F.M. Fisher, Y.B. Wang, F.T. Turner and M.F. Jund, 1992, Methane emission from rice fields: the effect of floodwater management, *Global Biogeochem. Cycles*, 6:249-262.

Schntz, H., A. Holzapfel-Pschorn, R. Conrad, H. Rennenberg and W. Seiler, 1989, A three-year continuous record on the influence of daytime season and fertilizer treatment on methane emission rates from an Italian rice paddy field, *J. Geophys. Res.*, 94:16,405-16,416.

Schntz, H., W. Seiler and R. Conrad, 1990, Influence of soil temperature on methane emission from rice paddy fields, *Biogeochemistry*, 11:77-95.

Seiler, W., 1984, Contribution of biological processes to the global budget of methane in the atmosphere, *in:* "Current Perspectives in Microbial Ecology," M.J. Kleig and C.A. Reddy (eds.), American Society of Microbiology, Washington DC, pgs. 468-477.

Sextone, A.J. and C.N. Mains, 1990, Production of methane and ethylene in organic horizons of spruce forest soils, *Soil Biol. Biochem.*, 22:135-139.

Strayer, R.F. and J.M. Tiedje, 1978, Kinetic parameters of the conversion of methane precursors to methane in hypereutrophic lake sediment, *Appl. Environ. Microbiol.*, 36:330-340.

Svensson, B.H., 1984, Different temperature optima for methane formation when enrichments from acid peat are supplemented with acetate or hydrogen, *Appl. Environ.Microbiol.*, 48:389-394.

Takai, Y., 1970, The mechanism of methane fermentation in flooded soils, *Soil Sci. Plant Nutr.*, 16:238-239.

Takai, Y., T. Koyama and T. Kamura, 1956, Microbial metabolism in reduction process of paddy soil (Part I), *Soil Plant Food*, 2:63-66.

Tsutsuki, K. and F.N. Ponnamperuma, 1987, Behavior of anaerobic decomposition products in submerged soils: Effects of organic material amendment soil properties and temperature, *Soil Sci. Plant Nutr.*, 33:13-33.

US-EPA (United States Environmental Protection Agency), 1990, "Greenhouse gas emissions from agriculture," Vol.1, Office of Policy Analysis, U.S. Environmental Protection Agency, Washington, D.C.

Vermoesen, A., H. Ramon and O. van Cleemput, 1991, Composition of the soil gas phase: Permanent gases and hydrocarbons, *Pedology*, 41:119-132.

Vogels, G.D., J.T. Keltjens and C. Van der Drift, 1988, Biochemistry of methane production, *in:* "Biology of Anaerobic Microorganisms," A.J.B. Zehnder (ed.), Wiley, New York, pgs. 707-770.

Wagatsuma, T., T. Nakashima, K. Tawaraya, S. Watanabe, A. Kamio and A. Ueki, 1990, Role of plant aerenchyma in wet tolerance and methane emission from plants, *in:* "Plant Nutrition – Plant Physiology and Application," M.L. van Beusichem (ed), Kluwer Acad. Publ., pgs. 455-461.

Wahlen, M., N. Takata, R. Henry, B. Deck, J. Zeglen, J.S. Vogel, J. Southon, A. Shemesh, R. Fairbanks and W. Broecker, 1989, Carbon-14 in methane sources and in atmospheric methane: The contribution from fossil carbon, *Science*, 245:286-290.

Wang Z.P., C.W. Lindau, R.D. Delaune and W.H. Patrick Jr., 1992, Methane production from anaerobic soil amended with rice straw and nitrogen fertilizers, *Fertilizer Research*, 33:115-121.

Wang, Z.P., R.D. Delaune, P.H. Masscheleyn and W.H. Patrick Jr., 1993a, Soil redox and pH effects on methane production in a flooded rice soil, *Soil Sci. Soc. Am. J.*, 57:382-385

Wang, Z.P., C.W. Lindau, R.D. Delaune and W.H. Patrick Jr., 1993b, Methane emission and entrapment in flooded soils as affected by soil properties, *Biology and Fertility of Soils*, in press.

Wang, M.X., A. Dai, R.X.Shen, H.B. Wu, H. Schntz, H. Rennenberg and W. Seiler, 1990, Methane emission from a chinese paddy field, *Acta Meteorologica Sinica*, 43:265-275.

Wang, Z., 1986, Rice-based systems in subtropical China, *in:* "Wetlands and Rice in Subsaharan Africa," A.S.R. Juo and J.A. Lowe (eds.), IITA Ibadan Nigeria, pgs. 195-206.

Ward, D.M. M.R. Winfrey, 1985, Interactions between methanogenic and sulfate-reducing bacteria in sediments, *Adv. Aquatic Microbiol.*, 3:141-179.

Watanabe, I. and P.A. Roger, 1985, Ecology of flooded rice fields, *in:* "Wetland Soils: Characterization Classification and Utilization," International Rice Research Institute, P.O. Box 933, Manila, Philippines, pgs. 229-246.

Whittenbury, R., K.C. Phillips and J.K. Wilkinson, 1970a, Enrichment isolation and some properties of methane-utilizing bacteria, *J. Gen. Microbiol.*, 61:205-218.

Whittenbury, R., S.L. Davies, J.F. Davey, 1970b, Exospores and cysts formed by methane-utilizing bacteria, *J. Gen. Microbiol.*, 61:219-226.

Whitton, B.A. and J.A. Rother, 1988, Environmental features of deepwater rice fields in Bangladesh during the flood season, *in:* "1987 International Deepwater Rice Workshop," International Rice Research Institute, P. O. Box 933, Manila, Philippines, pgs. 47-54.

Winfrey, M.R., 1984, Microbial production of methane, *in:* "Petroleum Microbiology," R.M. Atlas (ed.), Macmillan, N.Y., pgs. 153-219.

Winfrey, M.R. and J.G. Zeikus, 1977, Effect of sulfate on carbon and electron flow during microbial methanogenesis in freshwater sediments, *Appl. Environ. Microbiol.*, 33:275-281.

Worakit, S., D.R. Boone, R.A. Mah, M.E. Abdel-Samie and M.M. El-Halwagi, 1986, Methanobacterium alcaliphilum sp. nov. an H_2-utilizing methanogen that grows at high pH values, *Int. J. Syst. Bacteriol.*, 36:380-382.

Yagi, K. and K. Minami, 1990, Effects of organic matter application on methane emission from some Japanese paddy fields, *Soil Sci. Plant Nutr.*, 36:599-610.

Yoshida, S., 1981, "Fundamentals of Rice Crop Science, International Rice Research Institute," P.O. Box 933, Manila, Philippines, 269 pp.

Yu, T., 1985, "Physical Chemistry of Paddy Soils," Springer-Verlag, Berlin.

Zeikus, J.G., 1977, The biology of methanogenic bacteria, *Bacteriol. Review*, 41:514-541.

POLAR ATMOSPHERE AND SNOW CHEMISTRY

Leonard A. Barrie[1] and Robert J. Delmas[2]

[1]Atmospheric Environment Service
4905 Dufferin St., Downsview, Ontario, Canada, M3H 5T4
[2]Laboratoire de Glaciologie et Geophysique de L'Environment
CNRS
BP 96, 38402 St. Martin, d'Hères-Cedex, France

ABSTRACT

In the last fifteen years, considerable progress has been made in understanding the occurrence, origin, pathways, history and relevance to global change of natural and anthropogenic substances in the polar troposphere. In addition, glacial snow and ice have provided historical records of tropospheric composition of greenhouse gases and of snow deposited in the polar regions. One of the most remarkable features of polar studies is the extreme geographical contrast between the Arctic and Antarctic. The Arctic troposphere is underlain by an active ocean surrounded by pollutant-emitting, industrialized continents while the Antarctic troposphere lies over a massive, 4 km thick, ice sheet surrounded by the pollution-free southern ocean. The Arctic troposphere is much more polluted (Arctic haze) than its southern counterpart and has different exposure to compounds of natural origin such as wind blown dust or marine gases and particles. The pollution has an impact on both physical and chemical climatology. Anthropogenic Arctic haze aerosols of black carbon and sulphate have a net warming influence in the north in contrast to elsewhere on the globe where (with less black carbon) they tend to offset the impact of anthropogenic greenhouse gases. Biogeochemical cycles of many substances including carbon, sulfur and nitrogen are perturbed. Compounds potentially toxic to polar ecosystems also accumulate.

Phenomena of interest in the polar regions include chemistry associated with Arctic haze pollution, the destruction of lower tropospheric ozone over the Arctic ocean at polar sunrise induced by marine halogens in the presence of sulfuric acid aerosols, oxidant chemistry of both polar tropospheres, chemical control of clouds and atmospheric energy budgets, and the relationship between glacial and atmospheric composition.

Historical records of atmospheric composition are available from glaciers and instrumental records of atmospheric optical properties and atmospheric composition. They indicate that the Arctic region has been polluted particularly in the winter-half of the year since at least the turn of the century.

INTRODUCTION

The polar regions are important natural observatories from which to mount studies of global change. They play a key role in atmospheric chemistry and climate, for instance as major players in the global climate feedback system and as hosts of ozone depletion that increases ultra-violet radiation at the earth's surface. One of the most remarkable features of polar geography is the contrast between the Arctic and Antarctica in the land-ocean-ice configuration (Figure 1). The Arctic troposphere (0 to ~8 km) is underlain by a flat biologically-active ocean which is surrounded by pollutant-emitting, industrialized continents while the Antarctic atmosphere lies over a massive continental ice sheet (4 km thick) and is surrounded by the stormy southern hemispheric ocean. The Arctic has been heavily impacted by relatively short-lived (3 to 6 weeks) haze pollution from the northern continents (Barrie, 1986) while the Antarctic remains unaffected, protected by remoteness from pollution sources and by a stormy midlatitude atmosphere that cleanses the air of particles and soluble gases. Yet, the poles have many common chemical, climatological and geographical features. For instance, their atmospheres are both impacted by relatively long-lived (years), insoluble, greenhouse gases. They are cold, have a unique one-day/one-night per year light regime and massive glaciers that never melt in summer retaining chemical evidence of changes in the Earth's atmospheric composition over hundreds of millennia.

In IGAC's Polar Atmosphere and Snow Chemistry (PASC) Activity, a group of environmental scientists are studying global change in the polar regions. Atmospheric composition measurements are made at locations shown in Figure 1 while atmosphere-snow exchange studies and snow/ice composition measurements are studied on glaciers shown in Figure 2. The latter are situated at different altitudes and locations in the polar regions forming a network of glacial atmospheric sampling sites that represent time since at least the last interglacial period.

One of the tasks in PASC focuses on the transfer of chemicals from the atmosphere to snow to ice. The relationship between the concentrations of substances in the atmosphere and those in glacial snow/ice, called "the transfer function," depends on the scavenging of gases and particles from the atmosphere by wet and dry deposition processes followed by incorporation into the snowpack (Barrie, 1985, 1991), movement into the firn layer below (1 to ~70 m) and ultimately, incorporation into ice or air bubbles which form progressively under pressure at about 70 m depth (Neftel, 1991). Understanding the transfer function of a particular chemical constituent requires considerable research in a difficult environment, multi-disciplinary collaboration between meteorologists, atmospheric chemists, glacial chemists and glaciologists and simultaneous air, snow and snowpack observations. These studies are currently being undertaken at ATM-Summit Greenland and South Pole station (Figure 2). A second major goal of PASC is to understand the role of tropospheric chemistry in the polar regions in global change. For instance, studies have defined the nature, occurrence and potential effects of Arctic haze and led to the discovery of a lower tropospheric ozone hole at polar sunrise (Barrie *et al.*, 1988). What is the chemistry causing this ozone depletion which seems to involve halogen compounds, is it natural or man-made and how relevant is this chemistry to global tropospheric ozone are outstanding questions that need to be answered.

In this chapter, we will highlight some of the recent results obtained by the PASC group pertaining to global change. However, before presenting results, field activities will be summarized. Since 1983 there have been a number of internationally coordinated intensive studies in the Arctic atmosphere that have enhanced our awareness of the value of the lower troposphere as a unique natural photochemical laboratory and of the complexity of chemical, biological and physical processes occurring in it. These include: flights of the

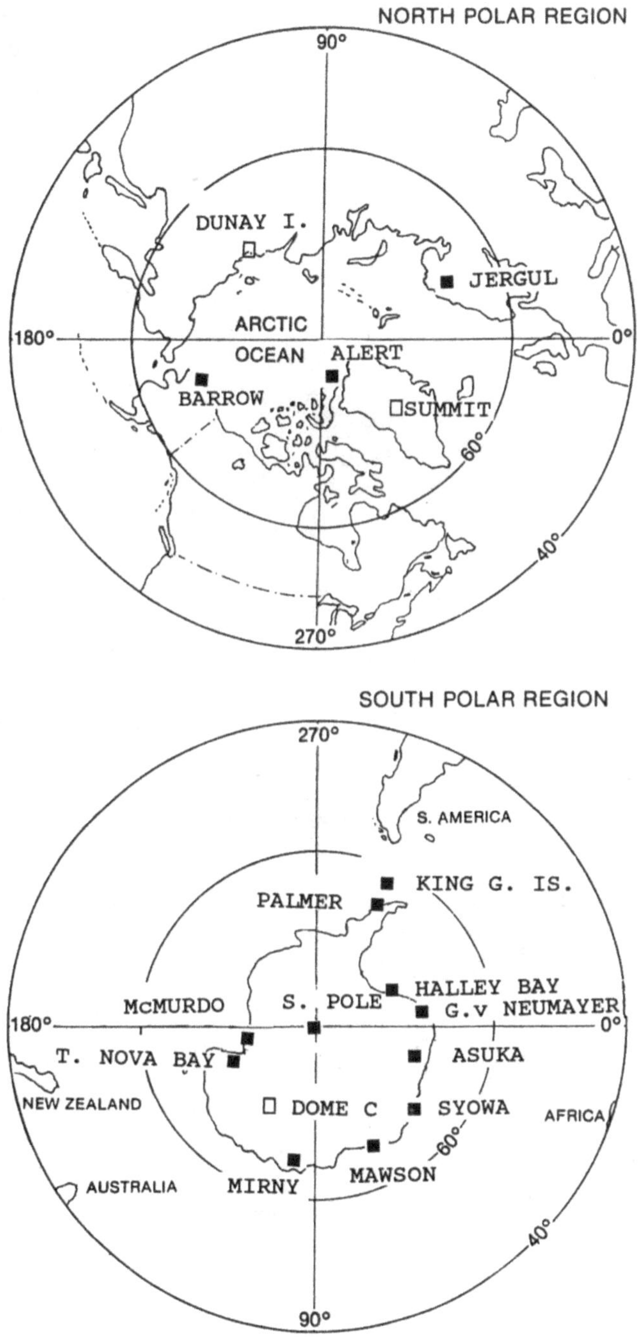

Figure 1. The polar regions and locations of tropospheric chemistry research stations.

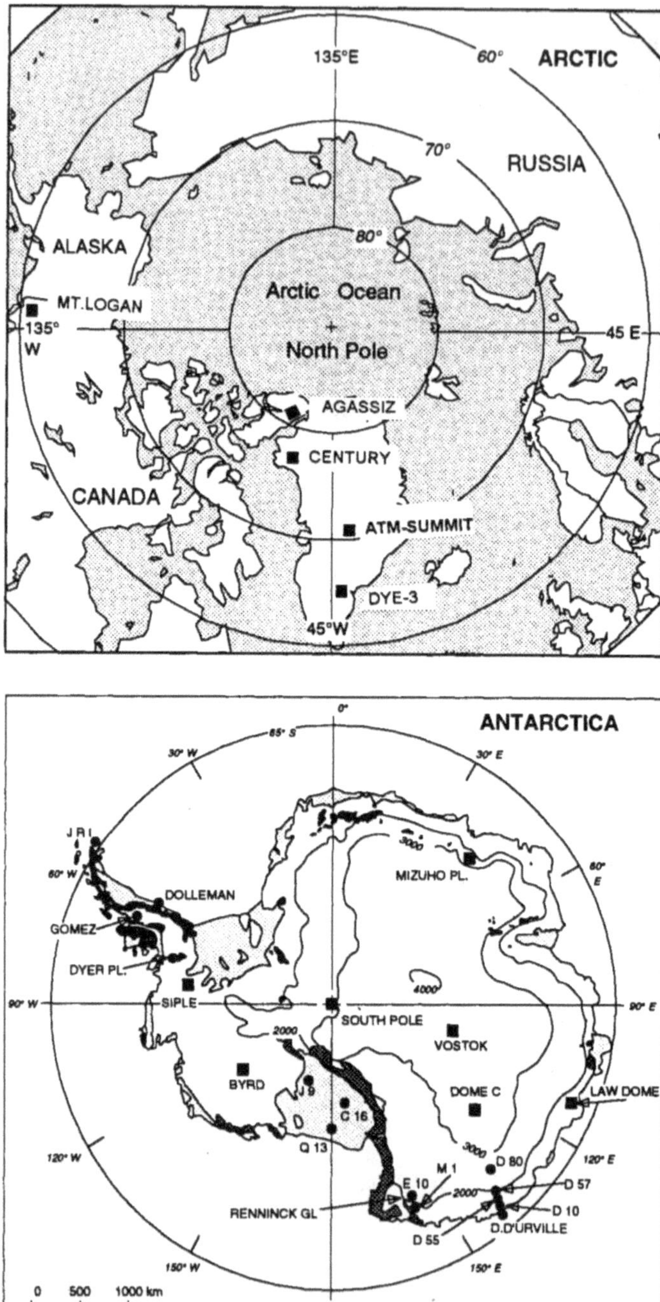

Figure 2. The northern and southern polar regions and locations of glaciers at which ice core chemistry and the atmosphere/glacier transfer function are being studied.

American Arctic Gas and Aerosol Sampling Program (AGASP) in 1983, 1986, 1989 and 1992 (Davidson and Schnell, 1993); a series of Canadian Polar Sunrise Experiments in 1985, 1986, 1988, 1989 and 1992 (Barrie and Bottenheim, 1993); and a number of studies conducted by the Norwegians around Spitsbergen Island in the mid-1980s (Ottar, 1989). Since the beginning of IGAC activities, field studies of the transfer function have taken place in both polar regions. In addition, shallow cores have been analyzed with the aim of documenting, in particular, the impact of anthropogenic pollution on global biogeochemical cycles.

At South Pole Station, fundamental research based on field observations have been conducted by a U.S. team, in particular on the deposition mechanisms of sulfur species. At more coastal sites at stations like Halley, G. von Neumayer and Dumont d'Urville, a coordinated research program has been initiated between European laboratories. Shallow core analyses and pit studies have documented the sulfur and nitrogen cycles at South Pole and in the Antarctic Peninsula.

In the North, a set of important field experiments have been carried out at Summit in Greenland (3.2 km above sea level [asl]), in parallel with the American GISP2 and European GRIP deep drilling projects to recover ice core down to bedrock (~3 km long). An in situ ice chemistry laboratory enabled the determination of most major ionic species in meltwater from snow pits, in particular, the temporal evolution of some unstable species supposed to be adsorbed on the snow surface. An air monitoring station (ATM) installed 30 km from the drilling site by U.S. scientists enabled multi-parameter air and aerosol sampling campaigns to be conducted every summer from 1989 onward. Finally, about 20 shallow firn cores (0 to ~70 m) covering the last 200 years have been recovered in the Summit area by U.S. and European glaciologists for detailed chemical investigations of global change in the northern hemisphere. Studies by Canadians at other northern glaciers such as at Agassiz (1.6 km asl, Figure 2) on Northern Ellesmere Island and at Mt. Logan (5.5 km asl) on the eastern rim of the North Pacific Ocean have added considerably to knowledge of regional differences.

RESULTS

Sulfur

Sulfur compounds originating from natural sources (e.g., as $(CH_3)_2S$ (DMS), H_2S) or anthropogenic activities (SO_2), after transformation in the atmosphere, form $SO_4^=$ and methane sulphonic acid (MSA) particles, which in turn can affect climate, acidify precipitation, host chemical reactions and reduce visibility. Records of these end-product sulfur compounds in glacial ice cores from South Pole and from Summit Greenland for this century (Figure 3) show clearly that $SO_4^=$ concentrations in snow are presently higher in Greenland than in Antarctica. They have also increased significantly in Greenland since the turn of the century but not in Antarctica. Cores from Agassiz show higher $SO_4^=$ concentrations than in Greenland but similar trends (Barrie *et al.*, 1985; Koerner, 1993). In Greenland, $SO_4^=$ reaches its climax (about 5 times the natural background level of 30 ng g^{-1}) in the 1970s. Atmospheric measurements of particulate chemistry in these regions since 1980 have shown that Arctic sulfates are mainly anthropogenic in origin (Li and Barrie, 1993) while those in Antarctica originate mainly from two types of marine sources: marine biogenic and sea spray aerosols as well as from volcanic sulfur emissions (Savoie *et al.*, 1993).

At the South Pole (Figure 3) during global volcanic events (e.g., Agung in 1965), $SO_4^=$ concentrations deviate significantly from a relatively constant background of 50 ng g^{-1}.

Indeed such deviations have been used to study the frequency of southern hemispheric volcanic events over the past 1000 years (Delmas *et al.*, 1992). MSA, which is an indicator of the marine biogenic sulfur source, exhibits large spikes (up to 4 times the background value of 6 ng g^{-1}) during the years of strong El Niño events in the southern hemisphere (Legrand and Fenlet-Saigne, 1991). This yields insight into variations in atmospheric circulation associated with El Niño since the source of biogenic sulfur to the 3 km high pole is the ocean on the periphery of Antarctica.

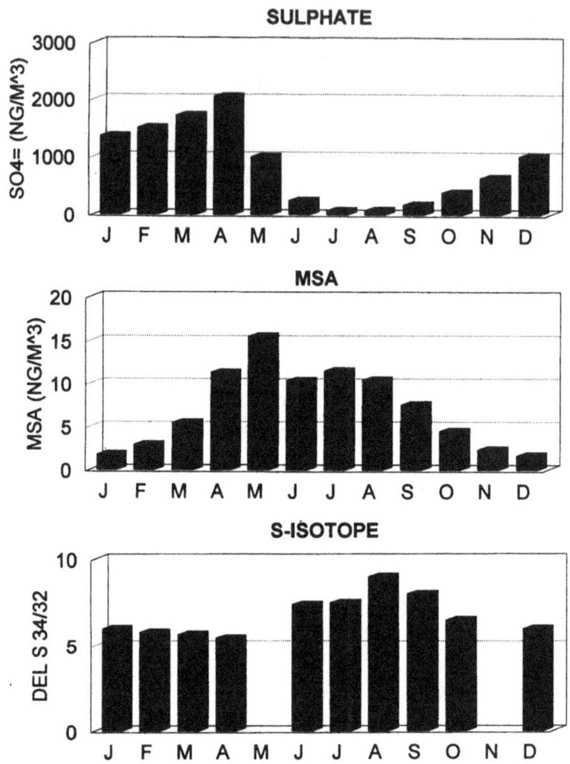

Figure 3. Sulfate concentration profiles in South Pole snow (Legrand and Feniet-Saigne, 1991) and at Summit, Central Greenland (Eurocore report, private communication, 1993) over the last decades. Methanesulfonic acid variations at South Pole are also shown; the peaks are linked to El Niño events. The years indicated in this figure correspond to the date of strong El Niños.

Seasonal variations of $SO_4^=$, MSA and S-isotopic composition in lower tropospheric aerosols at Alert (Figure 4) serve to illustrate several points. First, the winter Arctic aerosol whose main constituent is $SO_4^=$ is mainly anthropogenic in origin (low $\delta S^{34/32}$ ratio [5.5 to 6] similar to that in industrial sources [3 to 6]) while in summer it is a mixture of anthropogenic, sea salt and biogenic $SO_4^=$ ($\delta S^{34/32}$ intermediate between anthropogenic ratios [3 to 6] and marine ratios [15 to 21]). Second, Arctic marine biogenic sulfur, indicated by MSA, peaks twice in the year in spring and mid-summer. This is likely a result of long range transport from the north Atlantic in spring and local Arctic ocean sources in summer (Li and Barrie, 1993; Li *et al.*, 1993). Finally, in contrast to tropical areas, a major fraction of biogenic particulate sulfur mass in the polar atmosphere is comprised of MSA. In the high Arctic atmosphere, MSA/$SO_4^=$ ratios in biogenic sulfur range from 0.2 to 0.9 (Li and Barrie, 1993). In coastal Antarctic aerosols (Savoie *et al.*, 1993) and in snow (Mulvaney *et al.*, 1992), MSA concentrations can be higher than those of biogenic $SO_4^=$. This observation places limitations on the mechanism of oxidation of biogenic DMS in the atmosphere which must be explained by atmospheric chemists. A knowledge of this mechanism is critical in determining whether biogenic sources affect cloud reflectivity and hence climate.

The transfer function of sulfur from atmosphere to glacier is perhaps one of the better known ones. Even so, great uncertainties remain in such processes as dry deposition of particles to the snowpack (ventilation by wind) and reactions of gaseous sulfur compounds with snow.

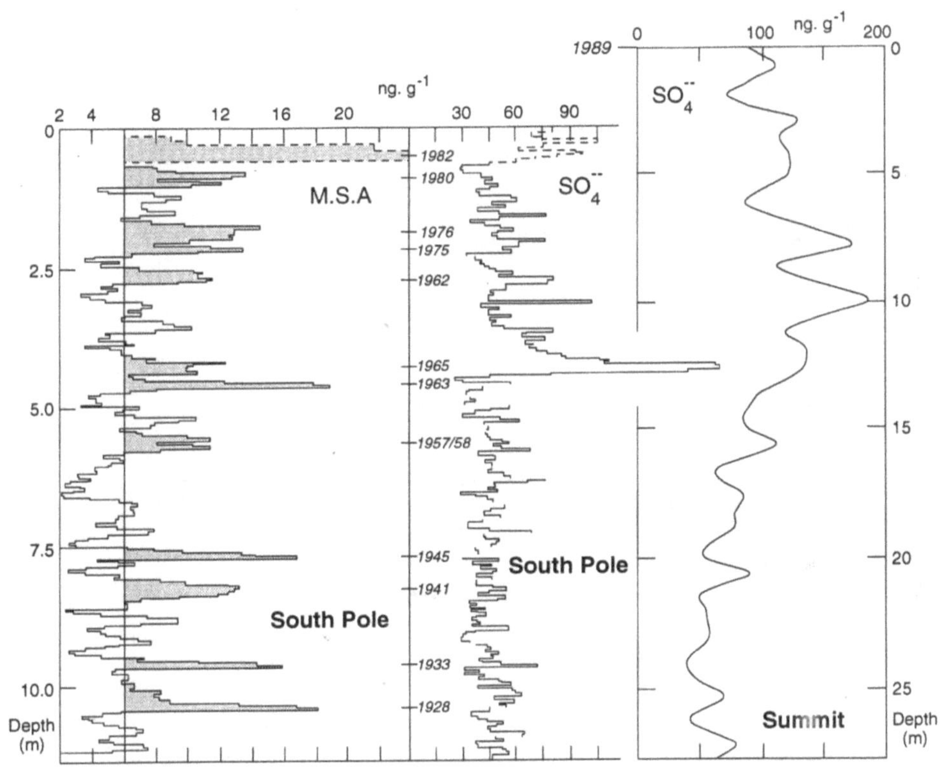

Figure 4. The seasonal variation of $SO_4^=$, MSA and isotopic composition of $SO_4^=$–S in aerosols at Alert, Canada (Li and Barrie, 1993; Nriagu *et al.*, 1991).

Nitrogen

Nitrogen compounds in the atmosphere can originate from anthropogenic (combustion processes) and natural sources (soil bacterial processes, lightning, stratospheric chemistry). They are important in the formation and destruction of tropospheric ozone (a major greenhouse gas) and of many other atmospheric trace constituents. When deposited to the Earth's surface, they provide nutrients to ecosystems.

There are limited observations in either polar troposphere. In the Arctic, it has been clearly shown from ground level observations at Alert (Bottenheim *et al.*, 1993) that in the polluted spring month of April, the bulk of atmospheric nitrogen oxides are comprised of peroxyacetylnitate (PAN) with alkyl nitrates second followed by HNO_3 and NO_2, a distant third. The seasonal variation of PAN and total inorganic NO_3^- (HNO_3 + particle-NO_3^-) at Alert (Figure 5) show a peak in late winter/spring and mid-winter, respectively. They are undoubtedly caused by anthropogenic sources. No extended observations of nitrogen oxides have been made in the middle troposphere at the surface of polar glaciers. At ATM-Summit on Greenland, Silvente (1993) observed that during summer months HNO_3 concentrations average about 8 pptv but, depending on the vertical mixing in the atmospheric boundary layer, sometimes peak at concentrations one order of magnitude

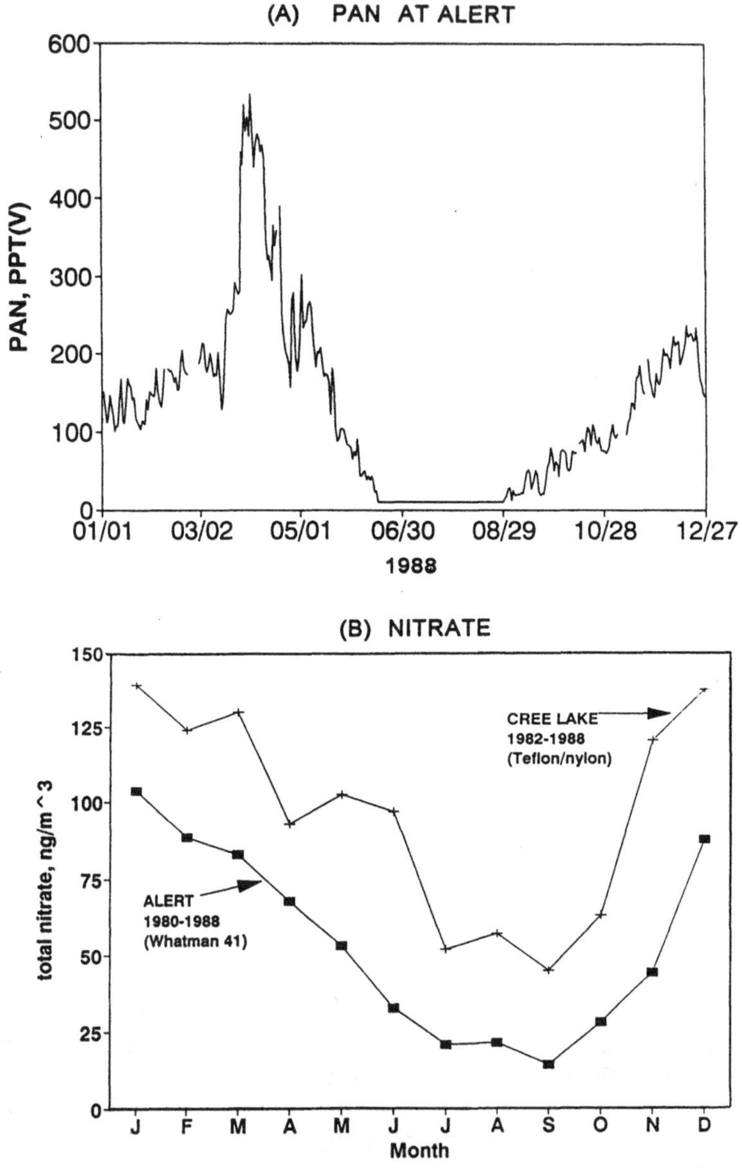

Figure 5. The seasonal variation of (A) PAN and (B) total inorganic NO_3^- at Alert and at a Cree Lake in northern Saskatchewan, Canada (Barrie and Bottenheim, 1991).

higher than average. This results from injection to the surface of free tropospheric air in which HNO_3 concentrations are three times higher than the average surface concentration at ATM-Summit (Talbot *et al.*, 1992).

The processes of incorporation of nitrogen into glaciers are poorly understood. Little is known about deposition of gaseous nitrogen compounds to snow. Most glacial observations have measured inorganic NO_3^-. There is evidence of post-depositional loss of NO_3^- from snowpacks after a snowfall (Neubauer and Heumann, 1988a, b; Mulvaney *et al.*, 1993). Thus, it is not surprising that glacial ice core records of NO_3^- such as those measured at South Pole and Dye 3, Greenland (Figure 6) must be interpreted cautiously. NO_3^- concentrations have risen in the northern glacier and not in the southern one. The rise in the north has occurred since World War II in parallel with the increase in nitrogen oxide emissions in the northern hemisphere. In the 1970s and 1980s, NO_3^- concentrations remained relatively constant at a mean value, twice that of the pre-industrial era. In south polar snow, concentrations are higher and natural fluctuations are large and irregular.

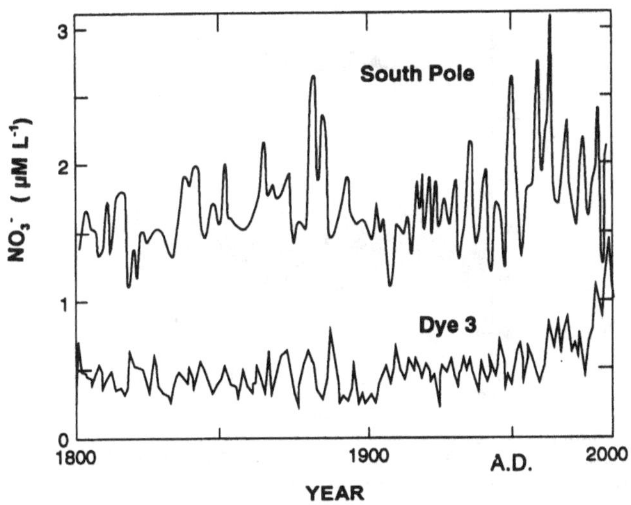

Figure 6. NO_3^- concentration profiles in snow at South Pole (Legrand and Kirchner, 1990) and at Dye 3, South Greenland (Mayewski *et al.*, 1990).

It has been suggested that long term variations in polar stratospheric temperatures over central Antarctica may be reflected in the nitrate record of Antarctic snow as a result of a winter/spring stratospheric denitrification by polar stratospheric ice crystals. The first measurements by Mayewski and Legrand (1990) in central Antarctic snow seem to support this proposal. More recent data from a firn core collected in an area within the polar vortex in the Weddell Sea sector and covering the last 25 years show the occurrence of a nitrate peak in the snow in spring or early summer only in this geographical area (Mulvaney and Wolff, 1993). No temporal trend of the size of the spring nitrate peak in association with the increasing amount of ozone depletion is observed. Further studies are still needed to account for such phenomenon as post-depositional loss and surface chemical production of NO_3^-.

Hydrogen Peroxide and Formaldehyde

H_2O_2 and HCHO are both linked with the oxidation capacity of the atmosphere. Long term records of their concentration in glacial snow from Summit have been measured (Figure 7). They are produced in the atmosphere by photochemically induced reactions involving CO, CH_4, nitrogen oxides (NO_x) and non-methane hydrocarbons.

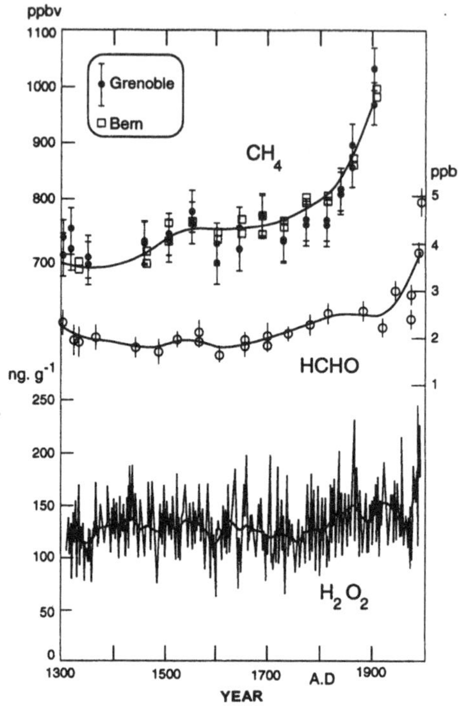

Figure 7. Methane (Blunier *et al.*, 1993), formaldehyde (Staffelbach *et al.*, 1991) and hydrogen peroxide (Sigg and Neftel, 1991) at Summit Greenland firn and ice.

Consider first H_2O_2 which is the most abundant trace impurity in snow. The mean concentration in Summit snow is about 140 ppb (Sigg and Neftel, 1991), which is about twice that found at Dye 3, in South Greenland. Its concentration undergoes a distinct seasonal variation which begins to disappear at 20 to 30 m depth as a result of vertical diffusion. The 700-year record exhibits two distinct parts. Up to 200 years from present, the concentrations fluctuate around a relatively stable mean value and then begin to increase gradually and then more rapidly over the past 15 years. This increase in high northern latitudes could be linked to the increase in hydrocarbon and nitrogen oxide anthropogenic emissions over this period.

HCHO was measured first on the Eurocore from Summit by Staffelbach *et al.* (1991). The concentrations which were consistent with independent snowpit samples exhibited no seasonal variations. The concentrations, relatively stable in the deepest part of the core, begin to increase slowly from about 1750 AD, then more rapidly in this century. They are currently about 1.9 times higher than at the beginning of the millennium. At the same time, methane concentrations have increased more than twofold (Figure 7).

The transfer function of H_2O_2 depends on co-condensation to an ice-surface with water vapor as well as evaporation and recrystallization processes in the snowpack (Neftel, 1990; Jacob and Klockow, 1993). The transfer function of HCHO depends on a balance between wet and dry deposition and post-depositional effects such as loss from the snowpack during reduction of specific surface area and evaporation of surface snow and photolysis of "fixed" HCHO molecules in the snow (Neftel, 1991). Consequently, the HCHO concentration of the firn is likely proportional to a multi-year average of the air concentration. Photolysis in the surface snow pack is poorly understood.

Greenhouse Gases (CO_2 and CH_4)

The concentrations of methane and carbon dioxide in the air bubbles from Summit cores have been measured for the first time in Bern and Grenoble laboratories by two different analytical methods. The methane results (Figure 7) from both laboratories are in good agreement over the last thousand years. The slow rise of the concentrations from 680 to 760 ppbv between 1500 and 1600 is apparently related to the early development of agriculture (Blunier *et al.*, 1993). The CH_4 increase linked to anthropogenic emissions begins to be markedly visible from about 1700 onward as already found from Antarctic ice core measurements at Siple and at D57 (Nagazawa *et al.*, 1993; Raynaud and Chappellaz, 1993). For CO_2, the agreement of the measurements is particularly satisfactory for the upper part of the record (i.e. from the 18th century) where the pre-industrial level is 280 ppmv. These Greenland observations complement, the previous Antarctic observations at Vostok (Reynaud *et al.*, 1993) by providing a comparison of hemispheric composition and more regional resolution. The transfer functions of these two relatively inert, ice-insoluble gases are well established (Neftel, 1991).

Lead

Lead is of interest for its potential toxicity to humans and wildlife as well as a tracer of the origin of air pollution. Isotopes of lead can be used to differentiate between sources. Lead has been an industrial air pollution product for centuries. Since the advent of the automobile, leaded gasoline has been the primary anthropogenic source and the smelting industry secondary. Since about 1970 as a fuel additive, lead has been phased out of gasoline until now in 1993, it is not used in North America and much less in Europe.

At Summit Greenland, ice core records show the historical use of lead and indicate that it peaked in concentration between 1950 and 1975 at a concentration (100 to 200 pg g^{-1}) several hundred times the pre-industrial level (Boutron *et al.*, 1991). Between 1970 and 1989, it decreased by at least an order of magnitude. Isotopic composition of this glacial lead (Rosman *et al.*, 1993) has shown that the relative contributions of U.S. sources has declined from more than 50% around 1970 to less than 29% in the late 1980s. The rest is believed to be contributed by Eurasian and to a lesser extent Canadian sources.

The contrast between the sources and trends of lead recorded on Greenland in the middle troposphere and at Alert (Figure 1) at sea level in the lower troposphere is remarkable. Trends in aerosol lead (Figure 8), atmospheric chemical transport modeling (Barrie *et al.*, 1989; Akerodolu *et al.*, 1993) and isotopic composition studies (Sturges and Barrie, 1989; Sturges *et al.*, 1993) indicate that the predominant source of lower tropospheric Arctic lead is Eurasian and that in the mid-1980s, western Europe, eastern Europe and the former Soviet Union contributed equally. This emphasizes that the middle tropospheric air-shed from which Greenland glaciers (3 km asl) receive their particulate-derived pollution is more global in extent than the air-shed of the lower tropospheric glaciers and monitoring sites such as Agassiz (1.6 km asl), Alert, Barrow, and Spitsbergen

(Figures 1 and 2) that are receiving Arctic haze pollution. The lower tropospheric air-shed for haze type pollution is predominantly affected by Eurasian sources. The contrast between middle troposphere and lower tropospheric circulation underlines the need and unique nature of a mid-tropospheric monitoring site at ATM-Greenland.

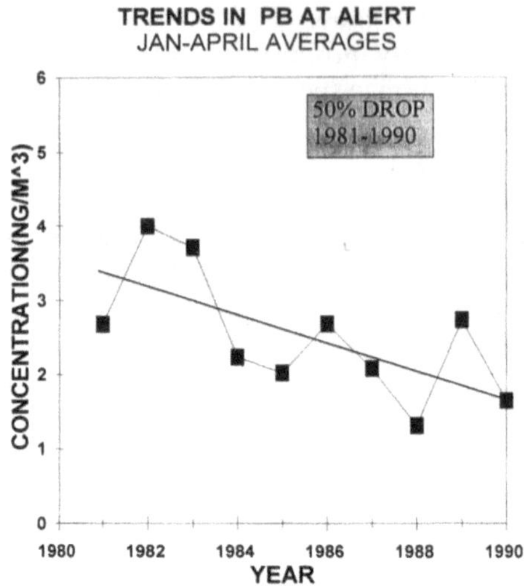

Figure 8. Trends in lower tropospheric lead concentrations at Alert, Canada in the 1980s.

Carboxylic Acids

Formic and acetic acids have been determined only recently in polar precipitation (Legrand and Saique, 1988). The extremely low concentrations (ppb level) encountered in these regions and the easy contamination of the samples make it difficult to obtain reliable data. Moreover, post-depositional effects are frequently observed in the firn layers for these compounds which are mostly deposited as gases (De Angelis, 1994).

In Antarctica surface snow, the concentration of acetate ion is extremely low (0.1 ng g^{-1}) while that of formate ion is about 10 times higher. In Greenland, the concentrations are found to be one to two orders of magnitude higher suggesting anthropogenic sources. The concentration range is 10 to 35 and 10 to 1000 ng g^{-1}, respectively. The very high formate values are linked with ammonium formate spikes from forest fires at sub-polar latitudes (Legrand *et al.*, 1992). Their duration is about one tenth of a year. They are observed always in summer layers and accompanied by peaks in oxalic acid. The contribution of carboxylic acids to the natural acidity of central Greenland snow is estimated to be about 20%.

Few studies of the nature, occurrence and origin of these compounds in the polar troposphere have been conducted. An investigation of the composition of the soluble mass of spring Arctic aerosol in Barrow Alaska (Li and Winchester, 1989) showed that formate and acetate are not negligible, forming about 20% of the soluble aerosol mass.

Tropospheric Chemical Changes At Arctic Polar Sunrise: Ozone Depletion

At polar sunrise, the Arctic troposphere (0 to ~8 km) is a unique chemical reactor influenced by human activity and the Arctic ocean, from which it is separated by a crack-ridden ice membrane 3 to 4 m thick. Ocean-atmosphere exchange of heat, water vapor, seasalt and marine biogenic gases influence the composition of the reactor. Superimposed on all this is the influence of industrial air pollution from the midlatitudes that is elevated by 10 to 20 times above summer levels in the winter months of December to April.

From 21 September to 21 December to 21 March, the region north of the Arctic circle goes from a completely sunlit situation to a completely dark one and then back to light. At the same time the lower troposphere is stably stratified. This hinders vertical mixing. In this environment, chemical reactions involving sunlight are much slower than farther south. Thus, it is not surprising to find the abundance of photochemically reactive non-methane hydrocarbons high at polar sunrise. Between complete dark in February and complete light in April, a number of chemical changes in the lower troposphere are observed. Perhaps the most sensational is the destruction of lower tropospheric ozone accompanied by the production of filterable bromine species (Barrie *et al.*, 1988). The latter are likely of marine origin although their production may involve anthropogenic compounds. Another change is the shift in the fraction of total sulfur in its end oxidation state (VI) from 50% to 90% (Barrie and Hoff, 1984). Several gaseous hydrocarbons disappear from the atmosphere at this time as a result of chlorine radical attack (Jobson *et al.*, 1993).

The Polar Sunrise Experiment 1992 was one of the most comprehensive international air chemistry studies every conducted in the Arctic troposphere aimed at understanding ozone depletion as the sun rises (Barrie and Bottenheim, 1993). The periodic disappearance of lower atmospheric ozone at polar sunrise associated with particulate Br production was confirmed at Alert (Figure 9). At the Arctic ocean surface 200 km north of Alert (Figure 1), the frequency of ozone loss was much higher than at Alert. BrO mixing ratios during the light-period of April averaged approximately 10 pptv and ranged from <3 to 17 pptv. This is sufficient to explain the ozone destruction. Chlorine was clearly implicated in polar sunrise chemistry by observations of selective destruction of hydrocarbons and alkylnitrates and production of formaldehyde and acetone. However, it cannot explain ozone depletion, leaving bromine as the prime candidate.

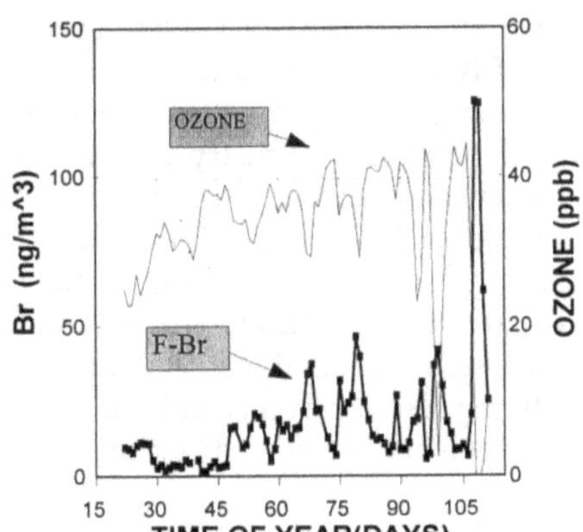

Figure 9. Observed anti-correlation between atmospheric ozone and filterable Br^- (F–Br) at ground level in Alert, Canada during Polar Sunrise Experiment 1992 (Barrie *et al.*, 1993).

The budget of total nitrogen oxide concentration (NO_y) was explained by measurements of individual compounds establishing that peroxyacetylnitrate (PAN) is the predominant constituent, at times comprising more than 70% of the total. In April, nitrogen dioxide was generally less than 50 pptv and was a minor fraction of NO_y. The presence of gaseous nitrite, possibly HONO, was detected in the dark winter at concentrations averaging 20 pptv and in the light period of April at approximately 8 pptv.

Gas phase Br^-, presumably HBr, HOBr or Br_2 was very episodic, peaking at approximately 30 to 80 pptv. In the dark of January and February, it averaged about 10 pptv in contrast to 5 pptv in the light period of April. The budget of atmospheric bromine was calculated from measurements of total bromine in air and of individual compounds. On average, $CHBr_3$, CH_3Br, and the sum of CH_2Br_2, CH_2ClBr, $CHClBr_2$, and $CHCl_2Br$ comprised 23, 44 and 5% of gaseous organic compounds, respectively. Bromoform and its chlorinated counterparts were correlated and peaked during ozone depletion episodes. Aircraft observations over the Arctic ocean ice at 400 m showed widespread destruction of ozone in the surface boundary layer in April.

FUTURE RESEARCH

Polar research is to some extent distinct from that in other parts of the globe. It has a strong individualistic component dictated, in part, by logistical requirements of work in remote areas of harsh climate and, in part, by strong national research programs for the north and south pole. Past efforts in atmospheric and snow chemistry have consisted of a large number of "small science projects" and a few loosely coordinated international efforts.

In the mid-1990s, PASC will encourage the integration of small science efforts into the larger picture through interdisciplinary specialty conferences and workshops as well as a detailed reporting of these activities through an active committee. Close contact with the Scientific Committee for Antarctic Research (SCAR) will be maintained. In parallel, coordinated international, interdisciplinary efforts on selected topics will be encouraged. These include: (1) a coordinated research program at ATM-Summit Greenland, where the deep coring of the U.S. and Europe has taken place, in order to better understand the exchange of gases and particles between the atmosphere and glacial snow/ice; (2) a northern tropospheric oxidants study (NOS) in collaboration with the North Atlantic Regional Experiment (NARE) of IGAC that will help to better define chemical controls on tropospheric ozone which has emerged as an important greenhouse gas; (3) a cooperative effort to develop an automated polar air chemistry measurement system; (4) ship and aircraft campaigns to better understand the role of polar oceans in atmospheric chemistry, stratospheric ozone depletion and global climate; (5) studies of the chemical processes involved in the accumulation of persistent organic pollutants (e.g., pesticides) in polar ecosystems via air and snow pathways; (6) in the coastal areas of Antarctic the aerosol composition will be studied to assess the role of sub-Antarctic marine biogenic activity in the sulfur cycle, particularly during El Niño events and finally; (7) in central Antarctic the hypothesis that glacial NO_3^- reflects the intensity of stratospheric ozone depletion will be tested.

In addition to major observational programs, efforts will be made to incorporate polar processes more accurately into global atmospheric chemistry and climate models. For instance, although globally anthropogenic aerosols have a net negative radiative forcing counteracting the greenhouse forcing of anthropogenic gases, they have a positive forcing regionally over snow covered polar regions. This creates large spatial gradients in radiative forcing by aerosols that have not been adequately represented in climate models to date. Close cooperation with the International Global Aerosol Program (IGAP) will be maintained in this regard.

REFERENCES

Akerodolu, F.A., L.A. Barrie, M.P. Olson, K.K. Oikawa, J. Keeler and J. Pacyna, 1993, The flux of anthropogenic metals into the Arctic from the midlatitudes in 1979/80, *Atmos. Envir.*, in press.

Barrie, L.A., 1985, Atmospheric particles: their physical and chemical characteristics and deposition processes relevant to the chemical composition of glaciers, *Annals of Glaciology*, 7:100-108.

Barrie, L.A., 1986, Arctic air pollution: an overview of current knowledge, *Atmos. Environ.*, 20:643-663.

Barrie, L.A., 1991, "Snow formation and processes in the atmosphere that influence its chemical composition," NATO ASI Series, Vol. G28, Seasonal Snowpacks, T. Davies et al. (eds.), pgs. 1-20.

Barrie, L.A. and R.M. Hoff, 1984, The oxidation rate and residence time of sulphur dioxide in the Arctic atmosphere, *Atmos. Environ.*, 18:2711-2722.

Barrie, L.A., D. Fisher and R.M. Koerner, 1985, Twentieth century trends in Arctic air pollution revealed by conductivity and acidity observations in snow and ice in the Canadian high Arctic, *Atmos. Environ.*, 19:2055-2063.

Barrie, L.A., J.W. Bottenheim, R.C. Schnell, P.J. Crutzen and R.A. Rasmussen, 1988, Ozone destruction and photochemical reactions at polar sunrise in the lower Arctic atmosphere, *Nature*, 334:138-141.

Barrie, L.A., M.P. Olson and K.K. Oikawa, 1989, The flux of anthropogenic sulphur into the Arctic from midlatitudes, *Atmos. Environ.*, 23:2502-2512.

Barrie, L.A. and J.W. Bottenheim, 1991, Sulphur and nitrogen pollution in the Arctic atmosphere, *in:* "Pollution of the Arctic Atmosphere," W. Sturges (ed.), Elsevier Press, 334 pp.

Barrie, L.A., R. Staebler, D. Toom, B. Georgi, G. Den Hartog, S. Landsberger and D. Wu, 1993, Arctic aerosol size-segregated chemical observations in relation to ozone depletion during Polar Sunrise Experiment 1992, *J. Geophys. Res.*, submitted.

Barrie, L.A. and J.W. Bottenheim, 1993, Polar Sunrise Experiment 1992, *J. Geophys. Res.*, in press.

Blunier, T., J. Chappellaz, J. Schwander, J.M. Barnola, T. Desperts, B Stauffer, D. Raynaud, 1993, Atmospheric methane, record from a Greenland ice core over the last 1,000 years, *Geophys. Res. Lett.*, 20:2219-2222.

Bottenheim, J.W., L.A. Barrie and E. Atlas, 1993, The partitioning of nitrogen oxides in the lower Arctic troposphere during spring 1988, *J. Atmos. Chem.*, 17:15-28.

Boutron, C., U. Görlach, J.P. Candelone, M.A. Bolshov and R.J. Delmas, 1991, Decrease in anthropogenic lead, cadmium and zinc in Greenland snows since the late 1960s, *Nature*, 353:153-156.

Davidson, C.I. and R.C. Schnell, 1993, Introduction to the special issue Atmospheric Environment: Arctic air, snow and ice chemistry, *Atmos. Environ.*, in press.

De Angelis, M., M. Legrand, and J.L. Steffensen, 1994, First investigation of post-depositional effects in polar snow, *in:* "Ice Core Studies of Global Biogeochemical Cycles," R.J. Delmas (ed.), NATO Advanced Studies Workshop, Annecy, France, April 1993.

Delmas, R.J., S. Kirchener, J.M. Palais and J.-R. Petit, 1992, 1000 years of explosive volcanism recorded at the South Pole, *Tellus*, 44:335-350.

Hopper, J.F., H.B. Ross, W.T. Sturges and L.A. Barrie, 1991, regional source discrimination of atmospheric aerosols in Europe using the isotopic composition of lead, *Tellus*, 43:45-60.

Jacob, P. and D. Klockow, 1993, Measurements of hydrogen peroxide in Antarctic ambient air, snow and firn cores, *Fresenius J. Anal. Chem.*, 346:429-434.

Jobson, B.T., H. Niki, Y.Yokouchi, J. Bottenheim, F. Hopper and R. Leaitch, 1993, Measurements of C_2–C_6 hydrocarbons during Polar Sunrise Experiment 1992, *J.Geophys. Res.*, submitted.

Koerner, R.M., 1993, Unpublished observations from on ice of 1989 to 1992 on Agassiz, Dept. Energy Mines and Resources, Ottawa, Canada.

Legrand, M. and C. Saigne, 1988, Formate, acetate and methansulfonate measurements in Antarctic ice: Some geochemical implications, *Atmos. Envir.*, 22:1011-1017.

Legrand, M. and S. Kichner, 1990, Origins and variations of nitrate in South Polar snow layers, *J. Geophys. Res.*, 95:3493-3507.

Legrand, M. and C. Feniet-Saigne, 1991, Strong El Niño revealed by methanesulphonic acid in South Polar snow layers, *Geophys. Res. Lett.*, 18:187-190.

Legrand, M., M. De Angelis, T. Staffelbach, A. Neftel, and B. Stauffer, 1992, Large perturbations of ammonium and organic acids content in the Summit-Greenland ice core: Fingerprint from forest fires?, *Geophys. Res. Lett.*, 19:473-476.

Li, S.M. and L.A. Barrie, 1993, Biogenic sulphur aerosols in the arctic troposphere I: Contribution to total sulphate, *J. Geophys. Res.*, 98:20,613-20,622.

Li, S.M., and J.W. Winchester, 1989, Geochemistry of organic and inorganic ions of late winter Arctic aerosols, *Atmos. Environ.*, 23:2401-2415.

Li, S.-M., R.W. Talbot, L.A. Barrie, R.C. Harriss, C.I. Davidson, and J.-L. Jaffrezo, 1993, Seasonal and geographic variations of methanesulfonic acid in the Arctic troposphere, *Atmos. Environ.*, in press.

Li, S-M., L.A. Barrie, and A. Sirois, 1993, Biogenic sulfur aerosols in the Arctic troposphere: II. Trends and seasonal variations, *J. Geophys. Res*., 98:20,623-20,632.

Mayewski, P.A., W.B. Lyons, M.J. Spencer, M.C. Twickler, C.F. Buck, and S. Whitlow, 1990, An ice-core record of atmospheric sulphate and nitrate, *Nature*, 346:554-556.

Mayewski, P.A. and M.R. Legrand, 1990, Recent increase in nitrate concentration of Antarctic snow, *Nature*, 346:258-260.

Mulvaney, R., E.C. Pasteur, D.A. Peel, E.S. Saltzmann and P.-Y. Whung, 1992, The ratio of MSA to non-sea salt sulphate in Antarctic Peninsula ice cores, *Tellus*, 44:295-303.

Mulvaney, R., G.F.J. Coulson and H.F.J. Corr, 1993, The fractionation of sea salt and acids during transport across the Antarctic ice shelf, *Tellus*, 45:179-187.

Mulvaney, R. and E.W. Wolff, 1993, Evidence for winter/spring denitrification of the stratosphere in the nitrate record of Antarctic firn cores, *J. Geophys. Res.*, 98:5213-5220.

Nakazawa, T., T. Machida, M. Tanaka, Y. Fijii, S. Aoki and O. Watanabe, 1993, Differences of the atmospheric CH_4 concentration between the Arctic and Antarctic regions in pre-industrial/pre-agricultural era, *Geophys. Res. Lett.*, 20:943-946.

Neftel, A, 1991, "Use of snow and firn analysis to reconstruct past atmospheric composition," NATO ASI Series, Vol. G28, Seasonal Snowpacks, T. Davies *et al.* (eds.), pgs. 385-416.

Neubauer, J. and K.G. Heumann, 1988a, Determination of nitrate at the ng/g level in Antarctic snow samples with ion chromatography and isotope dilution mass spectrometry, *Fresenius J. Anal. Chem*., 331:170-173.

Neubauer, J. and K.G. Heumann, 1988b, Nitrate trace determinations in snow and firn core samples of ice shelves at the Weddell Sea, Antarctica, *Atmos. Envir.*, 22:537-545.

Nriagu, J.O., R.D. Coker, and L.A. Barrie, 1991, Origin of sulphur in Canadian Arctic haze from isotope measurements, *Nature*, 349:142-145.

Ottar, B., 1989, Arctic air pollution: a Norwegian perspective, *Atmos. Envir.*, 23:2349-2356.

Raynaud, D. and J. Chappellaz, 1993, The ice record of methane, *in:* "The Global Cycle of Atmospheric Methane," NATO Advanced Studies Workshop, Oregon, Oct. 1991.

Raynaud, D., J. Jouzel, J.M. Barnola, J. Chappellaz, R.J. Delmas and C. Lorius, 1993, The ice record of greenhouse gases, *Science*, 259:926-934.

Rosman, K.J.R., W. Chisholm, C.F. Boutron, J.P. Candelone and U. Görlach, 1993, Isotopic evidence for the source of lead in Greenland snows since the late 1960s, *Nature*, 362:333-335.

Savoie, D.L., J.M. Prospero, R.J. Larsen, H. Fen, M. Izaguirre, H. Tao and T. Snowdon, 1992, Nitrogen and sulphur species in Antarctic aerosols at Mawson, Palmer Station and Marsh (King George Island), *J. Atmos. Chem.*, 17:95-122.

Sigg, A. and A. Neftel, 1991, Evidence for a 50% increase of H_2O_2 over the past 200 years from a Greenland ice core, *Nature*, 351:557-559.

Silvente, E., 1993, "Contribution à l'étude de la fonction de transfert air neige en régions polaires," Thesis, Univ. of Grenoble, 157 pp.

Silvente, E. and M. Legrand, 1993, Ammonium to sulphate ratio in aerosol and snow of Greenland and Antarctic regions, *Geophys. Res. Lett.*, 20:687-690.

Staffelbach, T., A. Neftel, B.Stauffer, and D. Jacob, 1991, Formaldehyde in polar ice cores; a possibility to characterize the atmospheric sink of methane in the past, *Nature*, 349:603-605.

Sturges, W.T. and L.A. Barrie, 1989, Stable lead isotope ratios in Arctic aerosols: evidence for the origin of Arctic air pollution, *Atmos. Envir.*, 23:2513-2520.

Sturges, W.T., J.F. Hopper, L.A. Barrie and R.C. Schnell, 1993, Stable lead isotopes in Alaskan Arctic aerosols, *Atmos. Envir.*, in press.

Talbot, R.W., A.S. Vijgen, and R.C. Harriss, 1992, Soluble species in Arctic summer troposphere: Acidic gases, aerosols and precipitation, *J. Geophys. Res.* 97:16,531-16,543.

TERRESTRIAL BIOSPHERE-ATMOSPHERE EXCHANGE IN HIGH LATITUDES

William S. Reeburgh[1], Nigel T. Roulet[2] and Bo H. Svensson[3]

[1]Department of Geosciences
University of California, Irvine
Irvine, CA 92717-3100
[2]Department of Geography
McGill University
Montreal, Quebec
Canada H3A 2K6
[3]Department of Microbiology
Swedish University of Agricultural Sciences
S-750 07 Uppsala, Sweden

ABSTRACT

This paper summarizes CH_4 flux measurements from high latitude tundra and wetland environments obtained during large field campaigns and a number of smaller local projects since 1986. Area-weighted time series CH_4 flux measurements at fixed sites in high latitude tundra and wetland environments suggest a CH_4 source strength of about 35 Tg yr^{-1}, which agrees reasonably with atmospheric model calculations. Individual flux measurements are subject to high spatial and temporal (daily to interannual) variability, and are strongly influenced by vegetation, soil temperature, thaw depth, and water table level, which influence the balance between CH_4 production and consumption. Field studies show that these subsurface parameters and CH_4 fluxes are often poorly correlated. Laboratory studies on varying water table level in tundra soil columns have shown long lag times and hysteresis in the fluxes under falling and rising water table conditions. These results indicate that the soil parameters are not independent variables and that a single parameter capable of predicting CH_4 flux is unlikely.

Boreal forest ecosystems appear to be a small net sink for CH_4. Microbially mediated oxidation of atmospheric CH_4 is the dominant process; wet bogs, fens, swamps, and beaver ponds are the only CH_4 sources. Microbial CH_4 oxidation is also an important modulator of tundra emission and has the potential to eliminate net CH_4 emission under warmer, drier conditions.

Global Atmospheric-Biospheric Chemistry
Edited by R.G. Prinn, Plenum Press, New York, 1994

Studies to date have emphasized CH_4, but future integrated studies along boreal forest and tundra transects will include measurements of other gases and carbon dioxide fluxes for estimates of carbon fixation and storage. Field manipulation experiments aimed at determining the sensitivity of high latitude trace gas emissions to temperature and moisture variations are also planned. High latitude biosphere-atmosphere studies to date have a strong North American bias. Collaborative studies in northern European and Siberian wetlands are essential to understanding the importance of high latitude ecosystems in trace gas budgets.

INTRODUCTION

Two circumpolar components of Arctic and sub-Arctic ecosystems, tundra and boreal forest, have accumulated and stored soil carbon since the last glaciation. Tundra (area: 7.3×10^{12} m^2; 7% of earth's land area) is the treeless Arctic and sub-Arctic plain populated with dwarf shrubs, graminoids and cryptogams, and is often underlain by permafrost. Tundra contains 13% of the earth's stored soil carbon. Boreal forest (11.1×10^{12} m^2) is a general term describing the mixed deciduous-coniferous forest biome located between the tundra and temperate forest or prairie. It contains about 14% of the stored soil carbon (Post *et al.*, 1982). Boreal wetlands (area: 1.5×10^{12} m^2) occupy regions between 45° and 60°N, and are counted as part of tundra under most vegetation classification schemes. Approximately 30% of all northern wetlands are contained in two continuous wetland complexes: the Hudson Bay Lowland in Canada (3.2×10^{10} m^2) and the Vasuugan peatland (5.4×10^{10} m^2) in Siberia. All of these systems are characterized by short (100 to 120 day) growing seasons, low temperatures, low decomposition rates, moist to waterlogged soils, and the presence of continuous or discontinuous permafrost.

Much of the stored soil carbon in high latitude systems is present as peat or frozen in permafrost, and is subject to limited biogeochemical degradation. The fate of this stored carbon under altered climate is a major question (Billings, 1987). Model projections (Mitchell, 1989) indicate that Arctic regions will be the first impacted by global warming. Permafrost temperature distributions (Lachenbruch and Marshall, 1986) are consistent with a decade-scale rise in temperature. We understand in broad terms that CH_4 emissions result from a balance between production (methanogenesis) and consumption (methylotrophy), and that this balance is governed by temperature, moisture/water table relations, and substrate quality, but which variables are the best predictors of methane flux, the importance of feedbacks, and generally, the role of tundra under altered climate remains unclear. Some projections (Khalil and Rasmussen, 1989; Lashoff, 1989; Oechel *et al.*, 1993) suggest that warmer climate will lead to warmer soil temperatures, thawing permafrost and increased CH_4 and CO_2 emission. Others (Whalen and Reeburgh, 1990a; Reeburgh *et al.*, 1993; Moore and Roulet, 1993; Roulet *et al.*, 1993) suggest that microbially mediated CH_4 oxidation, which is neglected in the above projections, has the potential to modulate CH_4 emission, and under appropriate conditions, consume atmospheric CH_4.

This paper is structured around progress on the tasks of the IGAC High Latitude Ecosystems as Sources and Sinks for Trace Gases (HESS) Activity, which include:

Estimating the magnitude of high latitude ecosystem trace gas emissions.

Understanding factors and processes controlling high latitude ecosystem gas exchange.

Understanding sensitivity of high latitude trace gas sources and sinks to climate change.

Since these tasks are sequential, this paper will emphasize recent work on trace gas emissions from high latitude ecosystems and our current understanding of the role of high latitude environments in the CH_4 budget (Task 1), spell out our present understanding of the important controls on CH_4 fluxes (Task 2), and finally, outline manipulation experiments planned to determine sensitivity of high latitude systems to climate change (Task 3).

ROLE OF HIGH LATITUDE ENVIRONMENTS IN THE GLOBAL METHANE BUDGET (Task 1)

Early Studies

The possible importance of high latitude environments in radiatively active trace gas budgets was recognized early, but prior to 1986 only three studies provided information on CH_4 fluxes from high latitude systems (Clymo and Reddaway, 1971; Svensson and Rosswall, 1984a, b; Sebacher *et al.*, 1986). High latitude wetlands have been prominent as a potentially large term in the global CH_4 budget (Cicerone and Oremland, 1988; Matthews and Fung, 1987; Fung *et al.*, 1991), but their contribution has been uncertain because of three problem areas: lack of basic flux data, lack of understanding of the processes controlling fluxes, and uncertainties in scaling up from sites to regional and global scales. Large-scale coordinated field campaigns and individual projects have led to rapid growth in the high latitude CH_4 flux data base. The wetland CH_4 flux work has been summarized and reviewed recently by Bartlett and Harriss (1993). Because of the almost total absence of CH_4 flux data, net CH_4 fluxes have received major emphasis, but measurements of other gas fluxes are being incorporated in current work.

Large Field Campaigns

ABLE 3A. The Arctic Boundary Layer Expedition (ABLE 3A) was conducted in the Arctic and sub-Arctic of North America during July and August of 1988. Studies included a comprehensive tower, chamber and aircraft campaign at a site in the Yukon-Kuskokwim Delta near Bethel, Alaska, and flights around Barrow, Alaska. A summary of the results appears in a dedicated 1992 issue (number D15) of the *Journal of Geophysical Research*; the following discussion will emphasize the CH_4 studies.

Methane flux measurements were made with methods covering three scales; aircraft boundary layer measurements (1000 km), tower measurements (103 m) and chamber measurements (1 m^2). Errors of approximately a factor of 2 between chamber and tower flux measurements resulted from poor resolution of CH_4 producing habitats in the tower footprint (Fan *et al.*, 1992). Mean aircraft fluxes were ~2 times higher than the tower fluxes. The aircraft data were obtained during afternoon in periods of good flying weather, and these conditions were associated with maximum tower fluxes. The tower and ground-based measurements were made in an area with small lakes, whose CH_4 flux varied with size and wind speed. A mean emission rate of 25 mg CH_4 m^{-2} d^{-1} for wet tundra, dry tundra and lakes, which is low compared to other sites, was made by areal weighting of vegetation types obtained from satellite imagery. Bartlett *et al.* (1992) extended this flux to a global emission rate for tundra of 11±3 Tg CH_4 yr^{-1}. Ritter *et al.* (1992) measured CH_4 source strengths of 25 to 85 mg CH_4 m^{-2} d^{-1} with aircraft boundary layer measurements, and established an averaged, seasonally adjusted source strength of 51 mg CH_4 m^{-2} d^{-1}. A global estimate for tundra CH_4 emissions using this rate was 44 Tg yr^{-1}.

NOWES/ABLE 3B. The Northern Wetlands study (NOWES), a joint experiment between the Canadian Institute for Research in Atmospheric Chemistry (CIRAC) and the NASA Atmospheric Boundary Layer Experiment 3B, was conducted in 1990 to assess the importance of a large continuous wetland, the Hudson Bay Lowland in Northern Canada (3.2×10^{10} m^2), as a source/sink of biogenic gases under current climate conditions (Roulet *et al.*, 1992a). Most of the measurements of CH_4 flux in high latitude systems have been from relatively small isolated systems, i.e. wetlands with an area <1000 km^2. A dedicated issue (number D1) similar to the ABLE 3A issue of the *Journal of Geophysical Research* appeared in 1994.

The experiment was conducted in two locations. The first location was a 120 km transect running from James Bay coast inland to the interior of the Hudson Bay lowland. Areal coverage of this transect by wetlands and lakes was 100%. The second location was near Schefferville, Quebec, in a region that is approximately 30% wetland. Fluxes of CO_2, CH_4, N_2O, NO_x, and NMHCs were measured at numerous sites with flux chambers. A micrometeorological flux tower at each study site was used to measure fluxes of CO_2, CH_4 and O_3. Periodic boundary layer flux measurements of CO_2, CH_4, CO and O_3 were made by aircraft during the experiment. In addition to flux measurements, the concentrations of other trace gases such as CO_2, VOC, NO, N_2O, NO_2, O_3 and some NMHC were made throughout the summer at Schefferville and an atmospheric observatory on the southern fringe of the lowland.

Low CH_4 fluxes for the Hudson Bay Lowland were the most surprising result of the campaign. Based on prior studies (Bartlett and Harriss, 1993) the Hudson Bay Lowland was expected to emit between 5 and 7 Tg CH_4 yr^{-1}, but field measurements gave an estimate of 0.5 ± 0.03 Tg CH_4 yr^{-1}. (Roulet *et al.*, 1994). These estimates were obtained by weighting chamber flux measurements by the fractional component of each major ecosystem determined by remote sensing. Fluxes measured by aircraft averaged 16 mg CH_4 m^{-2} d^{-1} (Ritter *et al.*, 1994), tower-derived fluxes averaged 16 mg CH_4 m^{-2} d^{-1} (Edwards *et al.*, 1994), and the modeled chamber fluxes averaged 20 mg CH_4 m^{-2} d^{-1}. Methane fluxes measured by these three independent methods agreed to within a factor of 2 at all times, and to within a factor of 1.2 most of the time (Roulet *et al.*, 1993). However, Moore *et al.* (1993) have shown that comparisons between tower and aircraft fluxes to greater than a factor of 1.5 are not possible for sensible heat due to errors in measurements at different scales. These errors are larger in comparison to those for trace gases. The highest daily fluxes of CH_4 were from the shallow ponds on the surface of the wetlands: 40 to 150 mg CH_4 m^{-2} d^{-1}. (Hamilton *et al.*, 1994). The fluxes from the peat surfaces were much smaller: <40 mg CH_4 m^{-2} d^{-1} (Moore *et al.*, 1994). Approximately 10% of the fluxes measured from the peat surfaces indicated uptake of CH_4. The NOWES CH_4 flux measurements were within 25% of those predicted by Fung *et al.* (1991).

The fluxes of N_2O were small, −3.8 to 7.9 mg m^{-2} d^{-1}, yielding a total source of ~1.2 GT yr^{-1}, an insignificant source in the global budget (Schiller and Hassie, 1994). Wetlands in general are NO_3^- depleted and limited, so their capacity to yield large amounts of N_2O is limited. The NMHC flux was measured in only a few locations, but was at least as large, and in many cases larger, than the CH_4 flux (Klinger *et al.*, 1994). The seasonal distribution of isoprene followed a pattern controlled by the source from vegetation and air temperature (Jobson *et al.*, 1994).

During NOWES there was little direct work on the controls of the fluxes of various trace gases. However, because the experiment was located along a transect in an area that has been isostatically rebounding since the last glaciation (Roulet *et al.*, 1992a), it presented an opportunity to study how 5000 years of peatland evolution influences the CH_4 flux. Moore *et al.* (1994) and Klinger *et al.* (1994) found clear relationships among various stages of ecological succession and CH_4 flux in individual wetland types, while Roulet *et al.* (1994) showed there was a significant increase in CH_4 flux from the region of younger wetlands (1,000 to 2,000 years old) near the James Bay coast to older wetlands (~5000 years old) in the interior of the Hudson Bay lowland. This was in large part due to the increase in the area of shallow ponds on wetlands inland from the coast.

Individual Studies

While the large campaigns have emphasized integration over large areas, individual studies have emphasized surveys, time series flux measurements, and processes controlling trace gas fluxes in a variety of high latitude wetland (Roulet *et al.*, 1992b; Roulet *et al.*, 1992c) and boreal forest (Whalen *et al.*, 1991) environments. These studies provide long term information on spatial and temporal variability of CH_4 fluxes; one study (Whalen and Reeburgh, 1992) has continued for 6 years, providing information on interannual variability. At least two of these studies have involved winter measurements (Dise, 1992; Whalen and Reeburgh, 1988). Winter fluxes are less than 5% of the annual emission in high latitude tundra systems, but can contribute as much as 20% of the annual emission in lower latitude peatlands with less extensive freezing. Scaling up site scale measurements to regional and global scales has involved areal weighting of fixed site time series flux data (Whalen and Reeburgh, 1988, 1992) and transect studies (Whalen and Reeburgh, 1990b). Measurements in an Alaska boreal forest system (Whalen *et al.*, 1991) showed that large areas of moist to dry soils actively consume atmospheric CH_4. Local bogs and fens are the only CH_4-emitting sites in boreal forest systems.

Importance of High Latitude Systems in the Global Methane Budget

Because little was known about the CH_4 source strength of high latitude environments, a global emission estimate has been made using flux data from virtually every study. Table 1 summarizes these high latitude CH_4 emission estimates. Most of these studies assume an active period of 100 to 120 days and a distribution of environments related to the flux data available. Morrissey and Livingston (1992) point out that high natural spatial and temporal variability in fluxes make this practice questionable. The earliest estimate (Svensson and Rosswall, 1976) is based on a CH_4 flux/moisture relationship. The estimates of Sebacher *et al.* (1986) and Whalen and Reeburgh (1990b) are from transect sampling campaigns; the latter campaign involved sampling at fixed intervals to obtain an areally representative distribution of samples. Time series flux measurements from fixed sites were used in the estimates of Crill *et al.* (1988), Whalen and Reeburgh, (1988, 1992), and Moore *et al.* (1990). The estimate of Bartlett and Harriss (1993) results from a comprehensive review of published CH_4 flux data. Fung *et al.* (1991) used the GISS global 3-D tracer model, a compilation of geographic and seasonal CH_4 sources and sinks, and CH_4 carbon isotope ratios to provide constraints on global CH_4 budget terms. Several budget scenarios were tested for their ability to reproduce the meridional gradient and seasonal variations observed at the GMCC stations. Their preferred scenario calls for CH_4 emissions of 35 Tg yr^{-1} from wetlands and tundra north of 50°. Reeburgh and Whalen (1992) pointed out that the high latitude CH_4 emission term seemed to be converging on about 35 Tg yr^{-1}. The additional global estimates since then, with the exception of the Bartlett *et al.* (1992) ABLE 3A estimate, lie between 20 and 45 Tg yr^{-1} and have large (±50%) error limits. High latitude terrestrial environments contribute slightly less than 10% to the global CH_4 budget (Fung *et al.*, 1991).

Table 1. High latitude CH_4 emission estimates.

Reference	Region	Location	Study Type	CH_4 Emission ($Tg\ yr^{-1}$)
Svensson & Rosswall (1976)	Arctic tundra	Stordalen, Sweden	summer fluxes extended with moisture relationships	1 – 25
Sebacher *et al.* (1986)	tundra and taiga	Alaska	flux transect, Arctic and boreal regions	46 – 106
Matthews & Fung (1987)	forested bog 50° – 70°N	—	estimated area × literature emission rates	20 – 62
	non-forested bog 50° – 70°N	—	estimated area × literature emission rates	42
Crill *et al.* (1988)	all bogs > 40°N	Minnesota peatlands	summer flux time series at permanent sites	72
Whalen & Reeburgh (1990)	Arctic tundra	U. of Alaska Arboretum	annual flux time series at permanent sites	19 – 32
Aselmann & Crutzen (1989)	wetlands 50° – 70°N	—	estimated area × literature emission rates	23
Moore *et al.* (1990)	northern fens	Quebec, Canada	summer flux time series at permanent sites	14
Whalen & Reeburgh (1990)	Arctic tundra	Alaska	flux transect, Alaska Pipeline Haul Road	38 – 53
	taiga	Alaska	flux transect, Alaska Pipeline Haul Road	15
Fung *et al.* (1991)	wetlands > 50°N	—	3-D tracer model, estimates sources (scenario 7)	35
Whalen *et al.* (1991)	moist taiga	Bonanza Creek Exptl. Forest, Alaska	thaw season flux time series at permanent sites	– 0.8
Whalen & Reeburgh (1992)	Arctic tundra	U. of Alaska Arboretum	annual flux time series (4 years of data)	42 ± 26
Bartlett *et al.* (1992)	ABLE 3A	Yukon-Kuskokwim Delta, Alaska	summer flux time series at permanent sites	11 ± 3
Ritter *et al.* (1992)	ABLE 3A	Yukon-Kuskokwim Delta, Alaska	aircraft boundary layer flux	44
Bartlett & Harriss (1993)	wetlands > 45°N	—	global summary	38

Present and Future Work

North America. *BOREAS.* The Boreal Ecosystem Atmosphere Study (BOREAS) is planned for 1994 at two locations in the boreal forest region of northern Canada. The objectives of the experiment are (1) to improve the understanding of the processes and states which govern the exchanges of energy, water, carbon and trace gases between boreal forest ecosystems and the atmosphere, with particular reference to those processes and states that may be sensitive to global climate change, and (2) to develop and validate remote sensing algorithms to extend this understanding of the processes from local to regional scales. The focus of the research will be two 600 km^2 sites: one dry, warm boreal forest site located near Prince Albert in central Saskatchewan, and the other, a moist, cool boreal forest site located near Thompson, Manitoba. A unique aspect of BOREAS is the concurrent measurement of trace gases, tentatively CO, CH_4, some NMHCs, N_2O and NO_x, and all important physical, biogeochemical and ecological variables at all the important ecosystems that comprise what would be considered the boreal biome. Using this approach, sites of both net emissions and uptake of trace gases can be quantified and the total influence of this portion on the boreal biome can be assessed. The ecosystems include various forest types (aspen, spruce, pine), wetlands, and beaver ponds. Measurements will consist of a full set of chamber, micrometeorological, and aircraft derived fluxes. The experiment will run from snowmelt to freeze-up in 1994, with three intensive measurement periods where all three types of flux measurements will be conducted simultaneously. One tower at each site will measure fluxes from August, 1993 to September, 1994.

ARCSS/LAII Flux Study. The NSF-sponsored Arctic System Science Land-Air-Ice Interactions Flux Study (ARCSS/LAII Flux Study) will be conducted along a transect in Alaska from Toolik Lake in the Brooks Range foothills to the Arctic Coastal plain from 1993 to 1996. The goal of this study is to understand the processes controlling carbon storage and emission as CO_2 and CH_4. This study will involve measurements of emission and consumption of CO_2 and CH_4 using micrometeorological tower and chamber studies, coupled with aircraft flux measurements at intervals during one field season. The gas flux studies are integrated with hydrologic and soil studies, vegetation mapping and spectral reflectance measurements for interpretation of satellite data in this multi-year project.

Scandinavia. Several programs examining trace gas fluxes are underway in Scandinavia and are projected to continue until 1998. A Swedish effort to quantify CH_4 fluxes from peatlands (Mikkelä *et al.*, 1992) also involves studies on the rates and controls on CH_4 production and oxidation (Svensson and Sundh, 1992). For example, Sundh *et al.* (1992) found that the oxidative capacity of a peatland is a direct function of the moisture content of the peatland. Other research is being conducted on the importance of peat quality on CH_4 flux (Nilsson, 1992) and the development of techniques such as FTIR for the measurement of the CH_4 flux.

Part of the Finnish Programme on Climate Change (SILMU) is examining the carbon balance of peatlands and climate change (Lainen and Päivänen, 1992). This experiment has been underway for three years and will continue for at least another three years. Chamber measurements of CO_2, CH_4 and N_2O flux are being made in areas subjected to various treatments and manipulations to determine the controls on trace gas emissions (Lien *et al.*, 1992; Martikainen *et al.*, 1992; Silvola *et al.*, 1992). The project includes experiments related to gas exchange from northern wetlands such as the emissions of CH_4 from virgin and drained peatlands, the effect of drainage intensity on carbon and nitrogen mineralization, rate of net peat accumulation, response of Sphagna to N deposition and increased CO_2 concentrations, and organic carbon in peatland waters. A particularly unique aspect of these experiments is the use of forest drainage sites as surrogates for climate change.

Russia. Several groups have initiated CH_4 flux studies in various parts of Russia. Panikov and Zelenev (1992) are working in Western Russia, and Samarkin *et al.* (1992) are involved in studies in the Kolyma lowland. A program of CO_2 flux measurements is underway near Chernski, Yakutia (S. Zimov, private communication), and a CH_4 flux measurement program is being established there. Established field stations and ongoing expeditions in these areas should facilitate collaborative flux time series studies.

CONTROLS ON HIGH LATITUDE TRACE GAS FLUXES (Task 2)

We understand in broad qualitative terms that CH_4 emissions result from a balance between production (methanogenesis) and consumption (methylotrophy). Physical factors that directly affect this balance are temperature, moisture, water table level, and substrate quality, but an understanding of the most important controls on CH_4 flux, the importance of feedbacks, and generally, the sensitivity of high latitude environments to altered climate remain unclear. Studies have concentrated on physical factors such as soil temperature because it is related to microbial activity, and the position of the water table, because it defines the zones of anaerobic (production) and aerobic (oxidation) processes. These are also the simplest variables to measure continuously.

Examples of relationships between CH_4 flux and soil temperature are summarized in Bartlett and Harriss (1993). A number of relationships have been reported for short term flux data sets using soil temperatures at fixed depths, but the number and variety of functions used (linear, polynomial, log-linear, and exponential), suggests that the relationship between temperature and CH_4 flux is not unique or straightforward (Whalen and Reeburgh, 1992).

Laboratory studies on the effect of water table level on CH_4 and carbon dioxide emission have been performed by Peterson *et al.* (1984) and Moore and Knowles (1987). Lowered wetland water tables are believed to control CH_4 fluxes by enhancing aerobic CH_4 oxidation. A lowered water table provides a more gas permeable surface zone that becomes aerobic through diffusive transport of atmospheric oxygen. This permits establishment of microbial populations capable of oxidation of CH_4 escaping from the deeper, waterlogged sediment layers. Whalen *et al.* (1993) showed that waterlogged tundra soils were anoxic only a few millimeters below the water table surface. Jar experiments with moist soils (Whalen and Reeburgh, 1990a; Whalen *et al.*, 1990) showed rapid equilibration with added tracers, suggesting gas phase diffusivities of $\sim 10^{-1}$ to $10^{-2}\ cm^{-2}\ sec^{-1}$, 10^4 times more rapid than aqueous phase molecular diffusion. The CH_4 oxidizing populations responded rapidly to changes in CH_4 concentration and consumed CH_4 to threshold values of 0.1 ppm (Whalen *et al.*, 1992). Methane oxidation rates were very sensitive to water content, apparently a result of limiting transport by sealing inter-grain boundaries. The initial moist soil work suggested that oxidation could provide a rapid response and a negative feedback to increases in CH_4 flux.

The above rapid responses have not been observed in wetlands with waterlogged soils. Subsoil CH_4 distributions and fluxes appear to be decoupled, as high standing stocks of soil CH_4 and steep soil water CH_4 gradients show poor correlations with surface fluxes (Whalen and Reeburgh, 1992). Correlations between CH_4 flux and water table level in wetlands are generally poor, and appear to be masked in whole-season correlations (Whalen and Reeburgh, 1992). One possible explanation is that the CH_4 emission response occurs too rapidly to appear in whole-season correlations. The magnitude of CH_4 standing stocks in soil waters compared with the surface CH_4 fluxes suggests residence times of days. However, laboratory water table manipulation experiments on peat cores give results that suggest much slower responses. Moore and Roulet (1993) showed very dissimilar CH_4 flux responses under falling and rising water table conditions, and similar studies by Kizer and coworkers (private communication) showed that the CH_4 fluxes of cores subjected to water table adjustments stabilize only after periods of weeks. In contrast to the rapid response by moist soils, saturated or waterlogged soils show responses that may be related to flooding history or frequency. For example, in a falling water table situation, the newly exposed soils have only a history of anoxic conditions, and some time may be required to repopulate the zone with aerobic oxidizers. Reconciling the results of wetland field observations with the laboratory core manipulations is an important problem that will require further study.

Whalen and Reeburgh (1992) noted that relationships between any single variable and CH_4 emission are site specific and of little value as predictors, emphasizing the fact that variables such as temperature and water table depth are not truly independent. They showed that parameters that integrate conditions over a season, such as thaw depth or centimeterdegrees, correlated best with CH_4 emission on a given date. Dise *et al.* (1993) showed that the position of the water table explained over 60% of the variance in CH_4 flux in Minnesota wetlands using multiple regression. Temperature explained 30% of the variance and a measure of substrate quality explained the remainder. Moore and Roulet (1993) showed that the response of CH_4 flux to a unit change in water table is similar among wetlands, but that the actual flux for a given water table varies considerably. Systems with small changes in water table level over the study period show no significant relationship between CH_4 flux and water table (Moore *et al.*, 1994). When this is the case, peat temperature dominates the system (Roulet *et al.*, 1992c).

The above attempts are directed toward a process-level understanding of the factors controlling CH_4 flux. Considering parameters like biomass and primary production rather than variables like temperature and water table level appears to offer a promising means of predicting CH_4 flux from some wetland systems. Recent work in diverse wetland systems shows strong linear correlations between CH_4 emission and biomass (Whiting *et al.*, 1991) and CH_4 emission and primary production (Aselmann and Crutzen, 1989; Whiting and Chanton, 1993). All sites considered were inundated and populated with plants capable of vascular transport. Concurrent measurements of net primary production and CH_4 flux showed that approximately 3% of daily net primary production was released as CH_4 (Whiting and Chanton, 1993). The correlations between biomass and primary production could result from either litter input or production of suitable methane-forming substrates below-ground by root exudation and root turnover, or they could be a result of the plants providing conduits for CH_4 emission. This relationship offers the potential of estimates of CH_4 emission through remote sensing of either photosynthetic activity or net primary production. Further work suggested by the net primary production/net CH_4 emission relationship includes determination of the origin and age of the methane-forming substrate, the role of below-ground CH_4 oxidation (Reeburgh *et al.*, 1993), and the applicability of the relationship to non-inundated wetlands free of vascular plants.

SENSITIVITY OF HIGH LATITUDE SYSTEMS TO CLIMATE CHANGE (Task 3)

Most climate models predict a large increase in winter temperature, and a moderate increase in summer temperature for northern regions. The predicted change in precipitation varies among models. Based on the output of the climate models, the summer climate of the northern wetland region is predicted to be warmer and slightly wetter, and it is not unreasonable to expect an increase in the annual emissions of CH_4 from northern wetlands. However, the biospheric feedbacks incorporating climate, hydrological and thermal regimes of wetlands, and the flux of CH_4 from wetlands are quite complicated and poorly understood. While an increase in summer precipitation would suggest an increase in the wetness of wetlands, the accompanying increase in temperature would also increase evaporative loss of water from wetlands. The direction of change in water storage in a wetland will be governed by the difference between precipitation input and evaporative and runoff outputs.

Models based solely on an increase in temperature suggest that the emissions of CH_4 from northern wetlands should increase in a $2 \times CO_2$ climate. Early models of Hameed and

Cess (1983) and Lashoff (1989) predicted an increase in CH_4 emissions with an increase in temperature. Harriss and Frolking (1992) developed a simple model based on the regressions of CH_4 flux and temperature developed by Crill *et al.* (1988) for peatlands in Minnesota. Using year-to-year temperature anomalies of ±2°C observed over the 20th century for five northern region wetlands, they estimate a ±5 Tg yr^{-1} variation in emissions. This deviation represents only 15% of the total emissions from northern wetlands for the expected range of temperature increase predicted by most GCMs for a 2 × CO_2 climate and suggests only a moderate feedback.

The CH_4 temperature feedback models assume that the moisture status of the wetland remains the same. However, Roulet *et al.* (1992b) using a simple hydrothermal model for northern fens, estimated a decrease in moisture storage of between 60 to 200 mm in a 2 × CO_2 climate. The same model predicted a 0.8 to 2.0°C increase in the peat temperature. The increase in peat temperature resulted in a 15% increase in CH_4 flux, a result very similar to that of Harriss and Frolking (1992). However, the decrease in water storage resulted in an 80% decrease in CH_4 flux, suggesting that the emissions of CH_4 from northern wetlands are more sensitive to change in moisture than temperature. A small decrease in water storage in a wetland can easily offset any temperature-induced increase in emissions. The models used to estimate the CH_4 biospheric feedback for wetlands are preliminary and crude. Several studies have used empirical approaches to examine the climate-CH_4 feedback in wetlands, either by attempting to correlate seasonal and inter-annual variability in the CH_4 with climate variables, or by measuring the CH_4 flux from wetlands that have undergone some drainage. The results of two studies conducted in Alaska suggest that the CH_4 flux from wetlands is extremely sensitive to change in moisture and temperature. A four-fold variation in annual CH_4 emissions from wetlands in the interior of Alaska was observed over a four year period (Whalen and Reeburgh, 1992). Livingston and Morrissey (1991) compared the difference in the CH_4 flux from Alaskan North Slope wetland with varying moisture and temperature regimes between two years with very different climates. They found a four-fold increase in flux with a 4°C increase in soil temperature and an increase in wetness. They also found a two-fold increase in flux with a 4°C increase in soil temperature under drier conditions. These results suggest a far stronger sensitivity of the CH_4 flux to the climate than was found in modeling studies.

A criticism of the use of normal climate variability as a surrogate for climate change is that ecosystems should be well-adjusted to this type of variability. However, with a persistent change in climate, ecosystems should be pushed beyond the range of normal variability. To find the effect of persistent change in the moisture, Roulet *et al.* (1993) measured the flux of CH_4 from several wetlands that have been drained. These wetlands experience the day-to-day natural variability in temperature and precipitation, but have a lower mean water table than natural wetlands. Their results show that even a slight drop in water table of 0.1 m effectively eliminates the flux. The flux decreased from 5 to 10 mg CH_4 m^{-2} d^{-1} with an average water table of −0.2 m to between 0.1 and −0.4 mg CH_4 m^{-2} d^{-1} when the average water table was below −0.3 m. An additional drop in water table to −1.0 m did not increase CH_4 uptake. Whalen *et al.* (1993) used kinetic data on CH_4 oxidizers in soil cores to estimate that lowering the water table 10 to 40 cm could result in a zero net CH_4 flux.

All of the studies of biospheric feedback of CH_4 assume that there is no change in the structure and function of the wetland ecosystem. This is clearly a limiting assumption. Vegetation changes were obvious at the forest drainage sites used by Roulet *et al.* (1993) only after 10 years of drainage (J. Bubier, private communication) With changes in structure, changes in substrate quality and transport mechanism should be expected. Without an understanding of how the plant communities of wetlands respond to changes in the climate it will be impossible to make any long-term predictions of future emissions from northern wetlands.

FUTURE STUDIES

A remarkable amount of data on trace gas emissions from high latitude systems has been obtained in a relatively short time, and the importance of these systems as global CH_4 sources is converging on a value between 20 and 45 Tg yr^{-1}. A first-order understanding of biogeochemical processes controlling the balance between CH_4 production and emission has resulted, and models dealing with sensitivity of emissions to various observed parameters are under development. Future work should emphasize the following.

Our current information on high latitude trace gas fluxes has a strong North American bias, and has placed almost total emphasis on CH_4 emissions. Studies in northern European and Siberian wetlands are beginning, and should be supported with collaborative studies and flux time-series measurements. Obtaining flux data from these systems, based on accepted site classification and nomenclature used in the other high latitude studies, is essential to understanding the global importance of high latitude systems. It is important that future work emphasize measurements of not only CH_4, but also of other gases, particularly CO_2.

Estimates of the sensitivity of high latitude systems to climate change are only beginning. Most use historical climate data (temperature, rainfall) and simple relationships to flux. We appreciate that none of the driving variables are independent and that a single process-level parameter capable of predicting CH_4 flux is unlikely. Feedbacks promise to be an important control and require more study. Realistic manipulation experiments on the 1 to 10 m scale are needed to confirm process-level controls and their role in feedbacks. Preliminary experiments involving manipulation of water table level and temperature are underway and should be repeated in a range of environments. The design and eventual results of these experiments are critical to scaling models of biogeochemical feedbacks to general circulation models and chemical tracer models.

Understanding microbially mediated CH_4 production and oxidation in wetlands and soils should be a high priority. Understanding the origin of the substrates important in CH_4 production are needed for a process-level understanding of the methane flux:primary production relationship. There are very few direct measurements of CH_4 oxidation, but the available information suggests that oxidation may be about half of the gross CH_4 budget (Reeburgh *et al.*, 1993)

REFERENCES

Aselmann, I. and P.J. Crutzen, 1989, Global distribution of natural freshwater wetlands and rice paddies, and their net primary productivity, seasonality and possible methane emissions, *J. Atmos. Chem.*, 8:307-358.

Bartlett, K.B., P.M. Crill, R.L. Sass, R.C. Harriss and N.B. Dise, 1992, Methane emissions from tundra environments in the Yukon-Kuskokwim Delta, Alaska, *J. Geophys Res.*, 97:16,645-16,660.

Bartlett, K.B., and R.C. Harriss, 1993, Review and assessment of methane emissions from wetlands, *Chemosphere*, 26:261-320.

Billings, W.D., 1987, Carbon balance of Alaskan tundra and taiga ecosystems: Past, present and future, *Quat. Sci. Rev.*, 6:165-177.

Cicerone, R.J. and R.S. Oremland, 1988, Biogeochemical aspects of atmospheric methane, *Global Biogeochem. Cycles*, 2: 299-327.

Clymo, R.S. and E.J.F. Reddaway, 1971, Productivity of sphagnum (bog moss) and peat accumulation, *Hidrobiologia*, 12: 181-192.

Crill, P.M. et al, 1988, Methane flux from Minnesota peatlands, *Global Biogeochem. Cycles*, 2: 371-384.

Dise, N.B., 1992, Winter fluxes of methane from Minnesota peatlands, *Biogeochemistry*, 17:71-83.

Dise, N., E. Gorham and E.S. Verry, 1993, Environmental factors controlling methane emissions from peatlands in northern Minnesota, *J. Geophys Res.*, 98:10,583-10,594.

Edwards, G.E., H.H. Neumann, G. den Hartog, G.W. Thurtell and G. Kidd, 1994, Eddy correlation measurements of methane fluxes using a tunable diode laser at the Kinosheo Lake tower during the Northern Wetlands Study (NOWES), *J. Geophys. Res.*, 99:1511-1518.

Fan, S.M. , S.C. Wofsy, P.S. Bakwin, D.J. Jacob, S.M. Anderson, P.L. Kebabian, J.B. McManus and C.E. Kolb, 1992, Micrometeorological measurements of CH_4 and CO_2 exchange between the atmosphere and subarctic tundra, *J. Geophys. Res.*, 97:16,627-16,643.

Fung, I., J. John, J. Lerner, E. Matthews, M. Prather, L.P. Steele and P.J. Fraser, 1991, Three-dimensional model synthesis of the global methane cycle, *J. Geophys. Res.*, 96:13,033-13,065.

Hameed, S. and R. Cess, 1983, Impact of global warming on biospheric sources of methane and its climate consequences, *Tellus*, 35B:1-7.

Hamilton, J.D., C.A. Kelly, J.W.M. Rudd and R.H. Hesslein, 1994, Flux to the atmosphere of CH_4 and CO_2 from wetland ponds on the Hudson Bay Lowland (HBL), *J. Geophys. Res.*, 99:1495-1510.

Harriss, R.C. and S. Frolking, 1993, The sensitivity of methane emissions from northern freshwater wetlands to global change, *in:* "Global Warming and Freshwater Ecosystems," P. Firth and S. Fisher (eds.), Springer-Verlag, New York, pg. 48, in press.

Jobson, B.T., Z. Wu, H. Niki and L.A. Barrie, 1994, Seasonal trends of isoprene, C2-C5 alkanes, and acetylene at a remote boreal site in Canada, *J. Geophys. Res.*, 99:1589-1600.

Khalil, M.A.K. and R.A. Rasmussen, 1989, Climate-induced feedbacks for the global cycles of methane and nitrous oxide, *Tellus*, 41B:554-559.

Klinger, L.F., P.R. Zimmerman, J.P. Greenberg, L.E. Heidt, and A.B. Guenther, 1994, Carbon trace gases along a successional gradient in the Hudson Bay Lowland, *J. Geophys. Res.*, 1469-1494.

Lachenbruch, A.H. and B.V. Marshall, 1986, Climate change: Geothermal evidence from permafrost in the Alaskan arctic, *Science*, 34:689-696.

Lainen, J. and J. Päivänen, 1992, Carbon balance of peatlands and global climatic change: Summary, *in:* "Carbon Balance of Peatlands and Climate Change," Progress Report, SIMLU, Helsinki.

Lashoff, D.A.,1989, The dynamic greeenhouse: Feedback processes that may influence future concentrations of atmospheric trace gases and climate change, *Clim. Change*, 14:213-242.

Lein, T., P. Martikainen, H. Nykänen and L. Bakken, 1992, Methane oxidation and fluxes in two drained peat soils, *SUO*, 43:231-236.

Livingston, G.P. and L.A. Morrissey, 1991, Methane emissions from Alaskan arctic tundra in response to climatic change, *in:* "Role of Polar Regions in Global Change," G.L. Weller, C.L. Wilson and B.A.B. Severin (eds), University of Alaska, Fairbanks, pg. 372.

Martikainen, P., H. Nykänen, P. Crill and J. Silvola, 1992, The effect of changing water table on methane fluxes at two Finnish mire sites, *SUO*, 43:237-240.

Matthews, E. and I. Fung, 1987, Methane emission from natural wetlands: Global distribution, area, and environmental characteristics of sources, *Global Biogeochem. Cycles*, 1:61-86.

Mitchell, J.F.B., 1989, The greenhouse effect and climate, *Rev. Geophys.*, 27:115 139.

Mikkelä, C., I. Sundh, J. Eilertsson, B. Svensson and M. Nilsson, 1992, Methane emissions from a Swedish peatland area: Temporal and spatial variations, *in:* "Proceedings of the 9th International Peat Congress," D. Fredericksson (ed.), 3:152-165.

Moore, K.E., D.A. Fitzjarrald and J.A. Ritter, 1993, How well can regional fluxes be derived from small-scale estimates?, *J. Geophys. Res.*, 98:7187-7198.

Moore, T.R. and R. Knowles, 1989, The influence of water table levels on methane and carbon dioxide emissions from peatland soils, *Can. J. Soil Science*, 69:33-38.

Moore, T.R., N. Roulet and R. Knowles, 1990, Spatial and temporal variations of methane flux from subarctic/northern fens, *Global Biogeochem. Cycles*, 4:29-46.

Moore, T.R. and N.T. Roulet, 1993, Methane flux: water table relations in northern wetlands, *Geophys. Res. Lett.*, 20:587-590.

Moore, T.R., A. Heyes and N.T. Roulet, 1994, Methane emission from wetlands of the southern Hudson Bay lowland, *J. Geophys. Res.*, 19:1455-1468.

Morrissey, L.A. and G.P. Livingston, 1992, Methane emissions from Alaska arctic tundra: An assessment of local spatial variability, *J. Geophys. Res.*, 97:16,66116,670.

Nilsson, M., 1992, "Fungi and Bacteria in Peat and Peat-forming Plant Communities," Diss. Swedish Univ. Agric. Sciences, Dept. of Forest Site Research, Stencil #20.

Oechel, W.C., S.J. Hastings, G. Vourlitis, M. Jenkins, G. Reichers and N. Grulke, 1993, Recent change of arctic tundra ecosystems from a net carbon dioxide sink to a source, *Nature*, 361:520-523.

Panikov, N.S. and V.V. Zelenev, 1992, "Emission of CO_2 and CH_4 from Northern Wetlands to the Atmosphere: Dynamics, Controlling Factors, and Tentative Mechanisms," Russian Academy of Sciences, Puschino, Moscow Region.

Peterson, K.M., W.D. Billings and D.N. Reynolds, 1984, Influence of water table and atmospheric CO_2 concentration on the carbon balance of arctic tundra, *Arctic and Alpine Res.*, 16:331-335.

Post, W.M., W.R. Emanuel, P.J. Zinke and A.G. Stangenberger, 1982, Soil carbon pools and world life zones, *Nature*, 298:156-159.

Reeburgh, W.S. and S.C. Whalen, 1992, High latitude ecosystems as CH_4 sources, *Ecol. Bull. (Copenhagen)*, 42:62-70.

Reeburgh, W.S., S.C. Whalen and M.J. Alperin, 1993, The role of methylotropohy in the global methane budget, *in:* "Microbial Growth on C-1 Compounds," J.C. Murrell and D. Kelley (eds.), Intercept, Andover, pgs. 1-14.

Ritter, J.A., J.D.W. Barrick, G.W. Sachse, G.L. Gregory, M.W. Woerner, C.E. Watson, G.F. Hill and J.E. Collins, Jr., 1992, Airborne flux measurements of trace species in the arctic boundary layer, *J. Geophys. Res.*, 97:16,601-16,625.

Ritter, J.A., J.D.W. Barrick, C.E. Watson, G.W. Sachse, G.L. Gregory, B.E. Anderson, M.A. Worner and J.E. Collins, 1994, Airborne boundary layer flux measurements of trace gas species over Canadian boreal forest and northern wetland regions, *J. Geophys. Res.*, 99:1671-1686.

Roulet, N.T., L. Barrie, W.A. Glooschenko, R.C. Harriss, J.I. MacPherson, T.R. Moore, H. Neuman and H.I. Schiff, 1992a, Northern Wetlands Study (NOWES): An assessment of the role of northern wetlands in the exchange of atmospheric trace gases, *Ecol. Bull. (Copenhagen)*, 42:77-85.

Roulet, N.T., T.R. Moore, J. Bubier and P. LaFleur, 1992b, Northern fens: Methane flux and climatic change, *Tellus*, 42B:100-105.

Roulet, N.T., R. Ash and T. Moore, 1992c, Low boreal wetlands as a source of atmospheric methane, *J. Geophys. Res.*, 97:3739-3749.

Roulet, N.T., R. Ash, W. Quinton and T.R. Moore, 1993, Methane flux from drained northern peatlands: An analog for persistent water table lowering, *Global Biogeochem. Cycles*, 7:749-770.

Roulet, N.T., A. Jano, C.A. Kelly, T.R. Moore, R. Protz, J.A. Ritter and W.R. Rouse, 1994, The Role of the Hudson Bay Lowland as a source of atmospheric methane, *J. Geophys. Res.*, 99:1423-1428.

Samarkin, V.A., E.M. Rivkina and E.A. Bocharnikova, 1992, "Methane Biogeochemistry in Arctic Marsh Soils," Russian Academy of Sciences, Puschino, Moscow Region.

Schiller, C.L. and D.R. Hassie, 1994, Emissions of nitrous oxide from the Hudson Bay Lowland, *J. Geophys. Res.*, 99:1573-1588.

Sebacher, D.I., R.C. Harriss, K.B. Bartlett, S.M. Sebacher and S.S. Grice, 1986, Atmospheric methane sources: Alaskan tundra bogs, an alpine fen and a subarctic boreal marsh, *Tellus*, 38B:1-10.

Silvola, J., P. Martikainen and H. Nykänen, 1992, A mobile automatic gas chromatograph to measure CO_2, CH_4 and N_2O fluxes from soil in the field, *SUO*, 43:263-266.

Sundh, I., C. Mikkelä, M. Nilsson and B.H. Svensson, 1992, Potential methane oxidation in a Sphagnum peat bog: Relation to water table level and vegetation type, *in:* "Proceedings of the 9th International Peat Congress," D. Fredriksson (ed), 3:142-151.

Svensson, B.H. and T. Rosswall, 1976, Methane production in tundra peat, *in:* "Microbial Production and Utilization of Gases (H_2, CH_4, CO)," H.G. Schelgel, G. Gottschalk, N. Pfennig, E. Goltze (eds.), Göttingen, pgs. 135-139.

Svensson, B.H. and T. Rosswall, 1984, In situ methane production from acid peat communities with different moisture regimes in a subarctic mire, *Oikos*, 43:341-350.

Svensson, B.H. and I. Sundh, 1992, Factors affecting methane production in peat soils, *SUO*, 43:183-190.

Whalen, S.C. and W.S. Reeburgh, 1988, A methane flux time series for tundra environments, *Global Biogeochem. Cycles*, 2:399-409.

Whalen, S.C. and W.S. Reeburgh, 1990a, Consumption of atmospheric methane by tundra soils, *Nature*, 346:160-162.

Whalen, S.C. and W.S. Reeburgh, 1990b, A methane flux transect along the trans Alaska pipeline haul road, *Tellus*, 42B:237-249.

Whalen, S.C., W.S. Reeburgh and K.A. Sandbeck, 1990, Rapid methane oxidation in a landfill cover soil, *Appl. Environ. Microbiol.*, 56:3405-3411.

Whalen, S.C. and W.S. Reeburgh, 1992, Interannual variations in tundra methane emission: A four-year time series at fixed sites, *Global Biogeochem. Cycles*, 6:139-159.

Whalen, S.C., W.S. Reeburgh and V. A. Barber, 1992, Oxidation of methane in boreal forest soils: a comparison of seven measures, *Biogeochemistry*, 16:181-211.

Whalen, S.C., W.S. Reeburgh and C.E. Reimers, 1993, Processes controlling methane fluxes from tundra environments, *in:* "Landscape Function: Implications for Ecosystem Response to Disturbance. A case Study in Arctic Tundra," J.F. Reynolds and J.D. Tenhunen (eds.), Springer-Verlag, New York, in press.

Whiting, G.J., J.P. Chanton, D.S. Bartlett and J.D. Happell, 1991, Relationships between CH_4 emission, biomass, and CO_2 exchange in a subtropical grassland, *J. Geophys. Res.*, 96:13,067-13,071.

Whiting, G.J. and J.P. Chanton, 1993, Primary production control of methane emission from wetlands, *Geophys. Res. Lett*., submitted.

EXCHANGE OF TRACE GASES BETWEEN THE TERRESTRIAL BIOSPHERE AND THE ATMOSPHERE IN THE MIDLATITUDES

Keith A. Smith[1], G. Philip Robertson[2] and Jerry M. Melillo[3]

[1]Soil Science Dept., SAC, School of Agriculture
West Mains Road, Edinburgh EH9 3JG, UK
[2]W.K. Kellogg Biological Station and Dept. of Crop and Soil Sciences
Michigan State Univ., Hickory Corners, MI 49060-9516, USA
[3]The Ecosystems Center, Marine Biological Laboratory
Woods Hole, MA 02543, USA

ABSTRACT

Most terrestrial ecosystems of the midlatitudes have been subjected to human influence. Large areas of forests and grasslands have been converted to agriculture; conversely, reforestation is also extensive in many regions. These changes in land use, the use of fertilizers on agricultural land, and high precipitation inputs of nitrogen from industrial emissions to forests and other natural ecosystems, all have consequences for trace gas exchange, and thus for the atmospheric concentrations of trace gases that contribute to climate change.

The study of these interactions, on a comprehensive world-wide basis, is the task of the IGAC Activity Trace Gas Exchange in Midlatitude Ecosystems (TRAGEX). Although TRAGEX as a coordinated program is still at the planning stage, there is a substantial current research effort in progress in North America, Western Europe and Australia that has already provided some insight into the most important processes governing trace gas exchange in the midlatitudes. Some of the highlights of this work, and major outstanding questions, are outlined in this paper.

Soil carbon availability, temperature/moisture interactions, pH, and nutrient dynamics have been identified as key variables in methane emission from temperate wetlands, but measurement and modeling processes of transfer to the atmosphere are in their infancy. Inputs of nitrogen and restrictions in aeration of topsoils have been shown to reduce very substantially the soil's capacity to act as a sink for CH_4 by microbial oxidation to CO_2. High N inputs promote emissions of N_2O, but it has been shown that large effects on fluxes are caused by variations in soil physical conditions, the chemical form of the N, and soil pH. The very high spatial and temporal variability of N_2O and CH_4 fluxes have made representative flux estimates very difficult to make, and have stimulated major research efforts in development of improved methods of analysis applicable to areas of 10^3 to 10^4 m^2 and ultimately to the km^2 scale. The linking of process models with the Global Information System (GIS) for large-scale integration is a focus of current planning activity.

Increased CO_2 concentrations and N deposition, and the fact that much of the present forest area of the region is in a mid-successional stage, suggest that there may be a substantial vegetation-related sink for CO_2 in the midlatitudes. Changes in soil organic matter in areas of reforestation and in response to changing agricultural practices may also represent an important contemporary sink for atmospheric CO_2. Feedbacks associated with CO_2 uptake by vegetation, its release during decomposition, and the nitrogen cycle processes that affect plant and microbial growth—including trace gas production—must be better understood to appreciate fully the significance of changing CO_2 sink strengths.

High priority must now be given to three areas: (1) the establishment of flux measurement networks in important mid-altitude regions, especially in those such as the former Soviet Union, China, temperate South America, where comparatively little data have been obtained; (2) the development of adequate process-based models; and (3) the scaling-up of the models to predict fluxes over large regions. This should lead to a substantial improvement in the quality of the input to climate models.

INTRODUCTION

Increasing atmospheric concentrations of gases such as carbon dioxide (CO_2), chlorofluorocarbons (CFC), methane (CH_4) and nitrous oxide (N_2O) are believed to be causing climate change: global warming, increased UVB flux to the earth's surface due to depletion of stratospheric ozone, and changes in precipitation patterns (Houghton *et al.*, 1990, 1992). Efforts such as those by the Intergovernmental Panel on Climate Change to assess the likely extent of these changes by means of climate models are heavily dependent on the quality of information regarding the fluxes of these trace gases as model inputs.

Apart from chlorofluorocarbon compounds, which are entirely industrial in origin, significant proportions of the other main "greenhouse gases" CO_2, CH_4 and N_2O come from processes occurring naturally in the biosphere, and their net fluxes to the atmosphere are increased by activities associated with land use. Additionally there are indirect contributions to the greenhouse effect from carbon monoxide (CO), non-methane hydrocarbons (NMHC), and NO_x (all of which have natural and anthropogenic sources), through their role as precursors of tropospheric ozone.

These biospheric contributions to total emissions are of particular significance in the midlatitudes (30° to 60°). A large proportion of the world's population and the most industrialized societies are concentrated in this zone, mainly in the northern hemisphere. The impacts on the natural terrestrial environment have been correspondingly heavy. They include deforestation, the cultivation of virgin grassland, irrigation and wetland drainage, increasingly intensive use of mineral fertilizers, large-scale production of livestock (and livestock waste), urban waste disposal in landfills, and large contributions of nitrogen and sulfur compounds to atmospheric deposition. All of these processes affect the emissions of greenhouse gases (Andreae and Schimel, 1989).

Currently we have a modest understanding of the principal processes contributing to fluxes of trace gases between the land and atmosphere in the midlatitudes. In recent years we have been able to quantify fluxes under some conditions, to develop models of processes, and to begin the task of scaling flux measurements on a small areal scale to larger scales. However, even in those regions where work has been most intensive, we are not yet able to quantify fluxes well enough to satisfy fully the requirements of the climate models. In addition, there are large regions of the midlatitudes, including most of the former Soviet Union, China, and temperate South America, for which very little information on trace gas fluxes is available. Perhaps most importantly, we do not understand sufficiently well how global change will affect gas fluxes.

In this paper we highlight on going work in some of these fields, including the development of the IGAC-TRAGEX program. TRAGEX is intended to fill these gaps in our knowledge of trace-gas fluxes in midlatitude ecosystems, and to improve our capability of predicting changes in these fluxes.

CO_2 FLUXES

In the last century, the principal cause of atmospheric CO_2 increases was conversion of virgin forest and grassland, much of it in the midlatitudes, to agricultural use (Wilson, 1978). Today, the principal biogenic source is deforestation in the tropics; ecosystems in the midlatitudes appear to be having either no net effect on atmospheric CO_2 concentrations, or are acting as a weak sink. Different modeling approaches, dealing with the difficult problem of resolving relatively small differences between two large numbers—the terrestrial source of and sink for CO_2—have produced conflicting results for midlatitude ecosystems and much discussion.

The *potential* strength of a midlatitude CO_2 sink is, however, less contentious, although its magnitude will depend on the extent of climate change and the as yet unquantified effects of elevated CO_2 and nitrogen deposited from the atmosphere on forest growth and soil carbon storage. It will also depend on rates of reforestation in some regions, and increases in soil carbon storage brought about by changing agricultural practices, which may even today represent a net contemporary sink for CO_2.

N and CO_2 Fertilization of Midlatitude Ecosystems

The present-day response of carbon stores in midlatitude ecosystems to elevated CO_2 and land management changes is a topic of intense debate. Atmospheric boundary models (e.g., Tans *et al.*, 1990) suggest a major terrestrial sink for CO_2 in the northern hemisphere. This sink appears to total 2.0 to 3.4 Pg of C per year, depending on C sources in regions to the south and north of the midlatitudes. However, the mechanism for this sink is unknown, and a decade of terrestrial modeling has converged on present estimates of a 0.4 to 2.6 Pg *source* of atmospheric C globally (Dale *et al.*, 1991). With some significant exceptions (e.g., Harmon *et al.*, 1990; Melillo *et al.*, 1988) little of this contemporary source can be traced to the temperate latitudes of Europe and North America (Houghton *et al.*, 1987), but neither do these regions appear at present to be a significant net sink: even when accounting for upward revisions in estimates of recent midlatitude forest growth (e.g., Kauppi *et al.*, 1992 for European forests), both North America and Europe appear to be within 0.1 Pg yr^{-1} of no net emission (Figure 1).

While terrestrial models suggest no present sinks for CO_2 in the midlatitudes, there is general agreement among most models that *future* changes in atmospheric CO_2 concentration and climate will increase forest C storage. Rastetter *et al.* (1992), for example, predict a 4% increase in C storage in vegetation and soil in a deciduous forest of the eastern U.S., after 50 years of doubled atmospheric CO_2. Smith *et al.* (1992) predict a 0.4 to 9.5% increase in vegetation and soil C storage worldwide under different GCM climate scenarios based on Holdridge Life Zones (Holdridge, 1947); this increased storage derives mainly from northward shifts in the extents of tropical and boreal forests and in the midlatitudes from a decrease in the extent of deserts.

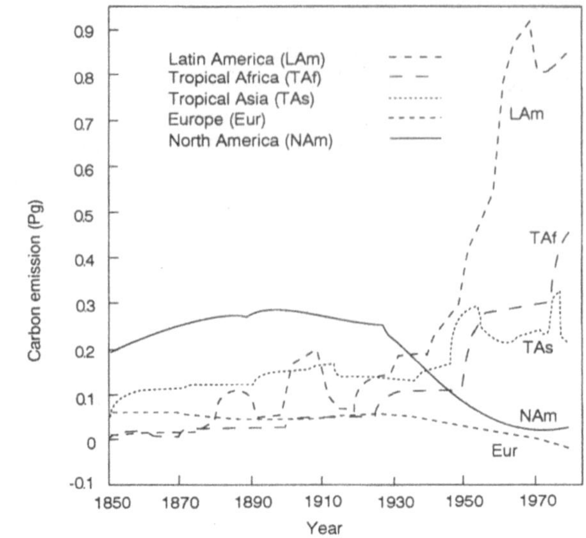

Figure 1. Estimated annual carbon emissions to the atmosphere from different continental regions due to land use changes (redrawn from Dale *et al.*, 1991).

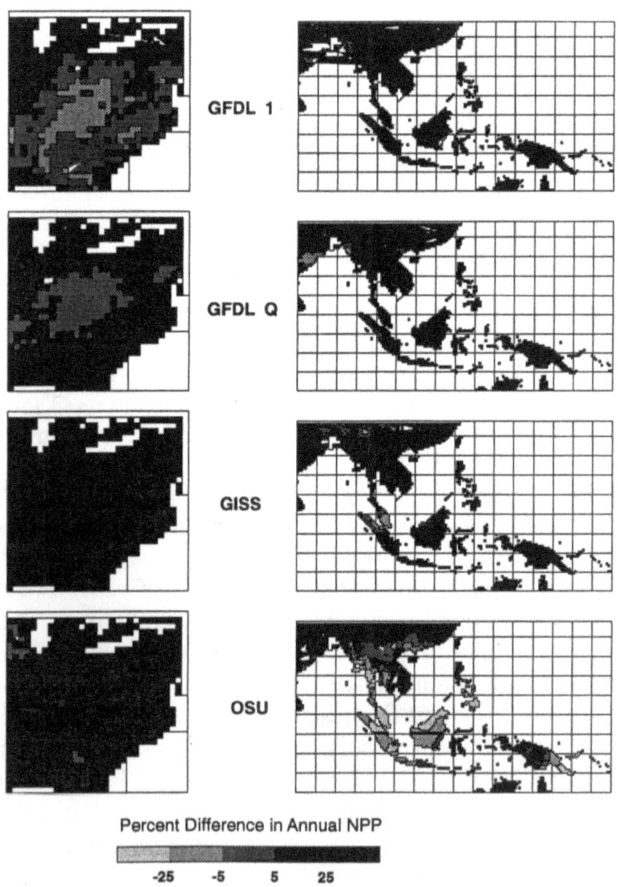

Figure 2. Percent difference in annual net primary production (NPP) between contemporary climate at 313 ppmv CO_2 and four GCM climates at 625 ppmv CO_2 as predicted by the Terrestrial Ecosystem Model for eastern North America (left) and southeast Asia (right) (Melillo *et al.*, 1993).

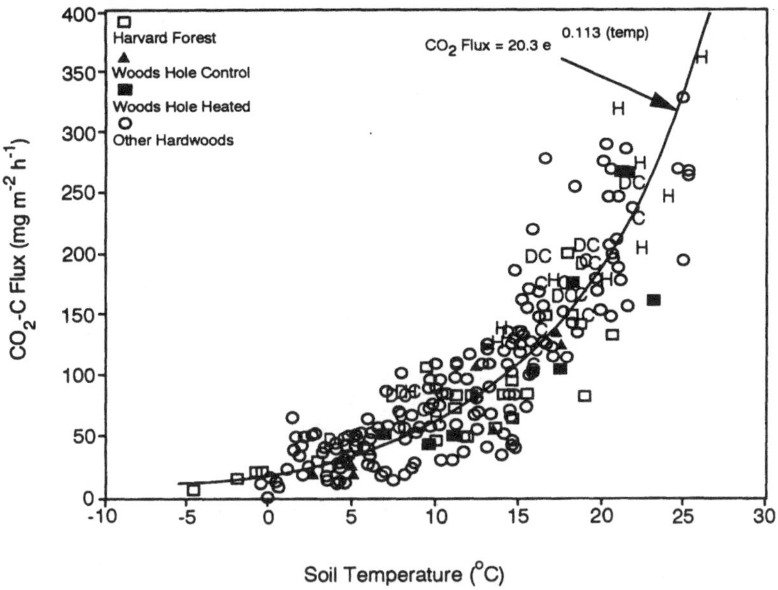

Figure 3. Relationship between CO_2 flux and soil temperature for hardwood forests around the world. Letters refer to Harvard Forest experimental plots (C = control; DC = disturbance control; H = heated plots) (redrawn from Peterjohn *et al.*, 1993).

Melillo *et al.* (1993) have used a process-level ecosystem model to estimate global patterns of net primary production (NPP) and soil nitrogen cycling following a doubling of CO_2 and associated climate changes. Doubling of CO_2 without climate change is predicted to result in a global NPP increase of 16.3%, while changes in both CO_2 and climate increase global NPP by 20 to 26% (Table 1). Predicted responses differ widely between different vegetation/climatic zones (Figure 2); in general, responses for northern and temperate ecosystems are less than 10%. The responses in tropical and dry temperate ecosystems, in their model, are dominated by the effects of elevated CO_2, but those in northern and moist temperate ecosystems reflect primarily the effects of temperature as it influences nitrogen availability and rates of CO_2 release by decomposition from soil organic matter. As Figure 3 shows, the latter effect is marked.

In the last few decades, many regions in the midlatitudes have experienced increased N deposition from the atmosphere as a result of industrial and agricultural emissions of NO_x and ammonia. About 18 Tg yr^{-1} of anthropogenic N is presently added to North American and European ecosystems from this source (Melillo *et al.*, 1989). This inadvertent fertilization will almost certainly increase carbon storage in these systems by stimulating net primary productivity, and may well affect the fluxes of other greenhouse gases such as CH_4 and N_2O (this is discussed further in following sections).

Carbon Storage in Agricultural Soils

In the mid-19th century, agriculture rapidly expanded across virgin grassland and forest regions of North America, temperate South America, Australia, New Zealand, and southern Africa. Native soil fertility and various social factors during this period discouraged the use of crop rotations, manuring, ley cropping and other cultural practices that might otherwise have maintained soil C stores. This led to a large pulse of CO_2 entering the atmosphere prior to 1900. $\delta^{13}C$ analysis suggests that this pulse may have been as high as 110 Pg (Wilson, 1978).

Table 1. Comparison of annual NPP (10^{15} g C) by vegetation type for experiment involving two levels of atmospheric CO_2 and five levels of climate. Climate scenarios correspond to contemporary climate and predictions of four separate global circulation models (GCMs) from the Geophysical Fluid Dynamics Laboratory (GFDL; models 1 and Q), the Goddard Institute for Space Studies (GISS), and Oregon State University (OSU) (Melillo *et al.*, 1993).

CO_2 Scenarios:	312.5 ppmv					625.0 ppmv				
Climate Scenarios†:	C	1	Q	G	O	C	1	Q	G	O
Polar desert/alpine tundra	0.4	0.4	0.4	0.4	0.4	0.5	0.5	0.5	0.5	0.5
Wet/moist tundra	0.6	0.7	0.7	0.7	0.7	0.6	0.8	0.7	0.7	0.7
Boreal woodland	1.1	1.4	1.3	1.3	1.3	1.1	1.6	1.4	1.4	1.4
Boreal forest	2.9	3.8	3.6	3.6	3.5	2.9	4.4	4.0	3.7	3.7
Temperate coniferous forest	1.1	1.1	1.1	1.1	1.1	1.2	1.3	1.3	1.4	1.3
Desert	0.6	0.6	0.5	0.6	0.6	0.9	1.0	0.9	1.0	1.0
Arid shrubland	1.8	1.8	1.8	1.9	1.9	2.3	2.5	2.5	2.7	2.6
Short grassland	1.0	1.2	1.2	1.2	1.1	1.1	1.4	1.3	1.4	1.2
Tall grassland	1.2	1.4	1.4	1.5	1.3	1.3	1.5	1.5	1.6	1.4
Temperate savanna	2.2	2.3	2.4	2.6	2.4	2.5	2.9	2.9	3.1	2.9
Temperate deciduous forest	2.2	2.0	2.1	2.5	2.3	2.3	2.4	2.6	2.8	2.6
Temperate mixed forest	3.3	3.4	3.5	3.7	3.6	3.6	4.0	4.1	4.2	4.0
Temperate broadleaf evergreen forest	2.2	2.3	2.2	2.2	2.2	2.6	2.8	2.8	2.8	2.7
Mediterranean shrubland	0.5	0.5	0.5	0.5	0.5	0.6	0.6	0.6	0.7	0.6
Tropical savanna	5.3	5.7	5.6	6.0	6.0	5.6	6.3	6.3	6.7	6.6
Xeromorphic forest	2.9	2.7	2.7	2.7	2.9	3.7	3.7	3.8	3.8	4.1
Tropical deciduous forest	3.8	3.4	3.5	3.3	3.5	4.5	4.5	4.6	4.5	4.6
Tropical evergreen forest	18.0	16.4	16.3	15.6	14.3	22.0	21.9	21.8	21.3	19.3
Total‡	51.0	51.1	50.8	51.5	49.8	59.3	64.3	63.8	64.2	61.2

†Column headings: C = Contemporary climate, 1 = GFDL-1 model, Q = GFDL-Q model, G = GISS model, O = OSU model.

‡Ecosystem-based estimates may not sum to totals because of the effects of rounding in reporting those estimates.

Farm mechanization and the intensive use of agrochemicals in the midlatitudes since World War II have further depleted soil C stores by the widespread replacement of mixed or animal-based farming systems with continuous monocultures. On the other hand, the higher productivity of intensively fertilized systems may be increasing soil C stores depleted by initial (pre-1900) cultivation where animal-based rotations were not previously important, e.g., most of the former North American prairie (Paustian *et al.*, 1993).

In most agroecosystems, annual cropping provides little or no scope for long-term C storage in vegetation, so soil C storage offers the principal potential for a C sink. Three mechanisms can increase soil C storage in these systems: (1) increased C inputs, (2) decreased decomposition rates, and (3) reduced amounts of CO_2 produced per unit of organic matter decomposed. Thus, where intensification of farming has resulted in higher productivity, C storage can be increased by the associated increased C inputs, so long as decomposition rates remain constant and crop residues are left on the field. Long-term soil C experiments bear this out (Paustian *et al.*, 1993), as do simulation models that show 20 to 30% increases in soil C with 40% increases in net primary production. In fact, Paustian *et al.* have shown that a modest increase in primary production (from increased yields or from incorporation of cover crops into a rotation) results in far more impact on soil C storage in soil than does a 2°C climate change (Figure 4).

Carbon storage may also increase in agricultural soils as a result of decreased decomposition rates, which may result from conservation tillage practices. Soil aggregation which protects organic matter from decomposition by restricting microbial access to substrates (Elliott *et al.*, 1980; Tisdale and Oades, 1982), is typically reduced by tillage. However, aggregate restabilization under conservation or reduced-tillage practices may—together with other effects of conservation tillage—lead to higher C storage under no-till systems (Figure 5). Given the widespread adoption of conservation tillage in North America (Crosson, 1981), this increase in C storage may be significant. Simulation modeling suggests that under optimal circumstances—high productivity, complete residue return, and no-till cultivation—soil carbon stores can largely regain and perhaps exceed pre-cultivation soil C levels under steady-state conditions (Figure 4).

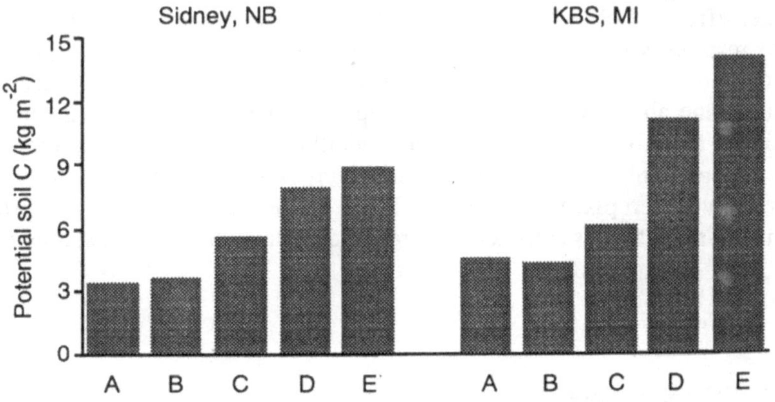

Figure 4. Simulated effects of agricultural management strategies under different climate scenarios on potential carbon storage in soil at a wheat fallow site in Nebraska, U.S., and at a maize site in Michigan, U.S. A = current management; B = A plus a 2°C temperature increase; C = B plus a 40% yield increase; D = C plus conservation tillage; E = D plus a 20% increase in residue lignin content (redrawn from Paustian *et al.*, 1993).

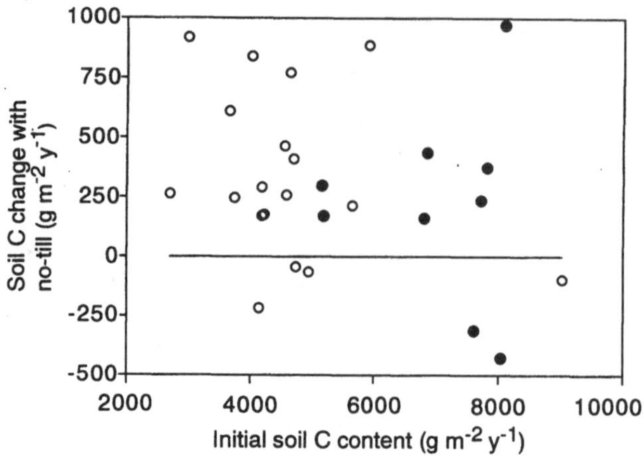

Figure 5. Effects of conservation tillage on soil carbon stores in paired long-term no-till (NT) versus conventional tillage (CT) row crop ecosystems of North America. Open circles represent light textured soils (e.g., sand and silt loams), closed circles represent soils of heavier texture (e.g., clay loams). Note positive effect of no-till on soil C even in soils of high OM content (redrawn from Paustian *et al.*, 1993).

Potential for Managing Carbon Storage

Present global inputs of C as CO_2 to the atmosphere from fossil fuels are circa 5.5 Pg C yr^{-1}. Even the most optimistic reforestation and soil organic matter management scenarios do not envisage the net capture of more than 1 to 2 Pg C yr^{-1} under present climate regimes (Houghton *et al.*, 1990), and once equilibrium is achieved (50 to 150 yr, depending on rates of tree growth and soil organic matter accumulation) this figure decreases to zero. Only under climate change scenarios that allow extensions of high productivity ecosystems into areas now occupied by ecosystems of lower productivity, e.g., the replacement of desert areas by grasslands and the extension of tropical and boreal forests polewards (Smith *et al.*, 1992) will substantial C capture significantly attenuate atmospheric loading from fossil fuels.

Any net effect of climate change on C storage will largely depend, then, on the differential responses of vegetation and soil organic matter to changes in temperature versus precipitation. Jenkinson *et al.* (1991) show via simulation modeling that increasing temperature in the absence of change in precipitation will accelerate soil organic matter decomposition, regardless of its effects on vegetation growth. Soil C stores will remain steady or increase only if: (1) precipitation increases sufficiently to allow enhanced vegetation growth to replace or exceed C lost from decomposition, or (2) precipitation decreases in the midlatitudes sufficiently to retard decomposition in mid-summer.

CH_4 FLUXES

Midlatitude CH_4 Emissions

Important biogenic sources of CH_4 from the midlatitudes include wetlands, paddy rice cultivation, landfills and sewage disposal sites, and ruminant livestock. Recent global estimates of the CH_4 balance (e.g., Tyler, 1991; Mathews and Fung, 1987; Cicerone and Oremland, 1988) suggest that midlatitude wetlands and paddy rice cultivation contribute little to the global flux relative to fluxes from high latitude and low-latitude regions.

However, wetland studies are continuing, and are providing information useful for process-level modeling and new measurement techniques that should also be applicable in other regions. For example, quadrupole mass spectrometry is being used to study the distribution of CH_4 with depth in flooded peatlands in the UK, in relation to the distribution of C sources, other nutrients, redox potential, pH, and temperature (Figure 6). Also, eddy accumulation, flux gradient and eddy correlation methods using a tunable diode laser, and aircraft-based methods are being used to obtain fluxes at different scales (Gallagher *et al.*, 1993; D. Fowler, private communication).

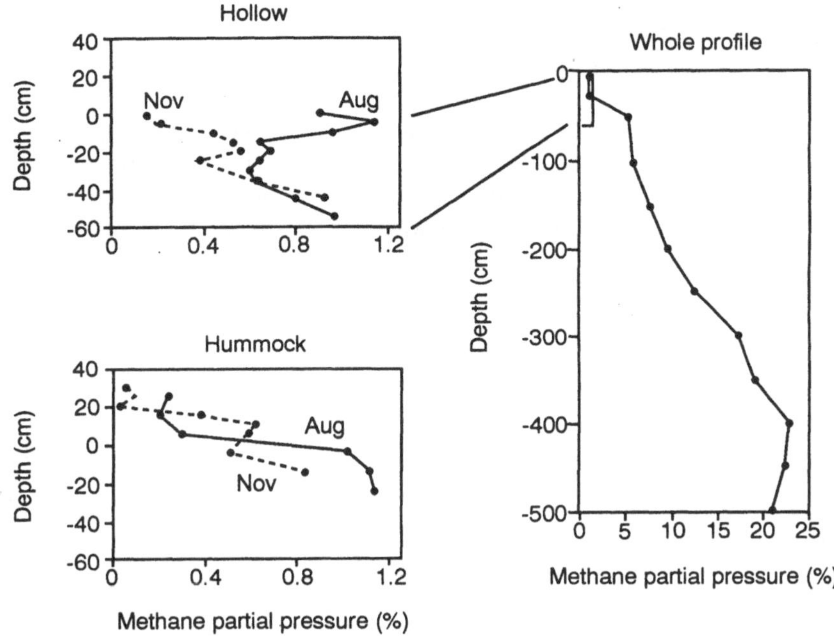

Figure 6. Measurements of methane partial pressure in waterlogged peat by quadrupole mass spectrometry (Clymo *et al.*, 1993).

In contrast with emissions from natural wetlands, CH_4 from midlatitude landfills may comprise a much higher proportion of the contemporary global landfill flux, estimated to be circa 30 Tg yr^{-1} (Houghton *et al.*, 1992; Bogner, 1992). This estimate is a reduction from previous estimates that ranged to 70 Tg yr^{-1} (Bingemer and Crutzen, 1987) because of the recognition of the very high oxidation rates occurring in landfill cover soils. Likewise, the midlatitudes are likely the most important global source of CH_4 from animal and poultry manures, which may produce an additional 25 Tg yr^{-1} CH_4 (Casada and Safley, 1990).

As for biogenic emissions of CO_2 from soils, the contemporary CH_4 flux from midlatitude systems and its contribution to current atmospheric loading rates should be balanced against historical fluxes. Region-wide, it is likely that biogenic methane emissions from soils today are less than emissions occurring prior to widespread wetland drainage in North America and elsewhere, and prior to the disappearance of large ruminant populations in midlatitude savanna regions. These potentials are, however, poorly estimated (Schimel *et al.*, 1992) and deserve much further evaluation.

The Soil as a Sink for Methane

The soil sink for CH_4, resulting from the microbial oxidation of CH_4 to CO_2, is estimated to be 30±15 Tg yr^{-1} (Houghton *et al.*, 1990; Whalen and Reeburgh, 1990). Although this value is only about 6% of total emissions, it is very similar to the estimated annual increase in the atmosphere (32±5 Tg; Houghton *et al.*, 1992).

Inputs of nitrogen have been shown to reduce this soil sink very substantially. The effect has been demonstrated in deciduous and coniferous forests (Steudler *et al.*, 1989; Figure 7) and in short-grass prairie (Mosier *et al.*, 1991) in the U.S. Results from a cropping systems experiment in the U.S. Midwest (Robertson *et al.*, 1993) suggest that the attenuation of CH_4 oxidation in cropping systems may be more related to soil disturbance and changes in organic matter dynamics than to changes in nitrogen availability (N fertilization) per se. Hansen *et al.* (1993) have further shown that compaction by farm equipment can also reduce CH_4 uptake.

The deposition of N from the atmosphere onto forests and natural grasslands and the use of fertilizers in agricultural ecosystems have increased greatly in the midlatitudes over the last few decades (Melillo *et al.*, 1989). If later results bear out early trends with respect to nitrogen attenuation of CH_4 oxidation, N deposition and land use changes may be reducing the CH_4 sink strength by more than 50% in areas already affected.

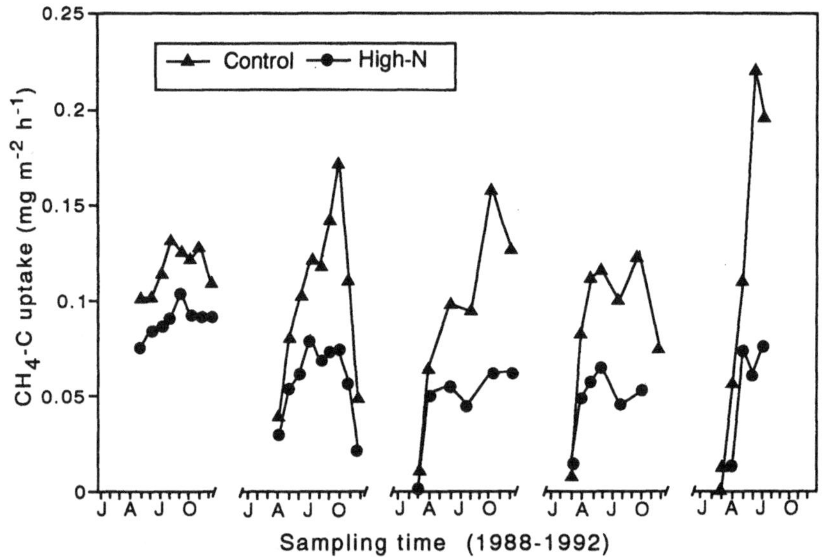

Figure 7. Effect of N fertilization on methane uptake in a red pine stand, Harvard Forest, MA, U.S. (J. Melillo, private communication).

N_2O FLUXES

Globally, soils are the major source of atmospheric N_2O (Houghton *et al.*, 1992; Robertson, 1993). Of the total emission rate of 8 to 18 Tg N yr^{-1}, 2.5 to 5.7 Tg are attributed to tropical soils, whereas the uncertainty ranges are much wider for temperate soils. Of particular note is the uncertainty associated with estimates of midlatitude agricultural activities.

Effects of N Inputs to Agricultural Land

In spite of the IPCC assessment, there is much indirect evidence that the flux of N_2O from agricultural land—much of which is in the midlatitudes—is in fact a major contributor to global emissions (Robertson, 1993). The emissions increase with increasing inputs of nitrogen in the form of mineral fertilizers, organic manures, and also N fixed biologically by legumes and other symbiotic systems.

The relationship between N inputs and emissions is not clearcut, however. In temperate grassland there is a variable but generally large response to N fertilizer, with short-term increases in flux ranging from factors of 2 to more than 10 when fertilizer is applied (Ryden, 1981; Bouwman, 1990). The chemical form of the N, as well as the amount applied, is important (Eichner, 1990; Figure 8). On the other hand, based on reported emissions from arable land (studies carried out predominantly in the northern midlatitudes), Bouwman (1990) derived the relationship

$$N_{em} = 1.879 + 0.0042N_f \tag{1}$$

where N_{em} is the N_2O emission and N_f is the fertilizer N applied, both in kg $ha^{-1}yr^{-1}$. Bouwman's regression suggests that an unfertilized arable field emits on average about 1.9 kg $ha^{-1}yr^{-1}$, and this only increases by about 0.4 kg ha^{-1} when 100 kg N ha^{-1} is applied.

Nitrogen fertilization should probably be regarded as a process that gives a major stimulus to the N dynamics of the system and consequently promotes N_2O production over a prolonged period, rather than simply as the addition of a transient source of mineral N which can serve directly as a substrate for N_2O production. Thus the high fluxes from unfertilized arable land may be due to the increased turnover of nutrients resulting from high net primary production (and attendant residue return) and soil disturbance by cultivation. Such factors increase the rates of N mineralization, nitrification, and denitrification compared with undisturbed environments, thus promoting N_2O production. The variability of the capacity of unfertilized grassland to emit N_2O may reflect management practices that influence these turnover processes.

The total area of cultivated land in the midlatitudes is of the order of 10^9 ha, giving rise, on the basis of Equation 1, to a *background* emission of circa 1.9 Tg N_2O–N yr^{-1}. Eichner (1990) suggests a smaller value based on a literature analysis of fertilizer-response experiments. The effects of fertilizer N inputs on global N_2O emission may, however, be far greater than either estimate. Robertson (1993), for example, points out that very little of the 80 Tg N added as fertilizer to croplands stays in these soil-plant systems very long; in general about 50% is either leached or denitrified within 6 months of application, 25% taken off as yield to end up eventually as sewage, and most of the 25% remaining in crop residue decomposes and is similarly lost over the next several years (soil organic matter rarely builds up under cultivation). Most of this 80 Tg N, then, will be likely eventually to be denitrified in order to close the global N cycle. Only a small portion of this fixed N will need to be denitrified to N_2O rather than to N_2 to have a substantial impact on the global N_2O budget.

The contribution of legumes to N_2O emissions has been little studied compared with the research devoted to conventionally fertilized systems. However, such evidence as there is suggests strongly that this source may be very significant. Duxbury *et al.* (1982) found annual emissions from alfalfa in New England about two and a half times those from a grass site. Similar observations have been made on leguminous pastures in Australia (I. Galbally, private communication); and recent work in Edinburgh showed a comparable increase for a clover/grass sward over a grass-only sward, with emissions an order of magnitude higher when the swards were plowed (Figure 9).

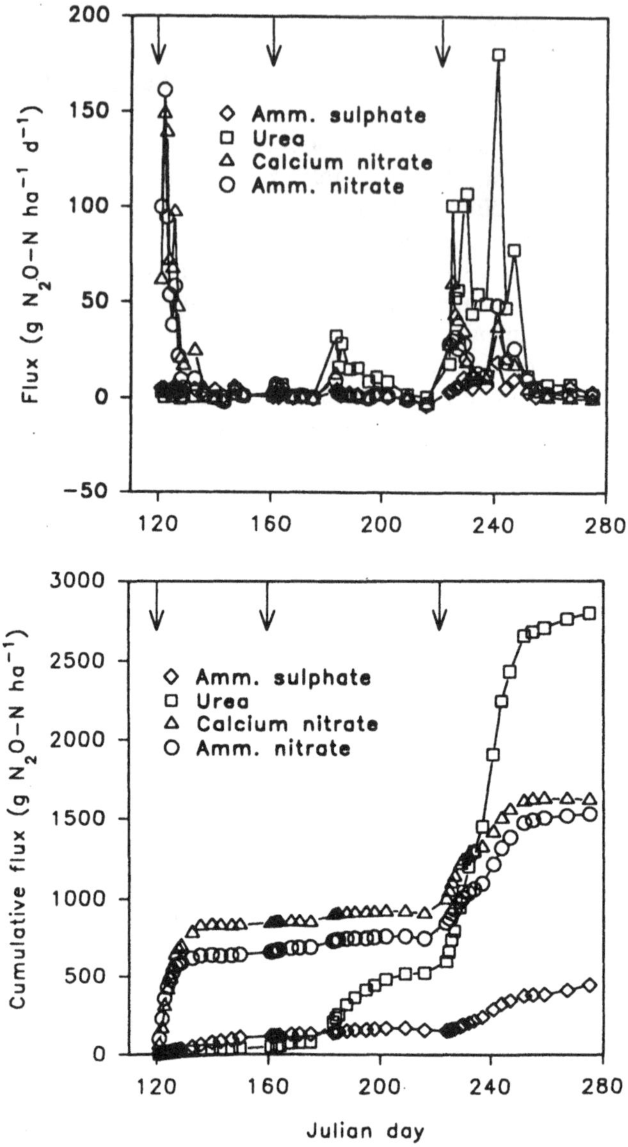

Figure 8. Temporal variations in fluxes (top), and cumulative fluxes (bottom) of N_2O from a grassland site near Edinburgh, fertilized with 4 forms of N fertilizer, 1992. Arrows indicate times of fertilizer applications (120 kg N ha^{-1} on each occasion) (H. Clayton, I.P. McTaggart and K.A. Smith, private communication).

The agricultural systems of temperate South America and Australasia are predominantly based on leguminous inputs of N. There is also a substantial movement towards greater use of leguminous green manures in the other continents, and the total global input of N from biological fixation is estimated to be roughly equal to that from industrially fixed N, circa 80 Tg N yr^{-1} (Isermann, 1993). The contributions from these sources to total global emissions of N_2O require further investigation.

A major challenge to future experimental programs is to obtain representative flux data for agricultural ecosystems not merely over whole annual cycles, but also over the lifetime of characteristic cropping rotations. This is necessary if we are to establish the total fluxes arising from the system as a whole and also to distinguish between the contributions directly associated with fresh N inputs and those from other processes. Such information is required if we are to have satisfactory predictive models to estimate the consequences of changes in land use practices on total emissions.

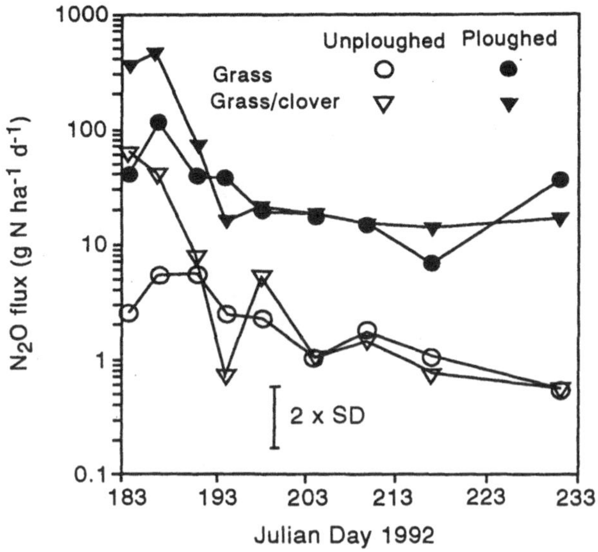

Figure 9. Effect of presence of clover and of plowing on emissions of N_2O from grassland (K.A. Smith, S. Bergos and M. Davies, private communication).

Effects of N Deposition

A major contributory factor to N_2O release in the midlatitudes is the deposition of N onto natural and near-natural ecosystems such as forests and moorlands. An acid beech forest soil receiving atmospheric N inputs of 40 to 60 kg ha^{-1} has been shown to emit 5.6 kg N_2O–N ha^{-1}yr^{-1} (Brumme and Beese, 1992). This rate is of the same order as those recorded from the most intensively fertilized agricultural land, and in fact additional fertilization of the beech soil only increased emissions by 40% (Figure 10). In contrast, in environments receiving less N input, additional fertilization can cause a dramatic increase in emissions (Figure 11). A significant part of N deposition in Europe is in the form of NH_4^+, and its acidifying effects may be a contributory factor to emissions (Nagele and Conrad, 1990); liming reduces the emissions considerably (Brumme and Beese, 1992; Figure 10).

PRECURSORS OF TROPOSPHERIC OZONE: NO_x, CO, NMHC

Tropospheric ozone is predicted to increase with increasing emissions of nitrogen oxides (NO_x), and with increasing emissions of carbon monoxide (CO), CH_4, and non-methane hydrocarbons (NMHC) when the atmospheric abundance of NO_x is greater than 20 to 30 pptv. These species also have complex effects on atmospheric OH, thus affecting the lifetime of other greenhouse gases (Houghton *et al.*, 1992).

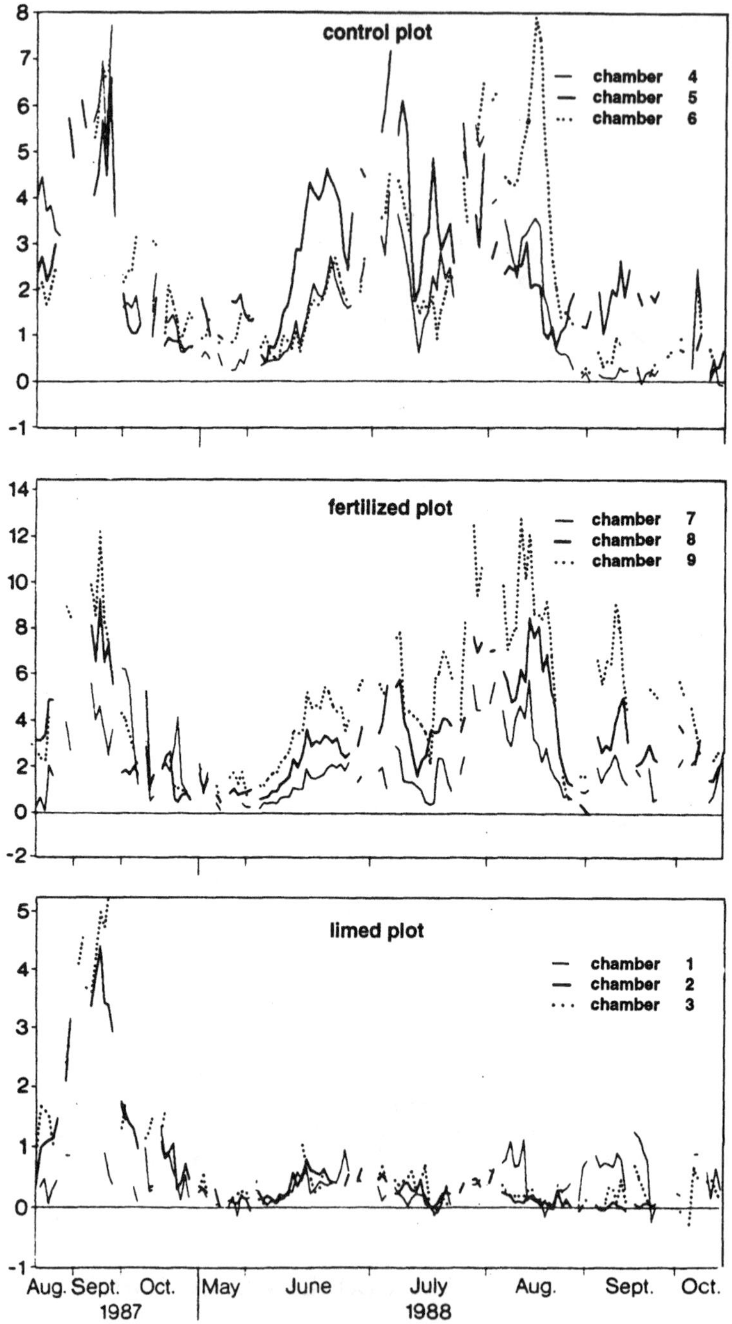

Figure 10. Daily mean values of N_2O emission from control, fertilized, and limed plots (data from three individual automated chambers per plot) (Brumme and Beese, 1992).

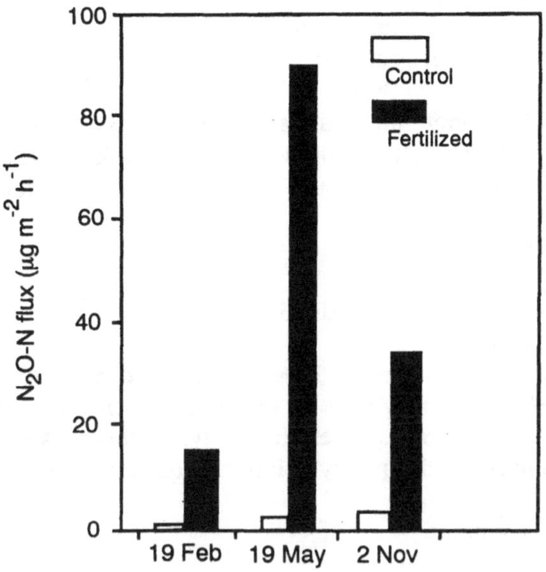

Figure 11. Effect of N fertilization on N_2O emission from soil in a slash pine plantation, Florida, U.S. in 1991 (Castro and coworkers, private communication).

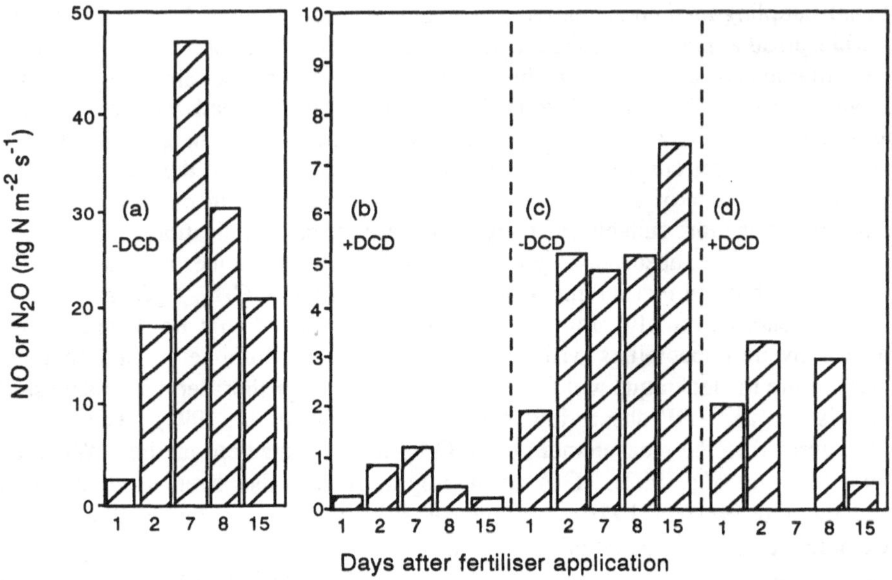

Figure 12. Emissions of NO (a, b) and N_2O (c, d) from sandy loam soil, following addition of ammonium sulfate with (a, c) and without (b, d) dicyandiamide (DCD). Note different y-axis (Skiba *et al.*, 1993).

Nitrogen oxides, CO and NMHC all have significant natural and anthropogenic sources, but budgets remain uncertain. NO is mainly emitted from industrial sources in the developed countries, but it is also a product of nitrification and denitrification. Its emissions by the former process can exceed those from N_2O by a substantial factor (Skiba *et al.*, 1993). Increasing use of ammonium-forming fertilizers and acid deposition is likely to increase NO release, but it has been shown that inhibition of nitrification is highly effective in reducing the emission (Figure 12).

The soil appears to be more important as a sink than as a source for CO, but too few data exist to produce a reliable estimate of the size of this sink. It may well be that the sink will be affected significantly by factors such as N inputs, as occurs for CH_4. More work is needed in this area.

NEW TECHNOLOGY FOR FLUX MEASUREMENT

Several encouraging developments have occurred recently in methods for trace gas flux measurement. We appear now to be on the verge of being able to obtain high-quality large-area flux measurements more or less routinely. If this prospect becomes reality, it should have a major impact on the quality of information being put into climate models.

Micrometeorological Methods

The flux gradient and the eddy correlation micrometeorological methods (Fowler and Duyzer, 1989) and the more recent eddy accumulation (or "conditional sampling") method (Businger and Oncley, 1990) permit flux measurements on the ha-km^2 scale, and thus overcome local spatial variability which is a problem with chamber methods. The conditional sampling method is relatively new and not widely tested, but seems very likely to find widespread application in the medium term. It depends on the determination of the difference in trace gas concentrations in sample bags used to collect air during updraughts and downdraughts. The sampling is controlled by high-speed valves linked to a 3-dimensional sonic anemometer. These need to be located at a height of at least 3 m above the ground. This corresponds to a fetch of 300 to 600 m, and the method integrates over areas of the order of 10 ha. For smaller areas the gradient method, using lower sampling heights, is more suitable (K. Hargreaves, private communication).

Tunable diode laser absorption spectroscopy (TDLAS) techniques now have sufficient sensitivity to permit the measurement of very small fluxes of CH_4, N_2O, and CO. The technique appears extremely promising, but it is recognized that its development for general use by non-specialists will occur only in the medium term, and that in the immediate future the technique will find application mostly in short-term intensive studies. The same is true for FTIR absorption spectroscopy (Galle *et al.*, 1993). These methods have been applied to the measurement of N_2O by the flux gradient method (Wienhold *et al.*, 1993; Galle *et al.*, 1993). Gas chromatography has also been used in this way (Hutchinson and Mosier, 1979), but the sensitivity is much lower than with the spectroscopic methods. It can be increased, however, by multiple replicate analysis of bulk gas samples, and thus used over a wider range of conditions (Arah *et al.*, 1993a; Hargreaves *et al.*, 1993). Figure 13 shows an intercomparison of these three analytical techniques applied to determination of N_2O fluxes by the gradient method, indicating good agreement between them.

Techniques which can directly measure concentration differences, such as dual path optical methods or the optical subtraction FTIR method under development in Sweden (L. Klemedtsson, private communication), and those such as TDL which can be adapted to make sequential analyses from large bag samples, are now being applied to the conditional sampling technique. Their use extends the range of fluxes measurable down to levels far lower than those attainable by GC.

Another integrating micrometeorological method which has only low technological requirements for gas measurement is that which exploits concentration buildups under nighttime inversions. For large relatively homogeneous source regions the method may augment enclosure studies. However, for both this and for micrometeorological measurements in general, it must not be forgotten that measurements of the depth of the boundary layer have to be made, requiring specialized equipment. Here, as in some of the other methods, there is a strong case for collaboration between scientists from several disciplines to provide the required expertise.

Recognizing the resource requirements for such flux measurement experiments, it has been recommended (IGAC, 1992) that certain of the TRAGEX network sites should be designated as lead sites where special resources could be deployed, and which could also serve as sites for intercalibration of instruments and methods.

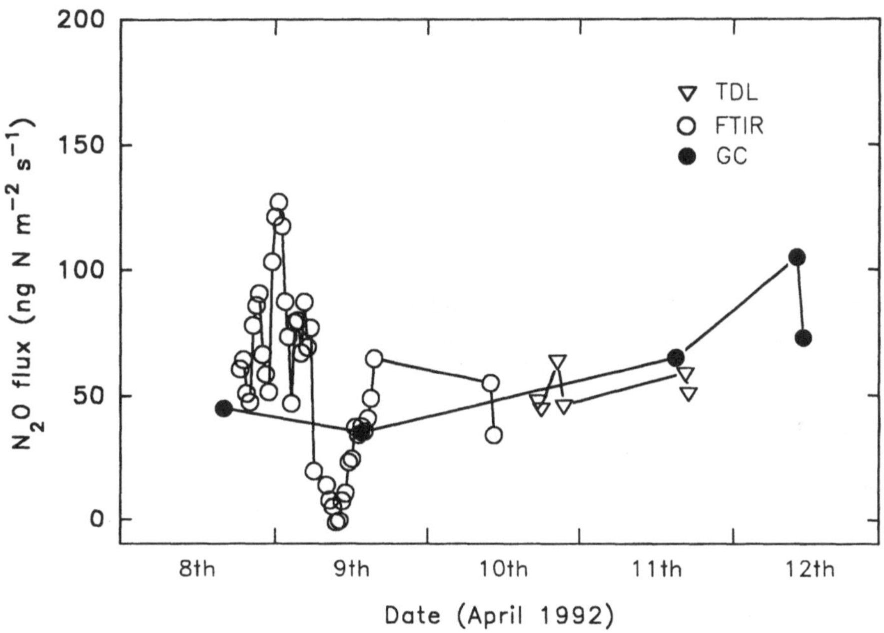

Figure 13. Measurements of N_2O flux from fertilized grassland, Stirling, Scotland, 1992, by flux gradient method, using three different analytical systems (Hargreaves *et al.*, 1993). TDL: tunable diode laser absorption spectrometry; FTIR: Fourier-transform infrared spectrometry; GC: automated gas-chromatography.

Chamber Methods

Large chambers. Where land use practices produce a small-scale mosaic of different crops, woodland and/or inputs, micrometeorological methods are difficult to apply, because of the resulting small fetch. This constraint applies especially to agricultural regions of Europe and Asia. For these regions the continued development of chamber methods is essential. One such development is the use of very large chambers together with long-path infrared absorption spectroscopy (Smith *et al.*, 1993). Such systems have two advantages:

they average emissions over areas 2 to 3 orders of magnitude larger than conventional small chambers, while still allowing flux measurements to be made at sites where the effect of different treatments on fluxes is being investigated, where micrometeorological methods could not be used.

Automatic (small) chambers. The recognition that longer time series of data are required to improve annual flux estimates has led to the development of several systems involving automation of chamber closure, gas sampling, and GC analysis. One such system (Brumme and Beese, 1992) is deployed in the measurement of CO_2 and N_2O fluxes in a temperate forest in Germany, and is being extended to include CH_4. Robertson (private communication) has a comparable system in a series of four cropping systems in Michigan, U.S. In Edinburgh, Scotland, automated chambers have been fitted to large soil monoliths to facilitate manipulation studies (Smith and Thomson, 1993) and a portable battery-powered system of automated field chambers and associated gas samplers has been developed (Smith and Scott, private communication) to take samples over a period at a remote site for return to the laboratory for automated analysis.

These developments reduce the labor requirement for field sampling and analysis enormously, and make much more feasible both long-term measurements and intensive short-term investigations required for process-related studies.

MODELING AT DIFFERENT SCALES

Carbon Dioxide

The modeling of CO_2 emission and uptake processes is more advanced than for the trace gases CH_4 and N_2O. For the assessment of CO_2 fluxes in the midlatitudes, modeling has a larger role to play than direct measurement. Some of the models for predicting changes in NPP and carbon storage in soil and biomass are referred to above in the section on CO_2 fluxes, and will not be considered further here.

Process Models for CH_4 and N_2O

The modeling of N_2O emissions is relatively well advanced at the process level, but scaling to field, landscape and regional levels is not so well developed. Firestone and Davidson (1989) have neatly represented N_2O formation/emission by a "hole-in-a-pipe" model, and Groffman *et al.* (1988) and Robertson (1989) have devised a conceptual model of the controls affecting nitrification and denitrification at different geographic scales. These models improve our capacity to evaluate the relative importance of controlling environmental variables for specific environments. They also serve to identify the key variables which need to be measured or recorded at experimental sites, in association with flux measurements, in order to make quantitative predictive flux models.

Parton *et al.* (1988) have developed a model of N_2O fluxes from nitrification, in which they have simply related fluxes to the rates of mineralization and nitrification occurring in soils of different texture in the semi-arid conditions of Colorado, U.S. The simulations were generally satisfactory in describing the fluctuations in soil mineral N, and accounted for some but not all of the fluctuations in N_2O emissions. This approach could be used to predict fluxes from other, generally well aerated soils in other regions, where nitrification is likely to be the dominant mechanism of production.

In wetter environments, such as northwest Europe and eastern and midwestern U.S., there is much evidence that denitrification is the dominant mechanism for N_2O production. Models of denitrification in soil have been developed that have been at least partly validated by experiment (Smith, 1980; Parkin and Tiedje, 1984; Arah and Smith, 1989; Smith and Arah, 1992; Arah *et al.*, 1993b). The fraction of the denitrified N emitted from the soil as N_2O varies dramatically with environmental factors, and modeling N_2O flux is thus even more of a problem than modeling denitrification per se.

The biochemical and physicochemical processes involved in CH_4 production, transport and oxidation in flooded soils are now being modeled. These processes are essentially similar in both natural wetlands and in rice paddies (cf., Neue and Sass, this volume), and models whose development may have arisen only in the context of one of these environments should be useful in both. The growing interaction between scientists involved in natural wetlands and rice research programs can be expected to promote this cross-fertilization. It is likely that we will see the outcome of present efforts within the next few years.

Scaling Up To Larger Areas

The capacity to extrapolate from process-based models applicable only to a tiny area or volume to landscape or catchment-sized areas and beyond is a major priority. Work in this area is still in its early stages, but it seems likely that the momentum will increase, with at least part of the stimulus coming from the TRAGEX program.

Scaling to larger areas requires a linkage between the controlling variables included in the process models and information contained in spatial databases, such as the Global Information System (GIS) database. Such linkages are most advanced for CO_2 fluxes. For example, Peterjohn *et al.* (1993) have shown that the flux from temperate forest soils is a function of soil temperature, which in turn is linearly related to air temperature. The latter parameter, but not the former, has already been incorporated into GIS, and Kicklighter *et al.* (1993) have modeled regional CO_2 fluxes using this information.

For the more complex relationships driving N_2O emissions, comparable "surrogate" parameters will need to be found. It should be possible to use information on soil texture, structure and water content, and air temperature, as an alternative to direct estimates of denitrification or nitrification. Similar considerations apply to CH_4 emission and uptake. It may thus be possible to score soils subjectively into a small number of classes in relation to each key factor, using soil and vegetation maps together with climatic data; such scoring, combined with an emission/uptake factor for each score class, will permit the aggregation of fluxes over larger areas with more confidence than can be achieved by direct extrapolation from point measurements.

THE IGAC-TRAGEX ACTIVITY

The need for a coordinated program to provide the necessary information is evident. Even where trace gas fluxes have been most intensively studied, data are rarely as comprehensive as modelers would desire. For example, in only a few instances have the estimates from a particular site been based on frequent measurements made over an entire year. As indicated above, there are whole regions for which there is little or no information, even from studies of short duration. Also, most studies have focused on small areas (e.g., individual fields or forest plots). As a result, many estimates of trace-gas fluxes in the temperate region are not very reliable at the regional scale.

The TRAGEX Activity (IGAC, 1992) is intended to remedy this situation; its purpose is:

- to document the contemporary fluxes of CO_2, CH_4, N_2O and the tropospheric ozone precursors, and
- to determine the factors controlling these fluxes and improve our ability to predict future fluxes.

Achievement of these objectives will involve three types of research activities:

- measurement of trace gas fluxes at an international network of representative sites;
- manipulation of what are perceived to be the controlling variables, to determine their impact on fluxes; and
- modeling of the processes responsible for trace gas fluxes.

Sampling networks are being planned to include the main temperate zone ecosystems in each continent. The networks are intended to produce the data necessary to quantify critical variables controlling fluxes of trace gases across each region. These networks must necessarily include the main ecosystem types, both natural and managed, within each geographical component of the temperate zone. Most importantly, the networks must embrace areas where current fluxes are not well known.

In this last category are several large midlatitude agricultural and forested areas of special importance, for which little data are available. These include areas of intensive agriculture in eastern Europe, the former USSR, and China, and the large region of temperate South America where there is a less intensive agricultural system based to an unusual degree on leguminous nitrogen inputs. Two major wetland areas that have been little studied, the Pripet marshes of Eastern Poland, Ukraine, and Belarus and the peatlands of western Siberia, are an additional priority.

The network also needs to include sites representative of the large areas of the midlatitudes that have been affected by atmospheric deposition of nitrogen, sulfur and acidity. The consequences for trace-gas fluxes of loading ecosystems with nutrients are just beginning to be understood, but more quantitative information is a priority within this program.

Human-driven biogenic fluxes that are unique to the temperate zone, such as those from landfills and agricultural wastes in anaerobic lagoons, also need to be quantified. The quantification of emissions from ruminants is not a specific remit of the TRAGEX Activity, but in view of the importance of this source it is an issue which needs to be addressed somewhere within the overall IGBP-IGAC Project.

Main Site Networks

The broad outline of a plan for a midlatitudes program was drawn up at the initial planning workshop of the Activity, held in Boulder, Colorado, U.S. in the autumn of 1991. The basic design calls for sites representing the major vegetation/land use types of the midlatitude zone: forests, rangeland, and agricultural land (Table 2). In addition, there is a need for data from a small number of specific temperate wetland types that are poorly characterized (Melillo *et al.*, 1992).

The first detailed plan for a network in onc region, building on the outline produced at Boulder, was that for the continental U.S. (excluding Alaska). This was devised at a workshop at Pingree Park, Colorado in October 1992 (Ojima *et al.*, 1993). The initial site

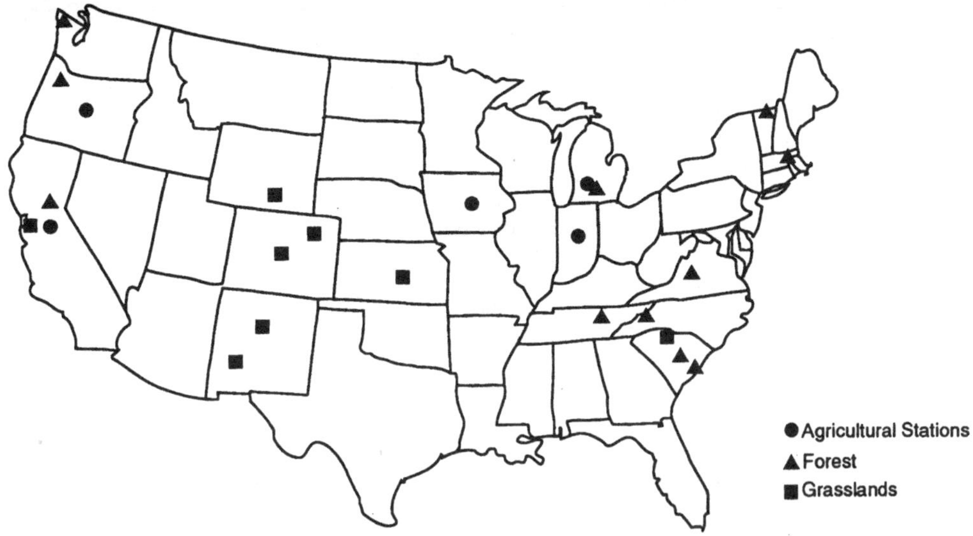

Figure 14. Locations of midlatitude sites for proposed U.S. trace-gas sampling network (Ojima *et al.*, 1993).

network planned for the U.S. is shown in Figure 14. Since then, a similar workshop to plan a network for Europe has been held in Munich, Germany in April 1993. Future meetings of the TRAGEX committee will have to address the problems associated with developing plans for other regions. It must be expected that outside Europe and North America the relative lack of scientific resources, infrastructure problems, and political and economic climates are all likely to make the task of developing and implementing a comprehensive plan a good deal more difficult.

Data Requirements

The Pingree Park meeting established protocols specifying the data that should be collected at each site (Ojima *et al.*, 1993). This had the primary objective of ensuring that the data obtained would all be compatible for modeling purposes, especially the extrapolation to larger regions. This approach is likely to set a precedent for the procedures to be adopted in other regions.

Many of the data collection activities within the monitoring program will need to be long-term, encompassing a wide range of temporal variation that will be helpful for understanding how fluxes will respond to climate change. Data acquired from manipulation experiments should complement those obtained from monitoring, and both will be used for the development of trace gas flux models incorporating the diverse factors that control these exchanges. These models will form the nucleus of a larger modeling activity capable of predicting trace gas exchanges at ecosystem scales. The linking of ecosystem models to atmospheric chemical transport models, and ultimately to general circulation models, will permit the necessary elucidation of trace gas fluxes at regional and global scales.

Table 2. Proposed locations of sampling site networks for midlatitude trace gas fluxes (IGAC, 1992).

Continent	Ecosystem/ Land Use Type	Key Ecosystems for Study	Existing Site Network
North America	Forest	Temperate rainforests (northwest coast)	LTER
		Mixed hardwoods (northwest, midwest)	LTER
		Southern coniferous	
		Montane coniferous	
	Rangeland	Desert	LTER
		Semi-arid shrubland	LTER
		Shortgrass steppe	LTER
		Tallgrass prairie	LTER
	Agricultural	Western irrigated	
		Western dryland	
		Corn belt	LTER
		Southeast	
		Intensive animal-based	LTER
Europe/ North Africa	Forest	Scots pine	NITREX
		Spruce	NITREX, TERN
		Beech	NITREX, TERN
	Rangeland	Desert (North Africa)	
		Semi-arid shrubland (Mediterranean)	
		Shortgrass steppe	
		Tallgrass prairie	
	Agricultural	Intensive pasture	TERN, TIGER
		Conventional arable	TERN, TIGER
		Intensive irrigated (horticultural)	
		Semi-arid	
Asia	Forest	Mixed hardwood	
		Coniferous	CERN
	Rangeland	Desert	
		Semi-arid shrubland	CERN
		Shortgrass steppe	CERN
		Tallgrass prairie	
	Agricultural	Irrigated	CERN
South America/ Australasia	Forest	Southern beech (*Nothofagus*)	
	Rangeland	Desert	
		Semi-arid shrubland	
		Shortgrass steppe	
		Tallgrass prairie	
	Agricultural	Temperate arable (including legume-based)	
		Intensive pasture (including legume-based)	

REFERENCES

Andreae, M.O. and D.S. Schimel (eds.), 1989, "Exchange of Trace Gases between Terrestrial Ecosystems and the Atmosphere," Wiley, Chichester, UK.

Arah, J.R.M. and K.A. Smith, 1989, Steady-state denitrification in aggregated soils: A mathematical model, *J. Soil Sci.*, 40:139-149.

Arah, J.R.M., I.J. Crichton, K.A. Smith, H. Clayton and U. Skiba, 1993a, An automated GC analysis system for micrometeorological measurements of trace gas fluxes, Proc. 1st IGAC Conf., Eilat, Israel; submitted to *J. Geophys. Res*.

Arah, J.R.M., K.A. Smith and I.J. Crichton, 1993b, Evaluation of a model of denitrification in soil microsites, *J. Soil Sci*., submitted.

Bingemer, H.G. and P.J. Crutzen, 1987, The production of methane from solid wastes, *J. Geophys. Res.*, 92:2181-2187.

Bogner, J.E., 1992, Anaerobic burial of refuse in landfills: Increased atmospheric methane and implications for increased carbon storage, *in:* "Trace Gas Exchange in a Global Perspective," D.S. Ojima and B.H. Svensson (eds.), *Ecol. Bull.* (Copenhagen) 42:98-108.

Bouwman, A.F., 1990, Exchange of greenhouse gases between terrestrial ecosystems and the atmosphere, *in:* "Soils and the Greenhouse Effect," A.F. Bouwman (ed.), Wiley, Chichester, pgs. 61-127.

Brumme, R. and F. Beese, 1992, Effects of liming and nitrogen fertilization on emissions of CO_2 and N_2O from a temperate forest, *J. Geophys. Res.*, 97:12,851-12,858.

Businger, J.A. and S.P. Oncley, 1990, Flux measurement with conditional sampling, *J. Atmos. Ocean. Technol.*, 7:349-352.

Casada, M.E. and L.M. Safley, 1990, "Global methane emissions from livestock and poultry manure," Report to Global Change Division, U.S. Environmental Protection Agency, Washington, D.C.

Cicerone, R.J. and R.W. Oremland, 1988, Biogeochemical aspects of atmospheric methane, *Global Biogeochem. Cycles*, 2:299-327.

Clymo, R.S., D.E. Pearce and W.E. Daulat, 1993, "Effects of environmental variation on methane and carbon dioxide dynamics of peatlands," Paper presented at UK TIGER Prog. Workshop, Grange-over-Sands, Feb. 1993.

Crosson, P., 1981, "Conservation tillage and conventional tillage: A comparative assessment," Soil Conservation Society of America, Ankeny, Iowa, U.S.

Dale, V.H., R.A. Houghton and C.A.S. Hall, 1991, Estimating the effects of land use change on global atmospheric CO_2 concentrations, *Can. J. Forestry Res.*, 21:87-90.

Duxbury, J.M., D.R. Bouldin, R.E. Terry and R.L. Tate, 1982, Emissions of nitrous oxide from soils, *Nature*, 298:462-464.

Eichner, M.J., 1990, Nitrous oxide emissions from fertilzied soils: Summary of available data, *J. Environ. Qual.*, 19:272-280.

Elliott, E.T., R.V. Anderson, D.C. Coleman and C.V. Cole, 1980, Habitable pore space and microbial trophic interactions, *Oikos*, 35:327-335.

Firestone, M.K. and E.A. Davidson, 1989, Microbiological basis of NO and N_2O production and consumption in soil, *in:* "Exchange of Trace Gases between Terrestrial Ecosystems and the Atmosphere," M.O. Andreae and D.S. Schimel (eds.), Wiley, Chichester, pgs. 7-21.

Fowler, D. and J.H. Duyzer, 1989, Micrometeorological techniques for the measurement of trace gas exchange, *in:* "Exchange of Trace Gases between Terrestrial Ecosystems and the Atmosphere," M.O. Andreae and D.S. Schimel (eds.), Wiley, Chichester, pgs. 189-207.

Gallagher, M.W., T.W. Choularton, K.N. Bower, I.M. Stromberg, K.M. Beswick, C. Dore, D. Fowler, K.J. Hargreaves and R. Storeton-West, 1993, Field measurements of methane at the landscape scale, Paper presented at UK TIGER Prog. Workshop, Grange-over-Sands, Feb. 1993.

Galle, B. and L. Klemedtsson, 1993, Application of an FTIR system for measurements of nitrous oxide fluxes using micrometeorological methods, an ultra-large chamber system and conventional field chambers, Proc. 1st IGAC Conf., Eilat, Israel; submitted to *J. Geophys. Res.*

Groffman, P.M., J.M. Tiedje, G.P. Robertson and S. Christensen, 1988, Denitrification at different temporal and geographical scales: proximal and distal controls, *in:* "Advances in Nitrogen Cycling in Agricultural Ecosystems," J.R. Wilson (ed.), CAB International, Wallingford, UK, pgs. 174-192.

Hansen, S., J.E. Maehlum and L.R. Bakken, 1993, N_2O and CH_4 fluxes in soil influenced by fertilization and tractor traffic, *Soil Biol. Biochem.*, 25:621-630.

Hargreaves, K.J., U. Skiba, D. Fowler, J. Arah, F. Weinhold, L. Klemedtsson and B. Galle, 1993, Measurement of nitrous oxide emission from fertilized grassland using micrometeorological techniques, Proc. 1st IGAC Conf., Eilat, Israel; submitted to *J. Geophys. Res.*

Harmon, M.E., W.K. Ferral and J.F. Franklin, 1990, Effects of carbon storage on conversion of old-growth forests to young forests, *Science*, 247:699-702.

Holdridge, L.R., 1947, Determinations of world formulations from simple climatic data, *Science*, 105:367-368.

Houghton, J.T., B.A. Callander and S.K. Varney (eds.), 1992, "Climate Change 1992: The Supplementary Report to the IPCC Scientific Assessment," Cambridge Univ. Press, Cambridge, UK.

Houghton, J.T., G.J. Jenkins and J.J. Ephraums (eds.), 1990, "Climate Change: The IPCC Assessment," Cambridge Univ. Press, Cambridge, UK.

Houghton, R.A., R.D. Boone, J.R. Fruci, J.E. Hobbie, J.M. Melillo, C.A. Palm, B.J. Peterson, G.R. Shaver and G.M. Woodwell, 1987, The flux of carbon from terrestrial ecosystems to the atmosphere in 1980 due to changes in land use: geographic distribution of the global flux, *Tellus*, 39B:122-139.

Hutchinson, G.L. and A.R. Mosier, 1979, Nitrous oxide emissions from an irrigated cornfield, *Science*, 205:1125-1127.

IGAC, 1992, Trace gas exchange: Midlatitude terrestrial ecosystems and atmosphere, IGAC, Cambridge, MA, U.S.

Isermann, K., 1993, Agriculture's share in the emission of trace gases affecting the climate and some cause-oriented proposals for sufficiently reducing this share, *in:* "Environmental Pollution 83," Elsevier, Amsterdam, Netherlands, in press.

Jenkinson, D.S., D.E. Adams and A. Wild, 1991, Model estimates of CO_2 emissions from soil in response to global warming, *Nature*, 351:304-306.

Kauppi, P.E., K. Mielikainen and K. Kuusela, 1992, Biomass and carbon budget of European forests, 1971 to 1990, *Science*, 256:70-74.

Kicklighter, D.W., J.M. Melillo, W.T. Peterjohn, E.B. Rastetter, A.D. McGuire and P.A. Steudler, 1993, Scaling field measurements to estimate regional carbon dioxide fluxes from temperate forest soils, *J. Geophys. Res*., submitted.

Mathews, E. and I. Fung, 1987, Methane emission from natural wetlands: Global distribution, area, and environmental characteristics of sources, *Global Biogeochem. Cycles*, 1:61-86.

Melillo, J.M., J.R. Fruci and R.A. Houghton, 1988, Land use change in the Soviet Union between 1850 and 1980: Causes of a net release of CO_2 to the atmosphere, *Tellus*, 40B:116-128.

Melillo, J.M., A.D. McGuire, D.W. Kicklighter, B. Moore, C.J. Vorosmarty and A.L. Schloss, 1993, Global climate change and terrestrial net primary production, *Nature*, 363:234-240.

Melillo, J.M., K.A. Smith, J. Bogner, P. Groffman, G. Harris, L. Klemedtsson, A.R. Mosier and W.H. Su, 1992, "Exchanges of N_2O, CH_4 and CO between terrestrial ecosystems and the atmosphere in midlatitudes," Report of IGAC Activity 7.2 Planning Workshop, Boulder, CO, U.S., Sept. 1991.

Melillo, J.M., P.A. Steudler, J.D. Aber and R.D. Bowden, 1989, Atmospheric deposition and nutrient cycling, *in:* "Exchange of Trace Gases between Terrestrial Ecosystems and the Atmosphere," M.O. Andreae and D.S. Schimel (eds.), Wiley, Chichester, UK., pgs. 263-280.

Mosier, A.R., D.S. Schimel, D. Valentine, K. Bronson and W. Parton, 1991, Methane and nitrous oxide fluxes in native, fertilized and cultivated grasslands, *Nature*, 350:330-332.

Nagele, W. and R. Conrad, 1990, Influence of pH on the release of nitric oxide and nitrous oxide from fertilized and unfertilized soil, *Biol. Fertil. Soils*, 10:139-144.

Ojima, D.S., E.A. Holland, J.M. Melillo, A.R. Mosier and G.P. Robertson, 1993, "Building a U.S. trace gas network (TRAGNET)," Report of Workshop, Pingree Park, CO, U.S., Sept. 1992.

Parkin, T.B. and J.M. Tiedje, 1984, Application of a soil core method to investigate the effect of oxygen concentration on denitrification, *Soil Biol. Biochem.*, 16:331-334.

Parton, W.J., A.R. Mosier and D.S. Schimel, 1988, Rates and pathways of nitrous oxide production in a shortgrass steppe, *Biogeochem.*, 6:45-58.

Paustian, K., G.P. Robertson, and E.T. Elliott, 1993, Management impacts on carbon storage and gas fluxes (CO_2, CH_4) in midlatitude ecosystems, *in:* "Greenhouse Gas Emissions and Carbon Sequestration, Advances in Soil Science," R. Lal, J. Kimble, E. Levine, and B.A. Stewart (eds.), Lewis Publishers, in press.

Peterjohn, W.T., J.M. Melillo, P.A. Steudler, K.M. Newkirk, F.P. Bowles and J.D. Aber, 1993, The responses of trace gas fluxes and N availability to experimentally elevated soil temperatures in an eastern US deciduous forest, *Ecolog. Applications*, submitted.

Rastetter, E.B., R.B. McKane, G.R. Shaver and J.M. Melillo, 1992, Changes in C storage by terrestrial ecosystems: How C-N interactions restrict responses to CO_2 and temperature, *Water, Air and Soil Pollut.*, 64:327-344.

Robertson, G.P., 1989, Nitrification and denitrification in humid tropical ecosystems: Potential controls on nitrogen retention, *in:* "Mineral Nutrients in Tropical Forest and Savanna Ecosystems," J. Procter (ed.), Blackwell, Oxford.

Robertson, G.P., 1993, Fluxes of nitrous oxide and other nitrogen trace gases from intensively managed landscapes: A global perspective, *in:* "Agricultural Ecosystem Effects on Trace Gases and Global Climate Change," ASA Special Publication No. 55, pgs. 95-108.

Robertson, G.P., W.L. Reed and S.J. Halstead, 1993, Methane consumption in agricultural ecosystems of the U.S. Midwest, *J. Geophys. Res*., submitted.

Ryden, J.C., 1981, N_2O exchange between a grassland soil and the atmosphere, *Nature*, 292:235-237.

Schimel, J., G.P. Robertson, D. Baldocchi, J.E. Bogner, E.A. Davidson, J. Duyzer, D. Ehhalt, D. Fowler, P. Groffman, K. Haider, V.A. Isodorov, L. Klemedtsson, J.M. Melillo, K.A. Smith, W.H. Su and W. Wieprecht, 1992, Impacts of trace gas fluxes in mid-latitude ecosystems, *in:* "Trace Gas Exchange in a Global Perspective," D.S. Ojima and B.H. Svensson (eds.), Ecol. Bull. (Copenhagen), 42:124-132.

Skiba, U., K.A. Smith and D. Fowler, 1993, Nitrification and denitrification as sources of nitric oxide and nitrous oxide in a sandy loam soil, *Soil Biol. Biochem.*, in press.

Smith, K.A., 1980, A model of the extent of anaerobic zones in aggregated soils, and its potential application to estimates of denitrification, *J. Soil Sci.*, 31:263-277.

Smith, K.A. and J.R.M. Arah, 1992, Measurement and modeling of nitrous oxide emissions from soil, *in:* "Trace Gas Exchange in a Global Perspective," D.S. Ojima and B.H. Svensson (eds.), *Ecol. Bull.* (Copenhagen), 42:116-123.

Smith, K.A. and P. Thomson, 1993, Investigation of soil and environmental processes responsible for production of nitrous oxide in soils, Paper presented at UK TIGER Prog. Workshop, Grange-over-Sands, Feb. 1993.

Smith, K.A., Scott, A., Galle, B. and Klemedtsson, L., 1993, Use of long-path infra-red gas monitor for measurement of nitrous oxide flux from soil, Proc. 1st IGAC Conf., Eilat, Israel; submitted to *J. Geophys. Res.*

Smith, T.M., J.F. Weishampel, H.H. Shugart and G.B. Bonan, 1992, The response of terrestrial C storage to climate change: modeling C dynamics at varying temporal and spatial scales, *Water, Air and Soil Pollut.*, 64:307-326.

Steudler, P.A., R.D. Bowden, J.M. Melillo and J.D. Aber, 1989, Influence of nitrogen fertilization on methane uptake in temperate forest soils, *Nature*, 341:314-316.

Tans, P.P., I.Y. Fung and T. Takahashi, 1990, Observational constraints on the global atmospheric CO_2 budget, *Science*, 247:1431-1438.

Tisdale, J.M. and J.M. Oades, 1982, Organic matter and water-stable aggregates in soils, *J. Soil Sci.*, 33:141-163.

Tyler, S.C., 1991, The global methane budget, *in:* "Microbial Production and Consumption of Greenhouse Gases," J.E. Rogers and W.B. Whitman (eds.), American Society for Microbiology, Washington, D.C., pgs. 7-38.

Wahlen, M. and W.S. Reeburgh, 1990, Consumption of atmospheric methane by tundra soils, *Nature*, 346:160-162.

Wienhold, F.G., H. Frahm and G.W. Harris, 1993, Measurements of N_2O fluxes from fertilized grassland using a fast response tunable diode laser spectrometer, Proc. 1st IGAC Conf., Eilat, Israel; submitted to *J. Geophys. Res.*

Wilson, A.T., 1978, The explosion of pioneer agriculture: Contribution to the global CO_2 increase, *Nature*, 273:40-42.

GLOBAL MEASUREMENTS OF PHOTOCHEMICALLY ACTIVE COMPOUNDS

Andreas Wahner[1], Franz Rohrer[1], Dieter H. Ehhalt[1], Elliot Atlas[2] and Brian Ridley[2]

[1]Institute for Atmospheric Chemistry
Forschungszentrum Jülich, Germany
[2]Atmospheric Chemistry Division
NCAR, Boulder, USA

ABSTRACT

An understanding of the distribution, variation, and photochemical interactions of reactive species in the global troposphere is fundamental to questions about the budgets and trends of tropospheric ozone and other trace gases. These are among the major objectives of the IGAC-GLOCHEM Activity. Recently, several field campaigns have measured reactive trace gases and radical species in the remote troposphere to test models of transport and chemical processes. The experimental programs described here are STRATOZ III, TROPOZ II, and MLOPEX II. Examples from each of these programs will highlight some of the results.

Global-scale distributions of trace gases were measured during June 1984 (STRATOZ III) and January 1991 (TROPOZ II). The measurements cover latitudes from 60°N to 60°S and altitudes from 0 to 12 km. A number of different chemical species were measured during the flights, but here we focus on measurement of nitrogen oxides, NO_x, in the troposphere. The oxides of nitrogen to a large extent control hydroxyl radical concentration and ozone production in the troposphere. Thus, accurate knowledge of the global NO_x distribution is required for the understanding of tropospheric chemistry.

The northern hemispheric summertime vertical profiles are "C-shaped," with high NO mixing ratios at 8 to 10 km altitude and a strong vertical gradient. In contrast, high concentrations were observed in the northern hemisphere during winter at all altitudes nearly without a vertical gradient.

A two-dimensional, zonal model with limited chemistry of NO, NO_2 and HNO_3, applied to the latitudinal band of 40 to 50°N, reproduces the measured vertical profiles reasonably well. Even the formation of a maximum of NO mixing ratio at 8 to 10 km altitude over Europe and North America during summertime can be explained by fast vertical transport of surface-emitted NO, and the NO emissions by aircraft in the upper troposphere.

The Mauna Loa Observatory Photochemical Experiment (MLOPEX II) was conducted to measure simultaneously a large number of related chemical species to characterize the photochemical state of the remote free troposphere and to test our current understanding of photochemical interactions and processes. The measurements at this location were carried out in 1991 and 1992 during four seasonal intensive experiments, each lasting 5 to 6 weeks. Aircraft studies during one intensive were made to relate measurements at the ground site to trace gas concentrations on a large, regional scale. In addition to trace gas composition, free radical species (HO_2, RO_2, HO) and photolysis rates (for NO_2, O_3) were measured.

Because the experiment was only recently completed, data have only begun to be integrated and analyzed. This paper presents examples of the different measurements obtained during the experiment, and emphasizes the variations and relationships observed over diurnal, synoptic, and seasonal time scales.

INTRODUCTION

The increasing emission of reactive gases into the troposphere has been clearly documented in recent years (WMO, 1993). Emission of radiatively active trace gases such as CO_2, CH_4, N_2O, etc. will have a direct impact on the earth's heat budget and have the potential for long-term climate effects. Combustion and related sources of SO_x, NO_x, CO, and hydrocarbons are known to produce photochemical smog and lead to acid deposition over large regions near source areas. However, transport of primary emissions must also impact the troposphere distant from source regions. Certainly, strong sources of hydrocarbons and halogenated hydrocarbons from the industrialized northern hemisphere are clearly reflected in the observed latitudinal profiles of these trace gases (e.g., Singh *et al.*, 1988; Montzka *et al.*, 1993).

Still, our understanding of the global distribution and seasonal variation of most reactive gases in the troposphere, and their photochemical products, is limited, as is our understanding of the oxidative rates and processes occurring in the global atmosphere. Knowledge of these rates and processes is fundamental to questions about the budgets and trends of tropospheric ozone and other trace gases. Reactive gases are central elements in atmospheric oxidation processes through their influence on hydroxyl (OH), hydroperoxy (HO_2) and organic peroxy (RO_2) radicals, and radical species are sensitive to small changes in reactive gas mixing ratios typical of the remote troposphere.

In recent years several field campaigns have begun to undertake coordinated measurement of reactive trace gases and radical species in the remote troposphere. The results of these campaigns are increasing the understanding of the distribution, variation, and photochemical interactions of reactive species in the global troposphere. This understanding is one of the major objectives of the IGAC Global Atmospheric Chemistry Survey (GLOCHEM) Activity.

In this paper we discuss selected results of two field measurement programs conducted recently. From the airborne TROPOZ II campaign we will focus on the global distribution of NO measured during January 1991; from the ground-based Mauna Loa Observatory Photochemistry Experiment (MLOPEX II) we will present preliminary observations on the temporal variation of a variety of species measured during the program and we will address some issues raised from the pilot MLOPEX study conducted in 1988.

TROPOZ II

The nitrogen oxides, NO and NO_2, play a dual role in tropospheric chemistry: they are a major controlling factor of the concentration of the hydroxyl radical, OH, and they are primarily responsible for the tropospheric formation of ozone, O_3. It is the anthropogenic emission of NO in various combustion processes that causes the increase in tropospheric O_3 concentration observed at northern midlatitudes. Because O_3 is a greenhouse gas, anthropogenic emissions of NO can thus also have an impact on climate. Moreover, since the temperature increase at the earth's surface is predicted to be particularly sensitive to the O_3 concentrations in the upper troposphere (Wang and Sze, 1980), relatively small emissions of NO_x into the upper troposphere, by air traffic for instance, can play an unexpectedly large role.

Considering the importance of the nitrogen oxides, $NO_x = NO + NO_2$, relatively few measurements exist to define their global distribution. Even less is known about the distribution of NO_x in the upper troposphere and its seasonal variation. A major fraction of the available measurements of NO in the upper troposphere were obtained during the Tropospheric Ozone (TROPOZ II) (Rohrer *et al.*, 1993) and during the Stratospheric Ozone

Figure 1. Flight track of the STRATOZ III (June 1984) and TROPOZ II (January 1991) campaign.

(STRATOZ III) (Drummond *et al.*, 1988) aircraft campaigns. The TROPOZ II campaign was conducted during January 1991. In contrast the STRATOZ III aircraft campaign, which took place in June 1984, sampled summer conditions. Both campaigns followed very similar flight routes mainly along the coast lines of North America, South America, western North Africa and Europe between 70°N and 60°S latitude (Figure 1). Vertical profiles of NO were measured from 0 km to 12 km altitude.

Figure 2 and Figure 3 present an integrated view of the NO mixing ratio measurements during TROPOZ II and STRATOZ III, respectively. All the data are averaged into a grid of 1 km altitude by 10° latitude boxes and filtered to remove excursions and gaps. The measurements in the boundary layer were excluded to prevent propagation of the high boundary layer values into the atmospheric layer between 1 km and 2 km altitude. The data in Figure 2 and 3 are shown within the same scale, so the seasonal differences in the global NO distributions can be extracted by direct comparison. The global distribution of NO obtained during January 1991 (Figure 2) shows three major features:

- The NO mixing ratios are considerably higher in the northern hemisphere than in the southern hemisphere. The northern-most latitudes seem to be loaded with high concentrations of NO.
- The NO mixing ratios increases with latitude north of 20°N.
- The vertical gradients are weak at all latitudes.

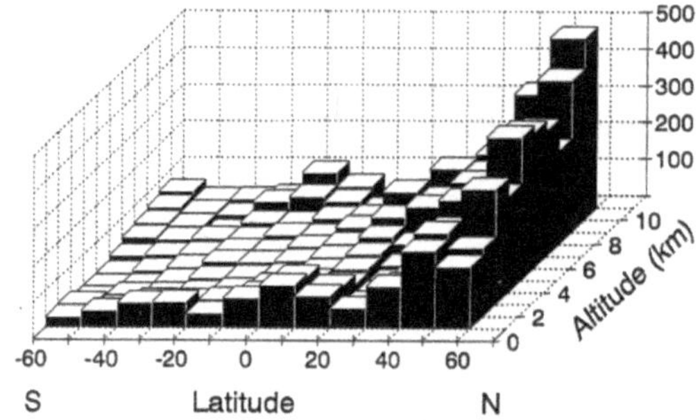

Figure 2. Meridional representation of the average NO mixing ratios measured during TROPOZ II (January, 1991) along the coast lines of eastern North America, South America, western North Africa and Europe.

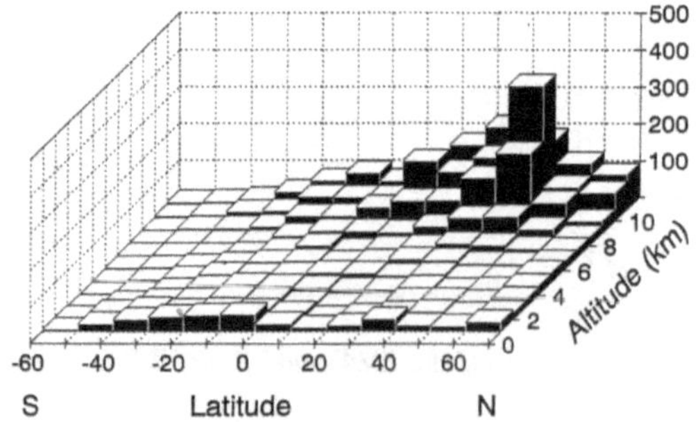

Figure 3. Meridional representation of the average NO mixing ratios measured during STRATOZ III (June, 1984) along the coast lines of eastern North America, South America, western North Africa and Europe.

The global distribution of NO obtained during June 1984 (Figure 3; Ehhalt *et al.*, 1992) is drastically different:

- There is a steep vertical gradient of the NO mixing ratio between 8 and 10 km altitude in all of the northern hemisphere reaching as far south as 20°S.
- In the middle troposphere at 3 to 7 km altitude the NO mixing ratio reaches a minimum of 10 to 20 ppt. Taking account of the high mixing ratios in the boundary layer the vertical profile of NO in the northern hemisphere during summertime appears strongly C–shaped.

What causes this drastic change in the vertical profile of the NO mixing ratio between summer and wintertime in the northern hemisphere? To answer that question we take a closer look at the NO profiles at 40° to 50°N latitude. In those latitudes, there are four different sources of tropospheric NO_x that have to be considered, namely, emissions by aircraft, production by lightning, stratospheric input, and sources at the earth's surface mainly from combustion of fossil fuel. Using these sources the summertime NO profiles were successfully simulated by a simple 2-D zonal model (Ehhalt *et al.*, 1992). Here we apply this model to explain the wintertime NO profiles. Before doing so we mention some of the model's salient features.

Only vertical transport is treated explicitly. Horizontal transport is introduced by allowing the NO_x source terms, surface boundary conditions, and vertical transport to change in time commensurate with a displacement in longitude by a 8 m s^{-1} westerly wind during summertime and 12 m s^{-1} westerly wind during wintertime. This is equivalent to moving the whole air column treated in the model zonally around the globe with a vertically uniform wind speed. The wind speed corresponds to the zonally and vertically averaged west wind component between 40° and 50°N latitude in June and January, respectively (Houghton, 1985). The model covers the altitudes from 0 to 14 km altitude for the June calculations and 0 to 12 km altitude for the January calculations with a 1 km height resolution except for the two lowest levels which are located at 0.1 and 0.5 km altitude. The tropopause is located at 12 km altitude in June and at 10 km in January. Temperature and density profiles are those of the U.S. Standard Atmosphere (1976). The model includes a fast vertical transport scheme, which injects air parcels from the planetary boundary layer directly into all tropospheric levels above 1 km altitude. The injection rates for the fast vertical transport are shown in Figure 4. They differ considerably between summer (Figure 4a) and winter (Figure 4b). The exchange rate of 0.1 day^{-1} at 8 to 10 km altitude in June over the continents drops to 0 day^{-1} in January. The exchange rates at 2 to 7 km altitude are larger, especially over the oceans, during January.

Chemistry is also kept to a minimum in the model. It includes the fast interconversion of NO and NO_2, the conversion of NO_2 to HNO_3 via the reaction with OH, and the conversion of NO_2 to nitrate via the reaction with O_3 to NO_3. The vertical OH and O_3 profiles are fixed to literature values. The vertical OH profiles for June and January at 45°N are taken from Volz *et al.* (1981), those of O_3 from Marenco and Said (1989). This results in a chemical lifetime for NO_x of about 4 days during June and 15 days during January at 10 km altitude. The NO_x chemical lifetime at 3 km altitude is 0.7 days during June and 2.5 days during January. The change is due to the lower OH mixing ratio during wintertime. Oxidation of NO_x to NO_3 contributes less than 20% during June and more than 80% during January to the loss of NO_x. NO_2 is allowed to dry deposit over the continents only, HNO_3 dry deposits over the whole surface. It is also removed by rainout. For details see Ehhalt *et al.* (1992).

The NO_x sources contributing to the atmospheric burden are listed in Table 1. Only the stratospheric input is zonally uniform. It is introduced by a downward flux of 1.5×10^{12} molecules $m^{-2}s^{-1}$ at 14 km altitude in June and at 12 km altitude in January. The surface sources of NO are represented by an upward flux boundary condition at 0 km. That flux vanishes over the oceans and varies with longitude over the continents to closely represent the natural and anthropogenic inputs. The surface flux is adopted from Hameed and Dignon (1988). The inputs by lightning and aircraft are volume sources and zonally non-uniform. The lightning input is concentrated over the continents and only present during summertime.

Table 1. NO_x emissions for June 1984 and January 1991 (10^6 t N/a) (% = fraction of the total source strength in the 40° to 50°N latitude band).

	40° – 50°N	June 1984	40° – 50°N	January 1991
Surface sources	8.63	95.5%	8.63	98.2%
Lightning	0.29	3.2%	0	0%
Aircraft (civil)	0.081	0.9%	0.124	1.4%
Stratosphere	0.037	0.4%	0.037	0.4%

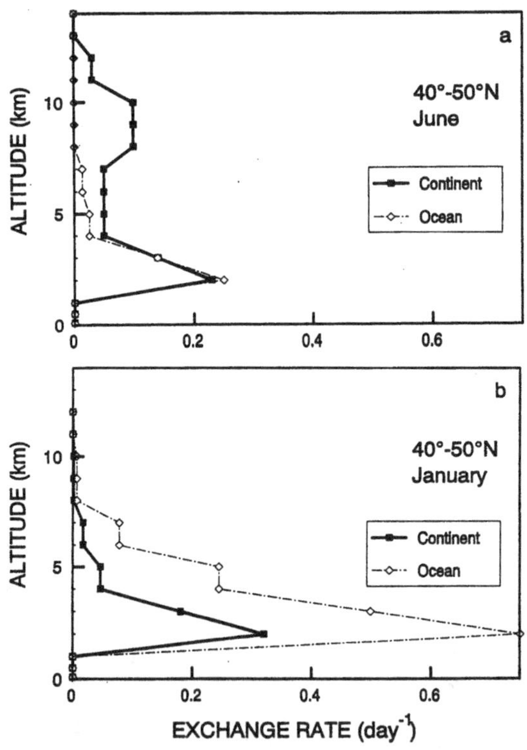

Figure 4. Vertical profiles of the exchange rates of the fast vertical transport over the ocean (dashed-dotted line) and over the continent (solid line) during June **(a, top panel)**, and January **(b, bottom panel)**.

The model yields a zonal NO distribution. The sections between 65°W and 5°W extending from the east coast of North America to the west coast of Europe are displayed as vertical profiles at 10° longitude intervals in Figure 5a and 5b. The calculated vertical profiles of NO for June describe the observed features: (1) a steep increase of the NO mixing ratio up to 200 ppt in the altitude range of 8 to 10 km; (2) a strongly C–shaped vertical profile at 65°W over the east coast of North America with surface sources present. The different shadings in Figure 5 identify the various contributions to the local NO mixing ratio. The aircraft emissions contribute about 30%, zonally averaged, during June 1984 to the upper tropospheric NO mixing ratio. The vertical gradient in the upper troposphere in June is additionally enhanced by the contribution of surface emitted NO, which is injected in the upper troposphere by fast vertical transport (see Figure 4a). Figure 5b, showing the calculated vertical profiles of NO for January also describes the observed features: the vertical profile at the North American coast shows hardly any vertical gradient between 1 and 10 km altitude and the mixing ratio of NO is about 200 to 300 ppt throughout the troposphere. The breakdown of the various contributions in Figure 5b at 55°W indicates that a possible vertical gradient that could be induced by the emission from aircraft in the upper troposphere is compensated by the fast vertical transport of surface emitted NO from the planetary boundary layer to the middle troposphere (2 to 7 km altitude) during wintertime (see Figure 4). The higher mixing ratio during January is a consequence of the longer chemical lifetime of NO_x during wintertime.

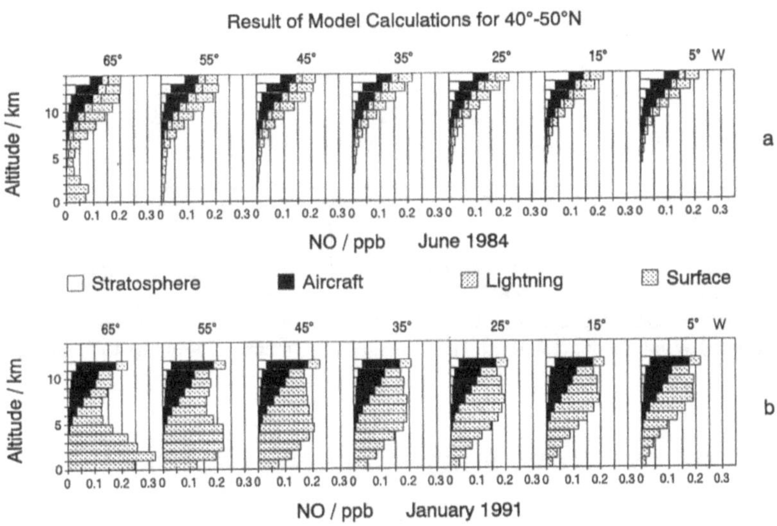

Figure 5. **(a, top panel)** Calculated vertical mixing ratio profiles of NO during June 1984, and **(b, bottom panel)** during January 1991 over the western North Atlantic. The different shadings indicate the contributions from the individual sources.

Figure 6 shows a direct comparison between the model calculated and measured vertical profiles of NO at 60°W longitude for June 1984 and January 1991 (Rohrer *et al.*, private communication). As pointed out before, the agreement between the shapes of the calculated and measured vertical profiles during summertime as well as during wintertime is surprisingly good. In addition the model reproduces the much higher mixing ratios observed during winter, although it tends to underestimate the mixing ratios in the upper troposphere and to overestimate the NO mixing ratio during summer at 6 to 9 km altitude. These deviations could be due to several reasons: the uncertainties of the various NO source strengths which are on the order of a factor of two, the model's use of climatological mean transport, radiation, temperature, and wind fields to simulate an episodic event, and the oversimplified chemistry. To remedy this, we are attempting to simulate the NO distribution with a global model.

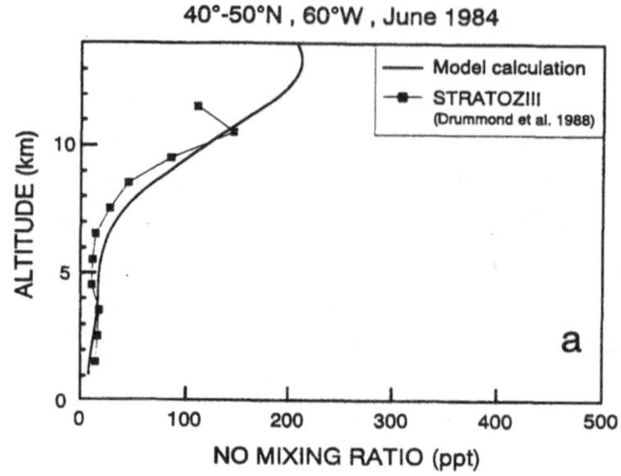

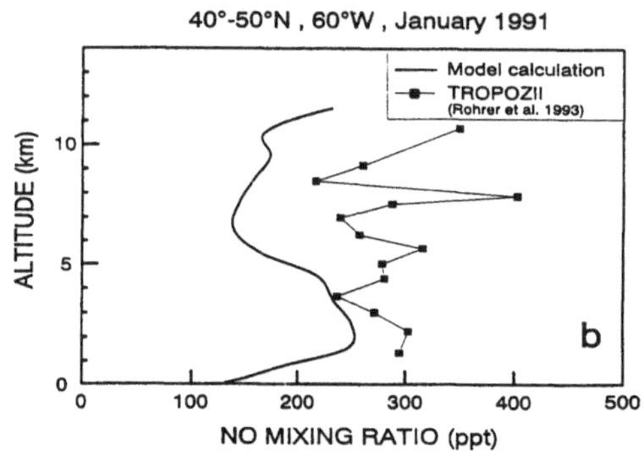

Figure 6. (a, top panel) Vertical profiles of the NO mixing ratio over the East coast of North America during June, 1984 on the south bound leg of STRATOZ III (Drummond *et al.*, 1988), and **(b, bottom panel)** during the January, 1991 south bound leg of TROPOZ II (from Rohrer, private communication, 1993). Solid line: model calculation 50°W longitude; closed squares: averaged measurements between 40°N and 50°N latitude.

MAUNA LOA OBSERVATORY PHOTOCHEMISTRY EXPERIMENT

As exemplified in the results from the TROPOZ experiments, atmospheric dynamics play a major role in shaping the distribution of trace gases in the remote troposphere. Meteorological factors and physical removal combine with chemical processes to modulate the variations observed in the composition of trace gases. Variations in the composition of a variety of trace species are then related to fast photochemical interactions, which determine radical abundance and oxidant balance. To understand better the complex factors that control the chemistry of reactive gases and radical species in the remote troposphere, a pilot study was conducted during the spring of 1988 at the Mauna Loa Observatory (MLO) on the island of Hawaii, and a second longer-term experiment took place over four seasons in 1991 and 1992. The emphasis of these studies was to measure a sufficient number of photochemically related species to be able to test and constrain models of the photochemistry of the remote atmosphere adequately. Budgets of odd-nitrogen and odd-oxygen were the main foci.

The ground-based approach at MLO offered several advantages to studying atmospheric chemistry. The site is located at 3.4 km in the remote north central Pacific ocean, and measurements at MLO can be representative of a significant fraction of the remote marine troposphere. With the combination of chemical and meteorological data, measurements made during downslope flow during the night and early morning can be filtered to be representative of free tropospheric conditions. The site at MLO, operated by the NOAA Climate Modeling and Diagnostics Laboratory (CMDL), has been measuring longer-lived species and atmospheric physical parameters over the last several decades so the site is well-characterized. Furthermore, since many of the reactive species in the free troposphere are found at low mixing ratios, measurements are not routine. Performance of ground-based instrumentation often is enhanced over that from aircraft missions, and when this improvement is combined with longer term measurement periods, data precision can be improved significantly.

Table 2. Participants in the Mauna Loa Observatory Photochemistry Experiment (MLOPEX II), 1991-92. J(*i*) denotes photolysis rate (sec^{-1} for species *i*).

Measurement	Technique	Experimenters	Institution
HOOH, CH_3OOH, H_2CO	EF	B. Heikes, B. McCully	URI
HOOH, CH_3OOH	HPLC	G. Kok, T. Staffelbach	NCAR
H_2CO	IEF	T. Campos	NCAR
HOOH, H_2CO	TDLAS	G. Mackay, D. Karecki	UNISEARCH
Carbonyls	HPLC	K. Mopper, X. Chen, C. Schultz	WSU
		X. Zhou, Y-N Lee	BNL
HNO_3, NO_3^-, Formic, Acetic Acid	Filter/IC; Denuder	R. Norton, S. Buhr	NOAA-AL, CIRES
HNO_3, NO_3^-	Denuder/IC	J. Lind, T. Campos, E. Atlas	NCAR
NMHC,CO	GC/FID	J. Greenberg, P. Zimmerman	NCAR
	GC/HgO	S. Schauffler, D. Helmig	NCAR
	GC/MS	A. Wedel, G. Brailsford, W. Pollock	NCAR
$J(NO_2)$, $J(O_3)$	Actinometer	R. Shetter, T. Gilpin	NCAR
RO_2	CA	C. Cantrell, J. Calvert,C. Gilliland	NCAR
$J(O_3)$	Radiometer	W. Junkerman	Fraunhofer Inst.
OH	CI/MS	F. Eisele, D. Tanner	GTRI
PAN, MN, O_3	GC, UV	B. Ridley, J. Walega	NCAR
NO, NO_2	CL	F. Grahek	NCAR
NO_y, SO_2	CL, PF	G. Hübler	NOAA AL, CIRES
Organic Nitrates	GC/ECD	E. Atlas	NCAR
Radon, Thoron		S. Whittlestone	ANSTO
PAM Station		J. Walega, D. Montzka	NCAR
Trajectories		P. Hess, K. Mosher, S. Madronich	NCAR
O_3, O_3 column, CFCs, CO		MLO-CMDL	NOAA
CH_4, N_2O, CO_2, Particles, Winds, Sondes, Black Carbon		R. Schnell	NOAA
AIRCRAFT			UW (KING AIR)
HOOH, CO, O_3, NOy, CN		T. Staffelbach	NCAR
Grab samples - NMHC		G. Kok, J. Walega, B. Ridley	NCAR
organic nitrates		E. Atlas, J. Greenberg	NCAR

Results from the 1988 pilot study were recently summarized in a series of papers in the *Journal of Geophysical Research* (Ridley and Robinson, and other articles in the same issue, 1992; Heikes, 1992). The experiment raised a number of issues which were the basis of the more extensive set of experiments in 1991-92. The MLOPEX II campaign was a coordinated effort involving many different investigators (Table 2), and interpretation of the data is only now beginning. In this section, we present some preliminary results of the different investigators, and it should be realized that more detailed interpretations will be made later. Below, we briefly discuss some of the observations and questions from the 1988 study and present preliminary data from MLOPEX II relevant to these issues.

Transport to the Mid-Pacific

During the 1988 study, no clear-cut signatures of direct transport from continental regions could be observed in measurements of different trace gases, in spite of the fact that springtime is the climatological maximum in transport of Asian dust to the region (Parrington *et al.*, 1983). Cyclical variations on the order of 5 to 7 days in some of the trace species were attributable to synoptic scale meteorological changes which provided a contrast between low altitude, equatorial air masses and those from more northerly latitude and higher altitudes. Ultimately the latitudinal gradients implied from the measurements are derived from continental and/or stratospheric sources, but the extent to which direct transport from continental sources impacts the trace gas chemistry in the mid-Pacific was unclear from this data set. Other measurements at MLO have indicated a more important role of direct transport to the site. Nitric acid measurements over several years at MLO indicate a summertime maximum, which is interpreted as transport from North America (Galasyn *et al.*, 1987; Moxim, 1990). Indications of rapid transport from continents could be seen in winter episodes using radon and methane as tracers of continental emission (Harris *et al.*, 1992; Whittlestone *et al.*, 1992). Finally, there is the evidence from the widespread springtime dust transport to the Pacific that strong seasonal variations in transport should be expected (Merrill *et al.*, 1989).

Data from MLOPEX II reveal a complex variation in trace gas composition during the year. As in the spring 1988 study, variations in trace gas mixing ratios were clearly related to the 5 to 8 day changes in synoptic meteorological conditions. An example of this variation from May, 1992, is shown for propane in Figure 7. In this figure, all data are included and a strong diurnal cycle from the upslope/downslope flow pattern also is

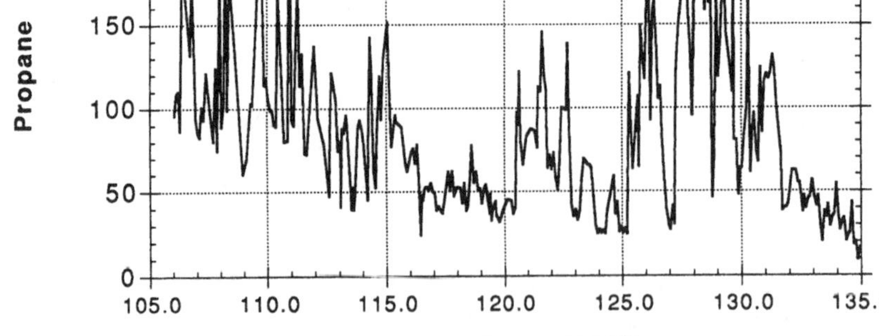

Figure 7. Variation of propane (pptv) during April/May, 1992 at the Mauna Loa Observatory. All data are included; no attempt has been made to filter the data for free tropospheric conditions. (Measurements made available courtesy of J. Greenberg and coworkers at NCAR).

apparent for much of the experimental period. Other hydrocarbons and tracers of continental air masses tended to show similar features. To relate the observed variations to transport from continental areas, careful calculation of individual air-mass trajectories to MLO are underway.

On a seasonal scale, pronounced variation was observed for different trace species, though the shorter cycle, synoptic variation often was larger than the seasonal differences. Trace gases with primary sources related to continental emissions or combustion/industrial activity tended to maximize during the January–February intensive, while the spring intensive seemed to be a transition period. Lowest mixing ratios of these species were observed in the summer and fall periods. Examples of this seasonal pattern are shown for

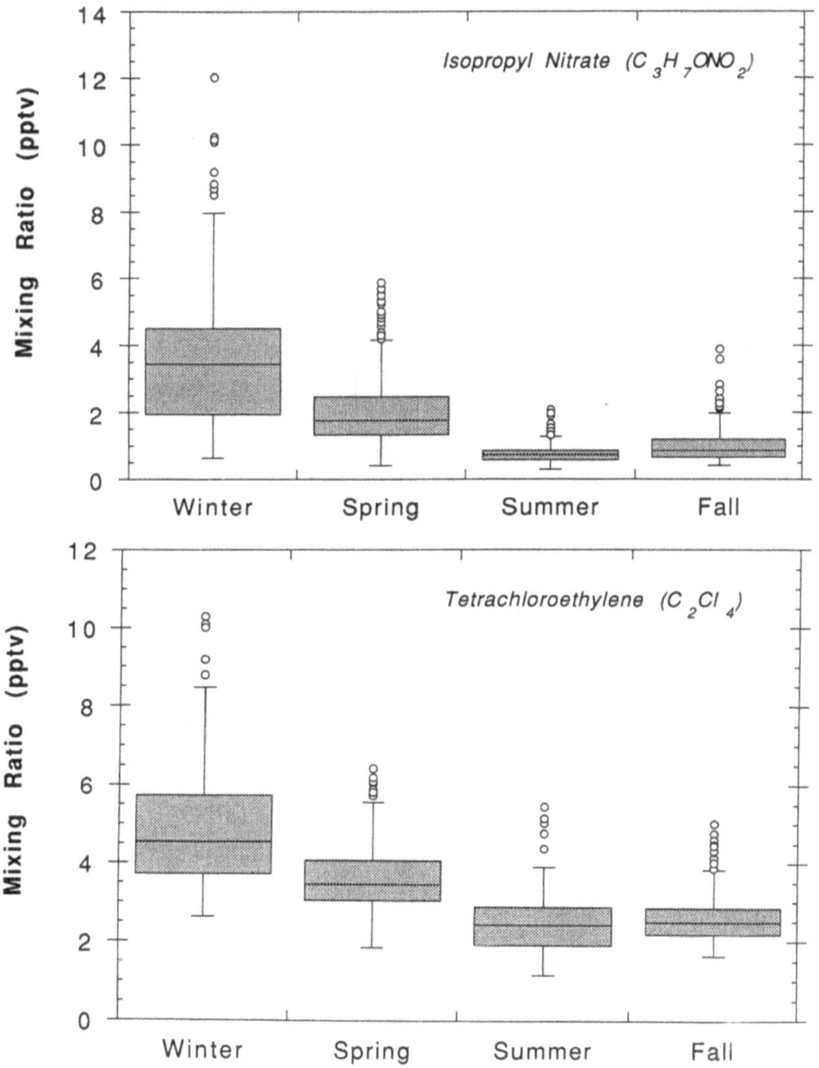

Figure 8. Seasonal variation of isopropyl nitrate **(top panel)** and tetrachloroethylene **(bottom panel)** during the Mauna Loa Observatory Photochemistry Experiment. The shaded box encloses 50% of the data, and the median is marked with a dark line in the shaded box. All data are included; no attempt has been made to filter the data for free tropospheric conditions.

tetrachloro-ethylene (C_2Cl_4), and isopropyl nitrate ($C_3H_7ONO_2$) (Figure 8). Most non-methane hydrocarbons followed a similar cycle. These observations are consistent with the large scale flow meteorology for MLO described by Miller *et al.* (1981) and Harris and Kahl (1990). Radon, which is a sensitive indicator of continental transport, had a slightly higher median value in the springtime compared to the winter, but maximum values of radon were observed during the winter period (Figure 9). These maxima should be related to the rapid transport of air masses from continental source regions, but detailed trajectory calculations are required to support this interpretation.

A springtime maximum was observed for a number of other trace gases. Besides the well-known maximum in ozone, highest median values were observed during the spring intensive for radon (as noted), peroxyacetyl nitrate (PAN), and nitric acid (HNO_3) (Figure 10). Along with the high ozone, these high median values during the spring suggest sources from the upper troposphere to MLO. However, given the relatively high levels of radon during the spring intensive, these upper level sources must also be linked to transport from continental regions.

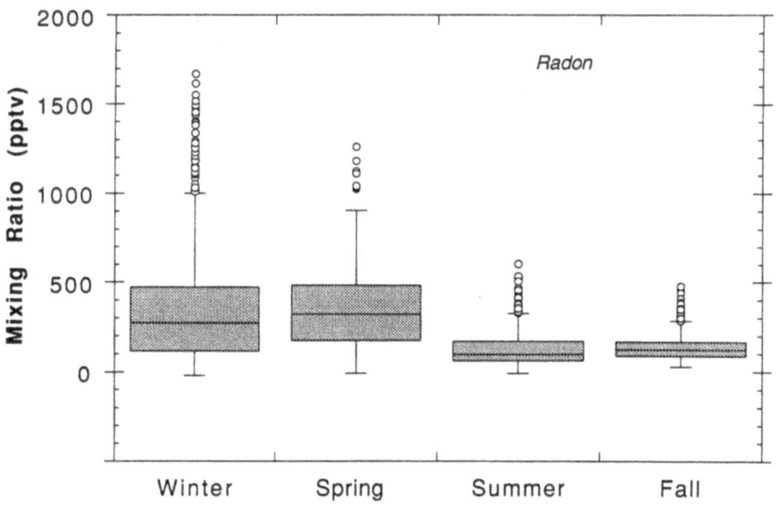

Figure 9. Seasonal variation of radon during the Mauna Loa Observatory Photochemistry Experiment. The shaded box encloses 50% of the data, and the median is marked with a dark line in the shaded box. All data are included; no attempt has been made to filter the data for free tropospheric conditions. (Measurements are made available courtesy of S. Whittlestone, private communication.)

Odd-Nitrogen Budget and Partitioning

One of the puzzling observations from a number of studies of the budget and partitioning of odd-nitrogen in the troposphere is the lack of agreement between measurement of total reactive oxidized nitrogen (NO_y) and the sum of the individually measured odd-nitrogen species. During MLOPEX I, the sum of the concentrations of measured odd-nitrogen species (HNO_3, PAN, NO_x, particulate nitrate, and alkyl nitrates) was approximately 25% less than total NO_y measured using a gold/CO catalytic conversion technique (Hübler *et al.*, 1992; Atlas *et al.*, 1992). The suggestion of a pool of "missing" odd-nitrogen has potential implications for the budget of NO_x in the remote troposphere if there actually are unknown species and if these species are reactive in the troposphere.

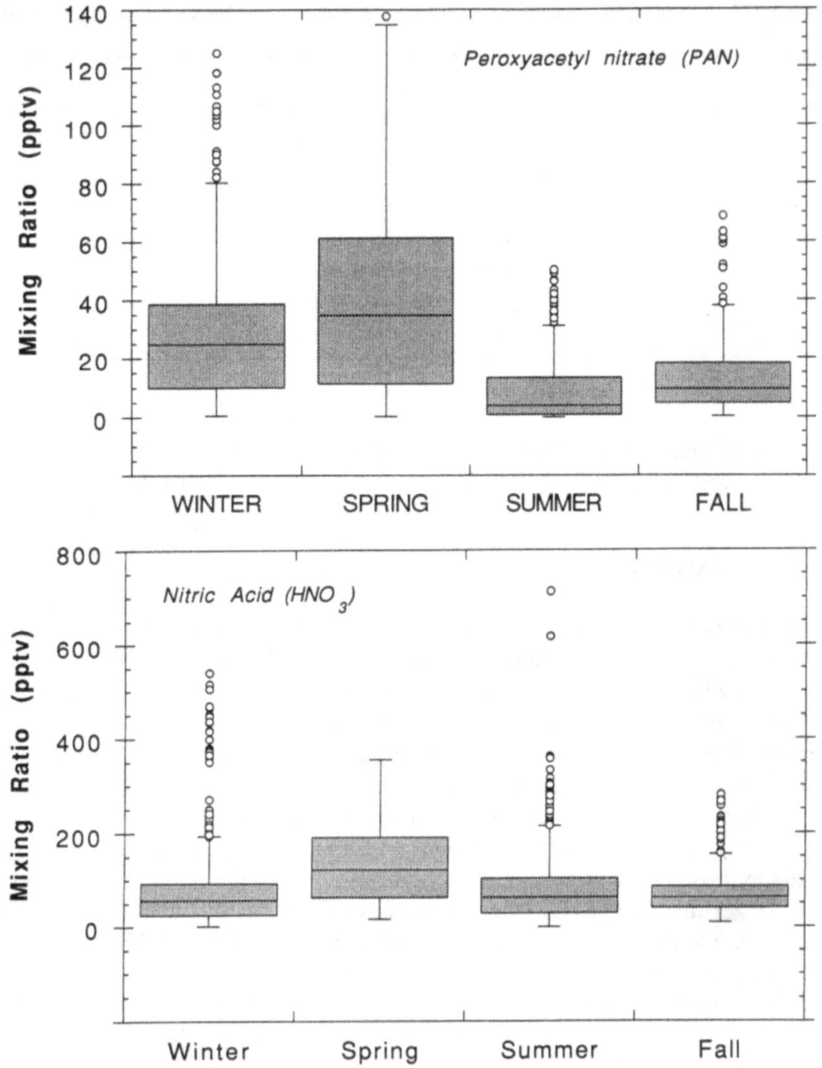

Figure 10. Seasonal variation of peroxyacetyl nitrate (PAN) **(top panel)** and nitric acid **(bottom panel)** during the Mauna Loa Observatory Photochemistry Experiment. Data are nighttime only data for PAN, and all data for nitric acid. No attempt has been made to filter the data for free tropospheric conditions. The shaded box encloses 50% of the data, and the median is marked with a dark line in the shaded box (data from experimenters listed in Table 2).

At Mauna Loa, nitric acid is the dominant odd-nitrogen species (Table 3), and during MLOPEX I a positive correlation was noted between the levels of nitric acid and the level of "missing" odd-nitrogen. For MLOPEX II, three separate techniques were employed to measure nitric acid, since this measurement may be the most critical in controlling the budget of odd-nitrogen. Preliminary evaluations of the data from MLOPEX II suggest reasonable agreement between the various techniques for measuring nitric acid. Still, preliminary data from the summer and fall intensives indicate that, on average, an imbalance from 20 to 40% in the odd-nitrogen budget may be found. Data are not yet available to make similar comparisons for the other intensive periods. During MLOPEX II

Table 3. Odd-nitrogen partitioning during MLOPEX I, 1988 (Atlas *et al.*, 1992).

Compound	All data	Median ratio to NO_y Upslope	Free troposphere
HNO_3	.448	.518	.431
NO_x	.145	.155	.139
NO_3^-	.111	.187	.052
PAN	.070	.088	.047
$MeONO_2$	.014	.015	.012
$RONO_2$ ($>C_3$)	.014	.019	.011
$\Sigma(NO_y)_i$	.844	1.108	.746

searches for additional complex organic nitrates that may help resolve the imbalance tended to reveal only traces (sub-pptv level) of different multifunctional organic nitrates.

Photochemical Modeling

Measurements of relevant photochemical parameters and comparison with model predictions for the spring 1988 study indicated areas of both good and poor agreement (Ridley *et al.*, 1992; Liu *et al.*, 1992). For example, Ridley *et al.* (1992) showed that measured NO_2:NO ratios were in good agreement with results from a detailed photochemical box model which was constrained by measured O_3, $J(NO_2)$ and other parameters. The abundance of peroxy radicals implied by the measurements were in good agreement with the model. Maximum daytime peroxy radical concentrations of about 60 pptv were predicted for average clear sky conditions during the spring, 1988 intensive. However, the hydroxyl radical concentration calculated from the photochemical box model was notably larger than suggested from earlier aircraft flights in the Pacific region. High sun maxima of 6.5×10^6 and 4.5×10^6 were estimated for 3.4 km at MLO and in the free troposphere, respectively .

Similarly, a comparison of measured and modeled CH_2O, HOOH, ROOH, and NO_x:HNO_3 showed significant discrepancies, which suggested that some fundamental inconsistencies exist. For example, Liu *et al.* (1992) also concluded that the NO_2:NO ratio measured at MLO was consistent with peroxy radical concentrations from their model, however, but there was some inconsistency between modeled and measured hydroperoxides. The measurement of hydrogen peroxide was consistent with HO_2 radical abundance predicted by the model, but organic peroxides measured at MLO were much lower than would be predicted from the RO_2 radicals in the model. This contradicts the conclusions from the analysis of NO:NO_2 ratios. Budget calculations by Liu *et al.* (1992) also suggested that measured OH radical concentrations may be overpredicted in models. For example, formaldehyde mixing ratios in the model averaged about 300 pptv compared to measured values of about 100 pptv. This difference suggests that OH radical (and/or CH_3O_2) is too high in the model or that additional sinks of formaldehyde are required. Similarly, the HNO_3:NO_x ratio is overpredicted in the photochemical model. Reduction of model OH or additional losses of HNO_3 are required to reconcile measurement with model prediction.

To address some of the issues raised in the model evaluation, MLOPEX II added several new measurement programs to help constrain photochemical model comparisons further. Since one of the critical indicators of hydroxyl radical chemistry is formaldehyde, additional independent methods were used to evaluate analytical biases and uncertainty. An early evaluation of the data indicates that the formaldehyde mixing ratios measured during MLOPEX I are comparable to those seen by the different investigators throughout the year during MLOPEX II. Formaldehyde mixing ratios in the range of 100 to 200 pptv appear to be characteristic of the marine free troposphere around Mauna Loa.

Measurements specifically related to radical formation rates and abundances were also added to the measurement program. A new measurement of $J(O_3)$ was developed to directly calculate the OH radical production rate from the photolysis of O_3. Initial comparisons suggest that $J(O_3)$ is reasonably well predicted from theoretical considerations, but significant effects may be caused by overhead aerosol abundance.

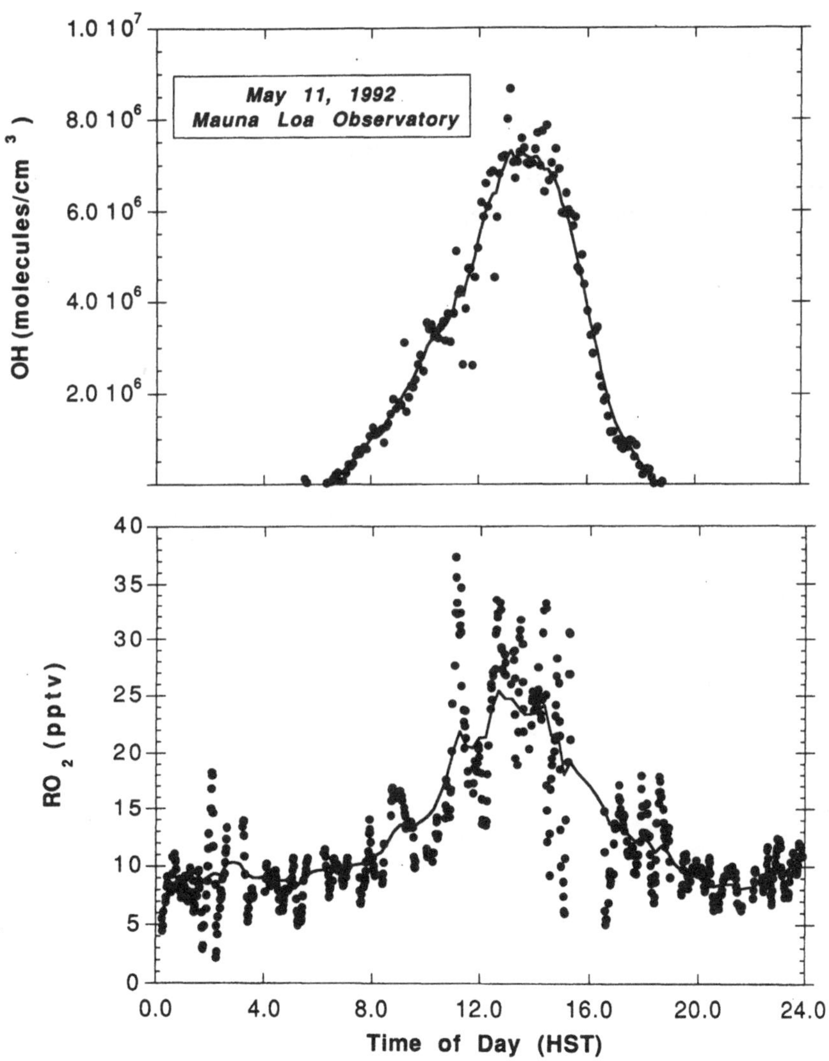

Figure 11. Diurnal variation of hydoxyl radical (OH) **(top panel)**, and total peroxy radicals (RO_2) **(bottom panel)** during 11 May, 1992 at the Mauna Loa Observatory. (Data made available by F. Eisele and D. Tanner [OH] and C. Cantrell and coworkers [RO_2]).

Radical species were also measured for the first time at MLO. A chemical amplifier technique was used to estimate total peroxy radical concentrations, and a chemical ionization/mass spectrometric technique was used for direct measurement of hydroxyl radical concentration. Figure 11 illustrates a "typical" cycle for these species for one spring day at Mauna Loa. Though model calculations have not been completed for MLOPEX II, the data are intriguing compared to predictions from the 1988 study. First, measured hydroxyl radical concentrations appear to be in the range of those predicted from photochemical models from MLOPEX I. Also, measured peroxy radical concentrations show a clear diurnal cycle, though maximum values may be lower than those predicted from the spring 1988 intensive. Evidence for possible nighttime peroxy radicals also appears in the data. At this point, though, further calculations and data evaluation are required to reach any conclusions regarding our ability to model the photochemistry of the remote troposphere accurately.

Regional Chemistry of the Mid-Pacific

As the MLOPEX data set is used to evaluate regional and global models of chemistry, it is useful to place the ground-based measurements in the context of the regional chemistry of the mid-Pacific atmosphere. Furthermore, it is necessary to show that the ground-based measurements at the Mauna Loa Observatory are, indeed, representative of the free

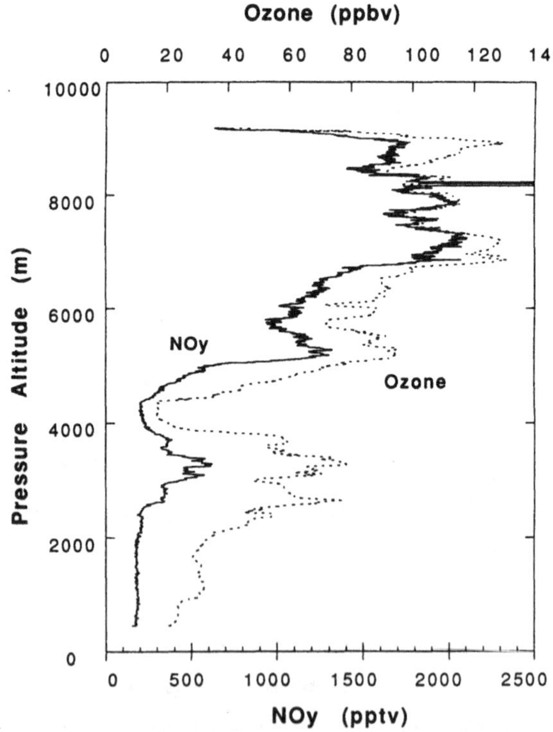

Figure 12. Vertical profiles of NO_y (solid line, bottom axis) and ozone (dotted line, top axis) during one spiral ascent well off the west coast of Hawaii. The Mauna Loa Observatory is located at about 3.4 km.

troposphere near 3.4 km. While this has been demonstrated by comparison of ground-based ozone measurements with ozone sondes (Oltmans, 1985), no evidence was available for more reactive species. To examine these questions in more detail, a limited aircraft campaign was conducted during the spring intensive experiment (see Table 2). As with many measurement programs, the observations are far more complex than imagined a priori. Measurements of different chemical species and tracers showed that the atmosphere in the region of Hawaii could be highly structured with multiple layering, and with it evidence during some flights of rapid transport of continental material in the upper troposphere. An example of measurements of ozone and NO_y from one flight is shown in Figure 12. High concentrations of NO_y in the upper troposphere are well beyond what is considered typical background data, either from other aircraft campaigns or from the MLOPEX data set. To combine the aircraft data into the framework of the ground-base data set is a challenge to be addressed, but it is clear that addition of the vertical and regional distributions of chemical species enhance the interpretation of the ground-based data set.

ACKNOWLEDGMENTS

The authors wish to acknowledge all participants in the MLOPEX program for making their data available for this manuscript. We also thank the staff and scientists of the NOAA CMDL station at Mauna Loa for their cooperative help and support. NCAR is supported by the National Science Foundation.

REFERENCES

Atlas, E.L., B.A. Ridley, G. Hübler, J.G. Walega, M.A. Carroll, D.D. Montzka, B.J. Huebert, R.B.Norton, F.E. Grahek and S. Schauffler, 1992, Partitioning and budget of NO_y species during the Mauna Loa Photochemistry Experiment, *J. Geophys. Res.*, 97:10,449-10,462.

Drummond, J.W., D.H. Ehhalt and A. Volz, 1988, Measurements of nitric oxide between 0-12 km altitude and 67°N – 60°S latitude obtained during STRATOZ III, *J. Geophys. Res.*, 93:15,831-15,849.

Ehhalt, D.H., F. Rohrer and A. Wahner, 1992, Sources and distribution of NO_x in the upper troposphere at northern midlatitudes, *J. Geophys. Res*, 97:3725-3738.

Galasyn, J.F., K.L. Tschudy and B.J. Huebert, 1987, Seasonal and diurnal variability of nitric acid vapor and ionic aerosol species in the remote free troposphere at Mauna Loa, Hawaii, *J. Geophys. Res.*, 92:13,105-13,113.

Hameed, S. and J. Dignon, 1988, Changes in the geographical distributions of global emissions of NO_x and SO_x from fossil-fuel combustion between 1966 and 1980, *Atmos. Environment*, 22:441-449.

Harris, J.M. and J.D. Kahl, 1990, A descriptive atmospheric transport climatology for the Mauna Loa Observatory using clustered trajectories, *J. Geophys. Res.*, 95:13,651-13,667.

Harris, J. M., P.P. Tans, E.J. Dlugokencky, K. A. Masarie, P.M. Lang, S. Whittlestone and L.P. Steele, 1992, Variations in atmsopheric methane at Mauna Loa Observatory related to long range transport, *J. Geophys. Res.*, 97:6003-6010.

Heikes, B.G., 1992, Formaldehyde and hydroperoxides at Mauna Loa Observatory, *J. Geophys. Res.*, 97:18,001-18,014.

Hübler, G., D.D. Montzka, R.B. Norton, P.C. Murphy, F.C. Fehsenfeld, S.C. Liu, B.A. Ridley, J.G. Walega, E. Atlas, F.E. Grahek, L.E. Heidt, J. Merrill, B.J. Huebert and B.A. Bodhaine, 1992, Total reactive oxidized nitrogen (NO_y) in the remote Pacific troposphere and its correlation with O_3 and CO: Mauna Loa Observatory Photochemistry Experiment 1988, *J. Geophys. Res.*, 97:10,427-10,447.

Houghton, D.D., 1985, "Handbook of Applied Meteorology," Wiley, N.Y.

Liu, S.C., M. Trainer, M.A. Carroll, G. Hübler, D.D. Montzka, R.B. Norton, B.A. Ridley, J.G. Walega, E.L. Atlas, B.G. Heikes, B.J. Huebert and W. Warren, 1992, A study of the photochemistry and ozone budget during the Mauna Loa Observatory Photochemistry Experiment, *J. Geophys. Res.*, 97:10,462-10,471.

Marenco, A. and F. Said, 1989, Meridional aand vertical ozone distribution in the background troposphere (70°N–60°S; 0–12 km altitude) from scientific aircraft measurements during the STRATOZ III experiment (June 1984), *Atmos. Environ.*, 23:201-214.

Merrill, J.T., M. Uematsu and R. Bleck, 1989, Meteorological analysis of long-range transport of mineral aerosol over the North Pacific, *J. Geophys. Res.*, 94:8584-8598.

Miller, J. M., 1981, A five-year climatology of back trajectories from the Mauna Loa Observatory, Hawaii, *Atmos. Environ.*, 15:1553-1558.

Montzka, S.A., R.C. Meyers, J.H. Butler, J.W. Elkins and S.O. Cummings, 1993, Global tropospheric distribution and calibration scale of HCFC-22, *Geophys. Res. Lett.*, 20:703-706.

Moxim, W.J., 1990, Simulated transport of NO_y to Hawaii during August: A synoptic study, *J. Geophys. Res.*, 95:5717-5729.

Oltmans, S., 1985, Tropsospheric ozone at four remote observatories, *in:* "Atmospheric Ozone: Proceedings of the Quadrennial Ozone Symposium," Reidel Publishing, Dordrecht, Netherlands, pgs. 796-802.

Parrington, J.R., W.H. Zoller and N.K. Aras, 1983, Asian dust: Seasonal transport to the Hawaiian Islands, *Science*, 220:195-197.

Ridley, B.A. and E. Robinson, 1992, The Mauna Loa Observatory Photochemistry Experiment, *J. Geophys. Res.*, 97:10,285-10,290.

Ridley, B.A., S. Madronich, R.B.Chatfield, J.G. Walega, R.E. Shetter, M.A. Carroll and D.D. Montzka, 1992, Measurements and model simulations of the photostationary state during the Mauna Loa Observatory Photochemistry Experiment: Implications for radical concentrations and ozone production and loss rates, *J. Geophys. Res.*, 97:10,375-10,388.

Singh, H.B., W. Viezee and L.J. Salas, 1988, Measurements of selected C_2–C_5 hydrocarbons in the troposphere: Latidudinal, vertical and temporal variations, *J. Geophys. Res.*, 93:15,861–15,878.

U.S., 1976, "Standard Atmosphere," U.S. Government Printing Office, Washington, D.C.

Volz, A., D.H. Ehhalt and R.G. Derwent, 1981, Seasonal and latitudinal variation of ^{14}CO and the tropospheric concentration of OH radicals, *J. Geophys. Res.*, 86:5163-5171.

WMO, 1993, "Scientific Assessment of Ozone Depletion, 1991," World Meteorological Organization, Global Ozone Research and Monitoring Project Report No. 25.

Wang, W.C. and N.D. Sze, 1980, Coupled effects of atmospheric N_2O and O_3 on the earth's climate, *Nature*, 286:589-590.

Whittlestone, S., E. Robinson and S. Ryan, 1992, Radon at the Mauna Loa Observatory: Transport from distant continents, *Atmos. Environ.*, 26A:251-260.

GLOBAL EMISSIONS AND MODELS OF PHOTOCHEMICALLY ACTIVE COMPOUNDS

Joyce E. Penner[1], Cynthia S. Atherton[1] and Thomas E. Graedel[2]

[1]Atmospheric and Geophysical Sciences Division
Lawrence Livermore National Laboratory
P.O. Box 808, Livermore, California 94550
[2]AT&T Bell Laboratories
600 Mountain Avenue, P.O. Box 636
Murray Hill, New Jersey, 07974

ABSTRACT

Anthropogenic emissions from industrial activity, fossil fuel combustion, and biomass burning are now known to be large enough, relative to natural sources, to perturb the chemistry of vast regions of the troposphere. A goal of the IGAC Global Emissions Inventory Activity (GEIA) is to provide authoritative and reliable emissions inventories on a $1^{\circ} \times 1^{\circ}$ grid. When combined with atmospheric photochemical models, these high quality emissions inventories may be used to predict the concentrations of major photochemical products. Then, comparison of model results with measurements of pertinent species will allow us to assess whether there are major shortcomings in our understanding of tropospheric photochemistry, the budgets and transport of trace species, and their effects in the atmosphere. Through this activity, we are building the capability to make confident predictions of the future consequences of anthropogenic emissions. This paper compares IGAC recommended emissions inventories for reactive nitrogen and sulfur dioxide to those that have been in use previously. We also present results from the three-dimensional LLNL atmospheric chemistry model that show how emissions of anthropogenic nitrogen oxides might potentially affect tropospheric ozone and OH concentrations, and how emissions of anthropogenic sulfur increase sulfate aerosol loadings.

INTRODUCTION

Two issues have received considerable attention with regard to the potential role of changing anthropogenic emissions to global tropospheric photochemistry. The first issue concerns the oxidizing capacity of the atmosphere and our potential to change this capacity

Global Atmospheric-Biospheric Chemistry
Edited by R.G. Prinn, Plenum Press, New York, 1994

through emissions of methane (CH_4), nitrogen oxides (NO_x), carbon monoxide (CO), and non-methane hydrocarbons (NMHCs). The oxidizing capacity of the atmosphere is determined mainly by the concentration of hydroxyl radicals (OH), although ozone (O_3) and hydrogen peroxide (H_2O_2) also play important roles. The hydroxyl radical reacts with a suite of species (Table 1), initiating photochemical reaction sequences that eventually lead to non-reactive products (e.g., CO_2) or to products that are removed by precipitation and dry deposition. If the OH concentration is decreased globally, the concentrations of gases removed by reaction with OH may increase even with no change in their sources. Hydroxyl radical formation is initiated by photolysis (at wavelengths shorter than about 310 nm) of O_3 which produces $O(^1D)$. $O(^1D)$ reacts with H_2O to form two hydroxyl radicals. Since the concentration of OH in the background atmosphere is determined mainly by the concentrations of CO, CH_4, NO_x, and H_2O, it is important to quantify the anthropogenic and natural emissions of these species to determine their effects on O_3 and OH. Anthropogenic emissions of each are tied to industrialization. Increases in NO_x emissions are particularly important in changing oxidant concentrations because the local concentration of NO_x determines whether photochemical sequences lead to net O_3/OH increases or decreases (Penner, 1989; Crutzen and Zimmerman, 1991). Besides its role as an oxidant, changes to tropospheric ozone are of interest because of its role in OH chemistry and because it acts as a greenhouse gas and has deleterious health and ecosystem effects (e.g., Fishman *et al.*, 1979; Tilton, 1989).

Table 1. Species which are removed or oxidized by reaction with OH. DMS is $(CH_3)_2S$, CCN is cloud condensation nucleus, NMHC is non-methane hydrocarbon, HCFCs are hydrochlorofluorocarbons, and CFCs are chlorofluorocarbons.

Species	Importance
CH_4	Greenhouse gas; Affects tropospheric O_3 and OH; Affects stratospheric O_3
CO	Urban pollutant; Affects tropospheric O_3 and OH cycles
CH_3CCl_3, CH_3Br, CH_3Cl	Greenhouse gases; Release chlorine or bromine in stratosphere which destroys O_3
HCFCs	Greenhouse gases; Replacements for CFCs which release chlorine in the stratosphere and destroy O_3
NMHCs	Urban pollutants; Enhance tropospheric O_3 in high NO_x areas; Sink for tropospheric OH
DMS, H_2S, SO_2	Form sulfate aerosols which contribute to acid rain and reflect solar radiation; Form CCN which alter cloud properties and cool the climate

The second major issue for tropospheric chemistry, which has emerged from a consideration of changing anthropogenic emissions, concerns the formation of anthropogenic aerosols and their potential to mask part of the warming expected from greenhouse gas emissions. Anthropogenic aerosols are composed of a variety of aerosol types (e.g., water-soluble inorganic species, organic condensed species, elemental or black carbon, and mineral dust), but the best-quantified and probably the largest component of the anthropogenic aerosol is that derived through photochemical interactions of anthropogenic SO_2 emissions (Charlson *et al.*, 1992). Global average climate forcing through direct reflection of solar radiation by anthropogenic sulfate aerosols may range from -0.3 to -1.1 Wm^{-2} (Penner *et al.*, 1993b). This forcing varies regionally and acts locally to mask anthropogenic greenhouse gas forcing. The magnitude of the calculated forcing varies depending on the underlying surface albedo, so it is important to estimate correctly the rate of the photochemical processes which convert SO_2 to $SO_4^=$ as the emissions of SO_2 are carried from source regions over areas with high albedo to ocean areas with low albedo. Because these processes are tied to the oxidizing capacity of the atmosphere, through the gas phase reaction of OH with SO_2 and through the aqueous phase reactions of H_2O_2 and O_3 with sulfur in solution, anthropogenic emissions of both NO_x and SO_2, as well as emissions of CH_4 and CO, are needed to calculate the climate forcing from sulfate aerosols accurately.

In this paper, we first describe the emissions inventories under development through the GEIA for anthropogenic NO_x and for anthropogenic SO_2, and we compare these to previously developed inventories. Then, we estimate the effects of anthropogenic NO_x and SO_2 emissions using the Lawrence Livermore National Laboratory (LLNL) three-dimensional (3-D) tropospheric photochemical model. While we have much more to learn in order to quantify the natural sources of sulfur and the nitrogen oxides, model studies in combination with comparison to measurements can be used to diagnose whether our understanding of the sources of these trace gases together with their transport, removal and photochemical processing in the atmosphere is adequate.

GEIA INVENTORIES FOR NO_x AND SO_2

The developing GEIA global inventories for anthropogenic SO_2 and NO_x have as their basis a country-level inventory of fuel combustion emissions compiled in 1985 by J. Dignon following techniques described in Dignon (1992). These emissions were allocated to a $1^\circ \times 1^\circ$ grid on the basis of population distributions and statistical relationships between fossil fuel use and emissions (Dignon and Hameed, 1985). Where available, the Dignon inventories have been replaced by 1985 regional and national inventories. In some cases the regional and national inventories also supply information on area and point sources such as motor vehicle miles traveled or smelter location and activity. The inventories from which information is being incorporated are as follows: Europe, inventory of the European Monitoring and Evaluation Programme (Sandnes and Styve, 1992); North America, inventory of the National Acid Precipitation Assessment Program (Saeger *et al.*, 1989; Wagner *et al.*, 1986); Southeast Asia, inventory of M. Kato and H. Akimoto (Kato and Akimoto, 1992; 1993); Former Soviet Union, inventory of J. Pacyna in consultation with FSU scientists (Pacyna, 1992); Australia, inventory of F. Carnovale (Carnovale, private communication, 1992). The GEIA team for the integrated NO_x and SO_x inventories consists of C. Benkovitz, J. Dignon, T. Graedel, J. Pacyna, T. Scholtz, L. Tarrason, and E. Voldner. While compilation of these new inventories is not yet complete, here we compare the updated GEIA inventories for North America, Europe, Asia, and Australia with those published by Dignon (1992) for 1980 for NO_x and with those published by Spiro *et al.* (1992) and Benkovitz (1982) for ~1980 for SO_2. These latter inventories have served as the basis for previous LLNL model runs to assess the importance of anthropogenic emissions of NO_x and SO_2 to tropospheric oxidant concentrations and to anthropogenic sulfate aerosol forcing of climate, respectively, and are used in the present study for those areas where IGAC inventories are not yet complete.

Figure 1a presents the 1985 IGAC inventory for NO_x for North America and Figure 1b shows differences from the 1980 inventory by Dignon (1992). North American emissions are smaller by at least a factor of 0.3 over much of the area, but larger by more than a factor of 1.5 in others. Summed over the entire area, the emissions from the two inventories are similar, however (6.2 and 6.7 Tg N yr^{-1} for IGAC and Dignon, respectively). Figure 2a presents the 1985 IGAC inventory for Europe while Figure 2b shows differences between the IGAC inventory and that of Dignon. As was the case in North America, this

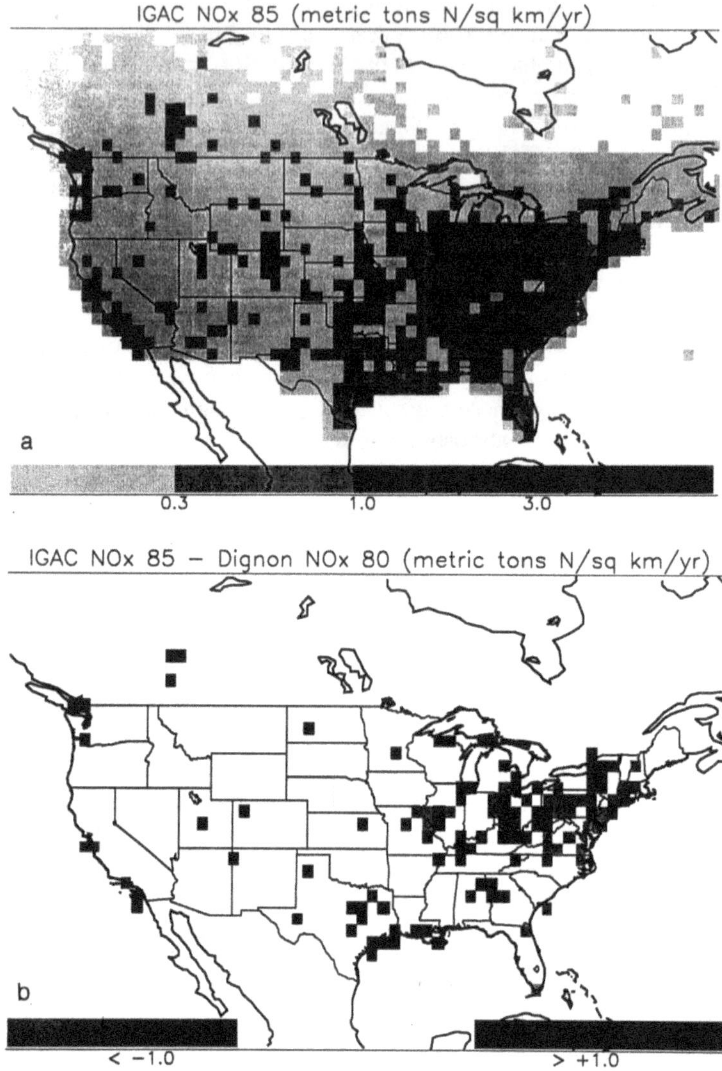

Figure 1. (a, top panel) The GEIA NO_x inventory for North America. The lightest shade shows areas with less than 0.3 tons N km^{-2} yr^{-1} and increasing darkness of shade delineates areas where emissions range from 0.3 to 1.0 tons N km^{-2} yr^{-1}, from 1.0 to 3.0 tons N km^{-2} yr^{-1}, and are greater that 3.0 tons N km^{-2} yr^{-1}, respectively. **(b, bottom panel)** Areas in North America where the difference between the GEIA inventory for NO_x and that developed by Dignon (1992) is greater than 1 ton N km^{-2} yr^{1} (black shade) or less than –1 ton N km^{-2} yr^{1} (lighter shade).

comparison demonstrates a redistribution of emissions, but overall emissions are again similar between the two inventories (5.5 Tg N yr^{-1} and 5.4 Tg N yr^{-1}, respectively). The differences noted in these figures may not cause large differences in photochemical response of the global LLNL model, because the resolution of the emissions used in the current version is prescribed by the resolution of the meteorological fields (which is currently ~4.5° × 7.5°). These emissions updates should be important to developing models that have significantly higher resolution, however. We note that in the new inventory, Asian emissions (not shown) are significantly larger than those from Dignon (1992), except

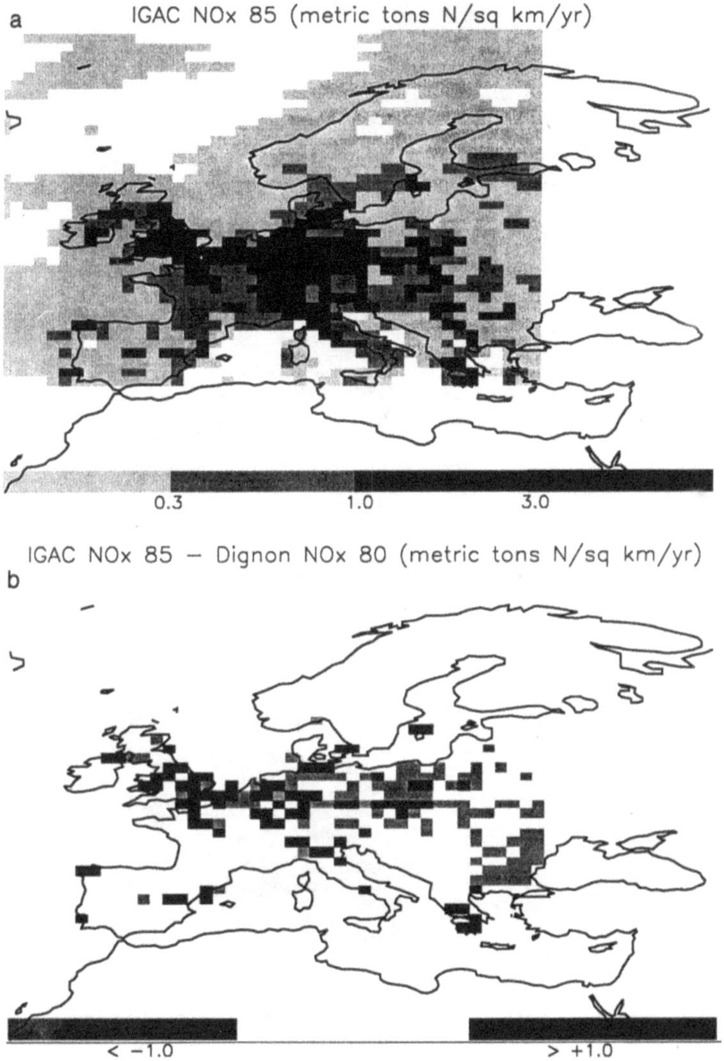

Figure 2. (a, top panel) The GEIA NO_x inventory for Europe. The lightest shade shows areas with less than 0.3 tons N km^{-2} yr^{-1} and increasing darkness of shade delineates areas where emissions range from 0.3 to 1.0 tons N km^{-2} yr^{-1}, from 1.0 to 3.0 tons N km^{-2} yr^{-1}, and are greater that 3.0 tons N km^{-2} yr^{-1}, respectively. **(b, bottom panel)** Areas in Europe where the difference between the GEIA inventory for NO_x and that developed by Dignon (1992) is greater than 1 ton N km^{-2} yr^{1} (black shade) or less than –1 ton N km^{-2} yr^{1} (lighter shade).

in Japan, where they have decreased by between 10 and 30%. Overall emissions in Asia increased from 3.6 to 4.1 Tg N yr^{-1}. Total emissions from Australia are also larger in the new inventory, having increased from 0.16 Tg N yr^{-1} to 0.24 Tg N yr^{-1}. As in other cases, however, the redistribution of emissions in the new inventory means that emissions actually decrease in some regions and increase in others.

Figures 3a and 4a present the IGAC SO_2 inventories for North America and Europe, while Figures 3b and 4b show areas of important differences between the IGAC inventories and those developed by Benkovitz (1982) for North America and by Spiro *et al.* (1992) for

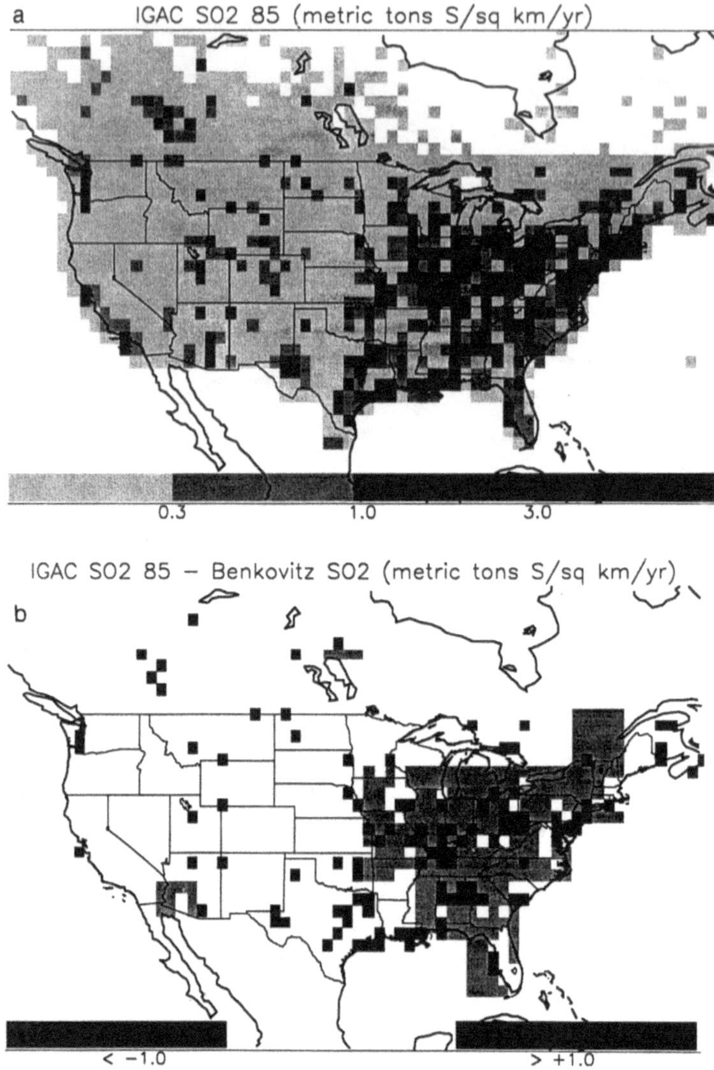

Figure 3. (a, top panel) The GEIA SO_2 inventory for North America. The lightest shade shows areas with less than 0.3 tons S km^{-2} yr^{-1} and increasing darkness of shade delineates areas where emissions range from 0.3 to 1.0 tons S km^{-2} yr^{-1}, from 1.0 to 3.0 tons S km^{-2} yr^{-1}, and are greater than 3.0 tons S km^{-2} yr^{-1}, respectively. **(b, bottom panel)** Areas in North America where the difference between the IGAC inventory for SO_2 and that developed by Benkovitz (1982) is greater than 1 ton S $km^{-2} yr^{-1}$ (black shade) or less than −1 ton S $km^{-2} yr^{-1}$ (lighter shade).

Europe. In Figure 3b, for SO_2, North American emissions in the Benkovitz (1982) inventory refer to calendar year 1975 for area sources and to 1977–1978 for point sources and have a resolution of 5° × 5°, whereas the IGAC inventory was derived from inventories developed by NAPAP for calendar year 1985 (Saeger *et al.*, 1989). The total SO_2 emissions in the Benkovitz (1982) inventory for ~1975 is similar to that developed by NAPAP for North America for 1980 (Spiro *et al.*, 1992). According to these inventory estimates, emissions of SO_2 may have decreased by 25%, from 16 Tg S yr^{-1} to 12 Tg S yr^{-1}, over this ~10 year time period. European emissions in the Spiro *et al.* (1992) inventory

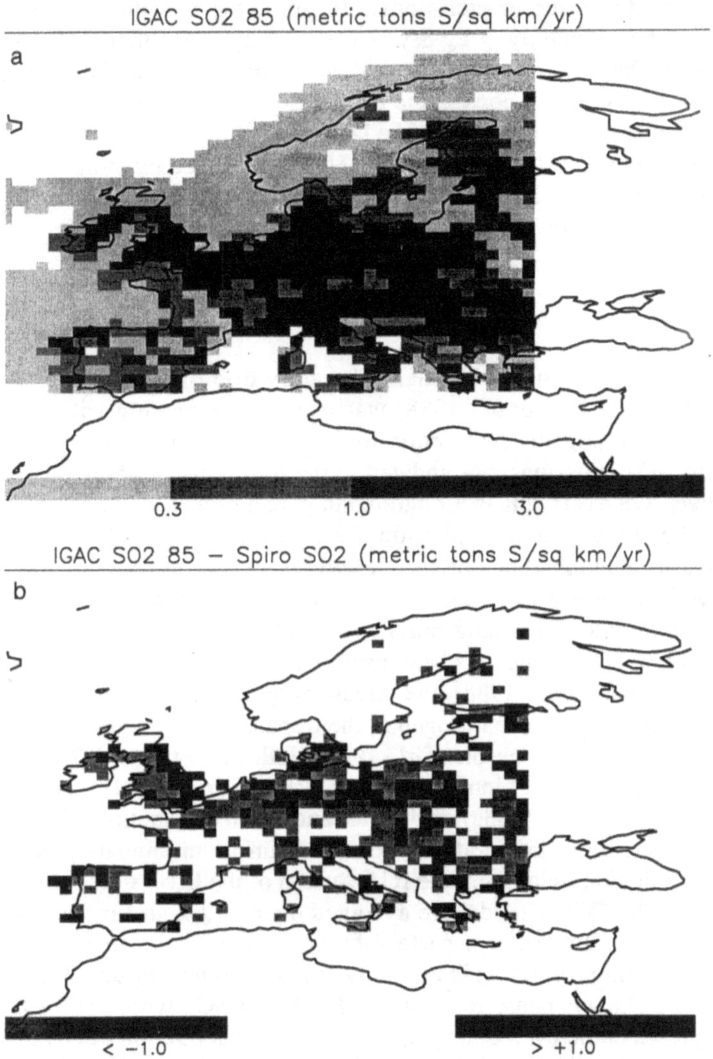

Figure 4. (a, top panel) The GEIA SO_2 inventory for Europe. The lightest shade shows areas with less than 0.3 tons S km^{-2} yr^{-1} and increasing darkness of shade delineates areas where emissions range from 0.3 to 1.0 tons S km^{-2} yr^{-1}, from 1.0 to 3.0 tons S km^{-2} yr^{-1}, and are greater that 3.0 tons S km^{-2} yr^{-1}, respectively. **(b, bottom panel)** Areas in Europe where the difference between the GEIA inventory for SO_2 and that developed by Spiro *et al.* (1992) is greater than 1 ton S km^{-2} yr^{1} (black shade) or less than –1 ton S km^{-2} yr^{1} (lighter shade).

were derived from an OECD inventory (Lubkert and De Tilly, 1989) for 1980, while those in the IGAC inventory were developed by the Cooperative Program for Monitoring and Evaluation of the Long Range Transmission of Air Pollutants in Europe (EMEP) for calendar year 1985 (Sandnes and Styve, 1992). According to these estimates, European SO_2 emissions may have decreased by 14%, from 21 Tg S yr^{-1} to 18 Tg S yr^{-1}, over this 5 year time period. The IGAC inventories for Asia and Australia also have decreased in the new inventory. The new emissions are 13 Tg S yr^{-1} and 1.0 Tg S yr^{-1} for Asia and Australia, respectively, smaller than those developed by Spiro *et al.* (1992) by 13% and 23%, respectively. The decreases for these regions may reflect differences in methodology rather than real decreases in emissions. Overall, the new inventories imply reduced emissions of approximately 9 Tg S yr^{-1} (or slightly more than 15%) from these four regions. It is not known whether decreased emissions should apply for other areas. Spiro *et al.* (1992) argue that emissions in the former Soviet Union decreased by almost 33% or 4.1 Tg S yr^{-1} between 1980 and 1989 (the latter year is represented in their inventory). We note that fossil fuel use in the former Soviet Union increased by more than 27% from 1980 to 1987 according to the statistics available through the United Nations. Total world wide fossil fuel use increased by almost 13% over this time period.

LLNL MODEL

The Lawrence Livermore National Laboratory tropospheric chemistry model (called GRANTOUR, cf., Walton *et al.*, 1988) was used in conjunction with the Dignon (1992) inventory for anthropogenic NO_x emissions and the Spiro *et al.* (1992) inventory for anthropogenic SO_2 emissions, but updated in the four regions where quality data for 1985 were available. GRANTOUR is a Lagrangian parcel model which may be run either off-line, using the wind and precipitation fields from a general circulation model, or interactively, in a mode that allows alterations of the wind and precipitation fields consistent with currently calculated species or aerosol concentrations. The model is typically run with 50,000 constant-mass air parcels whose dimensions average 100 mb × 330 km × 330 km. The model has been used to study the effects of smoke from a nuclear war on climate (Ghan *et al.*, 1988), the effects of an asteroid impact on climate (Covey *et al.*, 1988), the cycle of reactive nitrogen in the troposphere (Penner *et al.*, 1991), the sulfur cycle in the troposphere (Erickson *et al.*, 1991; Galloway *et al.*, 1992), climate forcing by smoke aerosols produced in biomass burning (Penner *et al.*, 1991), black carbon aerosol distributions (Penner *et al.*, 1993a), and ^{222}Rn and ^{210}Pb distributions (Dignon *et al.*, 1993).

At the present time, the model is normally run in an annual-cycle mode using the winds and precipitation fields from the R15 version of the Livermore/NCAR CCM1 model. The winds from the CCM1 model are averaged over a 12-hour time period for use by the LLNL GRANTOUR model. For surface-based sources, trace species are input into the model with a vertical profile that is normally assumed to be constant in mixing ratio in the lowest 100 mb. The gas phase reactions of DMS and SO_2 with OH for the model of the sulfur cycle presented below assumed a background concentration of OH that was specified according to the latitudinally- and seasonally-varying calculated concentrations from the LLNL 2-D model (cf., Penner *et al.*, 1991), while in our "oxidant" model these were calculated based on the reactions given in Tables 2 and 3. Additionally, in the oxidant model CO concentrations were fixed but varied monthly, with surface concentrations in remote areas taken from the literature (cf., Atherton, 1993) and with values increased in grid squares where fossil fuel emissions and biomass burning take place. The increase was proportional to the emissions of CO_2 from these activities in each grid square. The variation of CO with altitude and the maximum CO concentration were prescribed based on observations. Surface CH_4 concentrations were fixed according to the observations of Steele *et al.* (1987) and the variation of CH_4 concentration with altitude followed the predictions of the LLNL 2-D model. In addition, the concentration of O_3 above 200 mb was specified on a monthly basis according to results from the LLNL 2-D model scaled to simulate ozonesonde data (Komhyr *et al.*, 1989) more accurately. The H_2O concentration varied monthly according to the predicted monthly-averaged concentrations from the climate model. These latter concentrations have been compared to monthly observations of H_2O (Oort, 1983) and are, in general, between a factor of 1.0 to 1.5 high at the surface. They are up to a factor of 2 lower than observations in the middle and upper troposphere.

Table 2. Reactions and rate coefficients used in the model calculations ($cm^3 s^{-1}$).

Reaction	Rate coefficient at ground level and 298K
$O + O_2 \xrightarrow{M} O$	1.56×10^{-14}
$O_3 + NO \rightarrow NO_2 + O_2$	1.82×10^{-14}
$O + NO_2 \rightarrow NO + O_2$	9.72×10^{-12}
$O(^1D) + H_2O \rightarrow OH + OH$	2.20×10^{-10}
$O_3 + OH \rightarrow HO_2 + O_2$	6.83×10^{-14}
$O_3 + HO_2 \rightarrow OH + O_2 + O_2$	2.05×10^{-15}
$HO_2 + HO_2 \rightarrow H_2O_2 + O_2$	2.97×10^{-12}
$HO_2 + OH \rightarrow H_2O + O_2$	1.11×10^{-10}
$OH + NO_2 \xrightarrow{M} HNO_3$	1.17×10^{-11}
$OH + HNO_3 \rightarrow H_2O + NO + O_2$	1.62×10^{-14}
$H_2O_2 + OH \rightarrow H_2O + HO_2$	1.70×10^{-12}
$NO + HO_2 \rightarrow NO_2 + OH$	8.56×10^{-12}
$NO_2 + O_3 \rightarrow NO_3 + O_2$	3.23×10^{-17}
$CO + OH \xrightarrow{O_2, M} HO_2 + CO_2$	2.42×10^{-13}
$O(^1D) + XM \rightarrow O + XM$	2.89×10^{-11}
$NO_2 + NO_3 \xrightarrow{M} N_2O_5$	1.27×10^{-12}
$N_2O_5 \xrightarrow{M} NO_2 + NO_3$	3.73×10^{-02}
$N_2O_5 + H_2O \rightarrow HNO_3 + HNO_3$	5.00×10^{-22}
$HO_2 + HO_2 + H_2O \rightarrow H_2O_2 + O_2 + H_2O$	6.68×10^{-30}
$HO_2 + NO_2 \xrightarrow{M} HNO_4$	1.42×10^{-12}
$HNO_4 \xrightarrow{M} HO_2 + NO_2$	8.80×10^{-02}
$OH + HNO_4 \rightarrow H_2O + NO_2 + O_2$	4.65×10^{-12}
$OH + CH_4 \xrightarrow{O_2, M} CH_3O_2 + H_2O$	6.46×10^{-15}
$CH_3O_2 + NO \rightarrow CH_2O + NO_2$	7.68×10^{-12}
$CH_3O_2 + HO_2 \rightarrow CH_3OOH + O_2$	5.57×10^{-12}
$OH + CH_2O \xrightarrow{O_2} CO + HO_2 + H_2O$	1.00×10^{-11}
$OH + CH_3OOH \rightarrow CH_3O_2 + H_2O$	5.28×10^{-12}
$OH + CH_3OOH \rightarrow CH_2O + H_2O + OH$	2.15×10^{-12}
$HONO + OH \rightarrow H_2O + NO_2$	4.86×10^{-12}
$OH + NO \xrightarrow{M} HONO$	4.96×10^{-12}
$OH + HNO_3 \rightarrow H_2O + NO_2 + O$	1.31×10^{-13}
$DMS + OH \rightarrow SO_2$ (multi-step)	6.13×10^{-12}
$SO_2 + OH \rightarrow SO_4^{=}$ (multi-step)	8.89×10^{-13}

The reaction rate coefficients are given in Table 2 and are consistent with recent recommendations by DeMore *et al.* (1992) and Atkinson *et al.* (1992). They were varied monthly according to the average temperature and pressure at each location in the model. Factors which account for the diurnal averaging of reaction rates were taken from the LLNL 2-D model as were the photolysis rate coefficients for the reactions in Table 3. (A description of the LLNL 2-D model appears in Johnston *et al.*, 1989; Wuebbles and Kinnison, 1989; and Wuebbles *et al.*, 1993). Aqueous reaction of SO_2 to form $SO_4^=$ in clouds was treated in a simplified manner. This process was assumed to have an average e-folding lifetime of 30 hours at 40°N at the surface in summer. The lifetime was scaled proportionately to the square of the ratio of the OH concentration at 40°N and the locally specified concentration of OH at other locations. This formulation is a simplified attempt to account for the observed seasonal variations in sulfate concentrations and wet deposition in North America, by assuming that the rate of formation of $SO_4^=$ in clouds depends both on the frequency of interception of a cloud by an air parcel (average cloud-free periods vary from 10 to 80 hours; Lelieveld and Crutzen, 1990, 1991) and on the local concentration of H_2O_2 (assumed to be proportional to square of the specified concentration of OH).

In addition to the photochemical reactions in the model, each species may experience deposition if air parcels are in the lowest 100 mb. The deposition velocities are set by average observed values but may vary from ocean to land surfaces. In addition, trace species may be scavenged by precipitation with scavenging rates proportional to the amount of precipitation in the model (expressed as cm hr^{-1}) and scavenging coefficients set to reproduce measured washout ratios. The values used in the present simulations are shown in Table 4. Other details regarding the model formulation can be found in Walton *et al.* (1988), Penner *et al.* (1991), and Atherton (1993).

Table 3. Photolysis reactions in the model.

$O_3 \rightarrow O + O_2$	$CH_3OOH \xrightarrow{O_2} CH_2O + HO_2 + OH$
$O_3 \rightarrow O(^1D) + O_2$	$CH_2O \xrightarrow{O_2,M} CO + HO_2 + HO_2$
$NO_2 \rightarrow NO + O$	$CH_2O \rightarrow CO + H_2$
$HNO_3 \rightarrow OH + NO_2$	$HNO_4 \rightarrow OH + NO_2 + O$
$H_2O_2 \rightarrow OH + OH$	$HONO \rightarrow OH + NO$
$HO_2 \rightarrow OH + O$	$NO_3 \rightarrow NO + O_2$
$N_2O_5 \rightarrow NO_2 + NO_2 + O$	$NO_3 \rightarrow NO_2 + O$
$N_2O_5 \rightarrow NO_2 + NO + O_2$	$HNO_4 \rightarrow OH + NO + O_2$

MODEL PREDICTION OF TROPOSPHERIC SULFATE AEROSOL

Model calculations of the tropospheric sulfate aerosol abundance require reactive sulfur emissions from five main sources: production of (mainly) dimethylsulfide (DMS) in the oceans by phytoplankton, volcanic emissions of sulfur dioxide (SO_2), terrestrial emissions of DMS and hydrogen sulfide (H_2S) from soils, forests and crops, biomass burning emissions of SO_2, and fossil fuel and industrial emissions of SO_2. Table 5 gives a summary of the magnitude of the reactive sulfur sources that were used in the calculations reported here and the range in possible emissions magnitudes as reported by a number of authors (Graedel *et al*., 1993; Penner, 1993). In all cases, the inventories derived by Spiro *et al.* (1992) were used, except for the anthropogenic emissions which have been updated as noted above and except for the ocean emissions of DMS. For ocean emissions the inventory from Spiro *et al.* (1992) was doubled. The doubled emissions reflect the mid-range of estimates shown in Table 5 and also provide a better comparison with remote measurements of $SO_4^=$ (shown in Table 6).

As noted above, we expect that as emission inventories evolve and become more accurate, model comparison with data will provide a stringent test of whether the model correctly summarizes the effects of transport, photochemical transformation, and deposition. The adequacy of the current model may be judged from the comparisons presented below.

Table 4. Deposition velocities and precipitation scavenging coefficients used in the model†.

Species	Deposition velocity ($cm\ s^{-1}$)	Large-scale precipitation scavenging coefficient ($cm\ hr^{-1}$)	Convective precipitation scavenging coefficient ($cm\ hr^{-1}$)
O_3	0.6 (land)		
	0.06 (ocean)	0.0	0.0
NO	0.1	0.0	0.0
NO_2	0.5 (land)		
	0.1 (ocean)	0.0	0.0
H_2O_2	1.0	2.4	4.7
HNO_3	1.0	2.4	4.7
NO_3	0.4	0.0	0.0
N_2O_5	0.4	1.0	2.0
CH_2O	0.2	2.0	4.0
CH_3OOH	1.0	2.0	4.0
HONO	0.0	1.5	3.0
SO_2	0.8	0.8	1.5
$SO_4^=$	0.1	5.0	1.5

†These parameters were set to 0.0 for O, $O(^1D)$, OH, HO_2, HNO_4, CH_3O_2, and DMS.

Table 5. Global sources of reactive sulfur used in the model and an estimate of their possible range ($Tg\ S\ yr^{-1}$).

Source	Total	Range
Fossil fuel and industry	66	60 – 100
Biomass burning	2.2	1 – 4
Terrestrial soils, forests, crops	0.9	0.1 – 10
Oceans	23.7	12 – 40
Volcanoes		3 – 20
Degassing	3.4	
Eruptive	6.4	

Table 6. Observations and calculated concentrations of nss $SO_4^=$ (pptv).

Location	Observation	Model	Location	Observation	Model
Northern Hemisphere			Southern Hemisphere		
Canadian Arctic	210	22	Am. Samoa	84	75
Spetsbergen	168	95	New Caledonia	98	105
Bermuda	455	294	Norfolk Island	56	102
Midway	133	81	Cape Grim	21	112
Oahu	119	109	Mawson	21	25
Barbados	175	139	South Pole	20	6
Guam	119	86			
Belau	147	107			
Fanning	154	122			

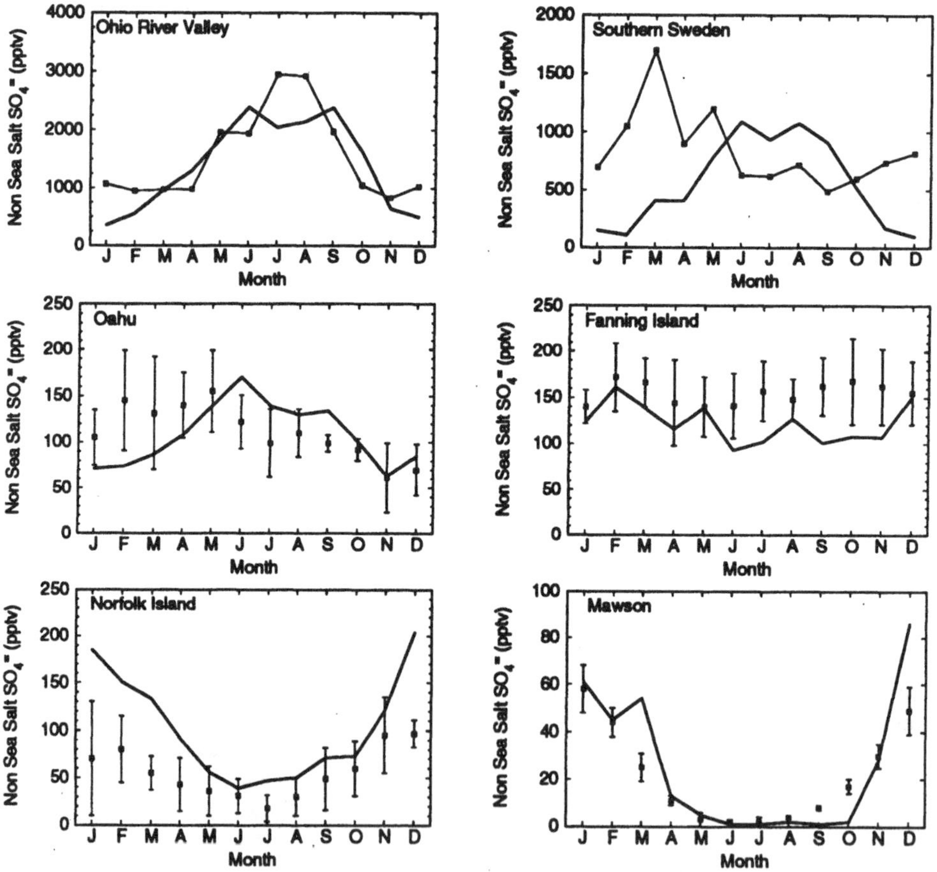

Figure 5. Model calculated (smooth line) and observed (squares) annual cycle of non-seasalt sulfate (in pptv) at selected locations throughout the year. Observations by several investigators have been summarized by Langner and Rodhe (1991).

Figure 5 shows a comparison of the predicted monthly averaged concentrations of $SO_4^=$ at selected locations with data summarized by Langner and Rodhe (1991). Note that some of the data appear to be represented quite well (e.g., $SO_4^=$ over the Ohio River Valley, U.S.), while other data are poorly predicted ($SO_4^=$ over southern Sweden). The seasonal changes seen for $SO_4^=$ are mainly a result of the seasonality of the $SO_4^=$ source. Both the source derived from reaction of SO_2 with OH and the source derived from the aqueous transformation rate of SO_2 to $SO_4^=$ vary strongly with season.

Table 6 compares predicted and observed concentrations of $SO_4^=$ at the set of remote locations compiled by Langner and Rodhe (1991; Table 5 in Langner and Rodhe, 1991). The predicted concentrations appear to be somewhat low in most remote locations except for three regions in the southern hemisphere, where concentrations are high due to poor resolution in the model near coastal areas. Table 7 compares predicted and observed deposition of sulfate at remote locations while Figures 6 and 7 compare predicted and observed deposition in the important source areas of North America and Europe. In July, the predicted wet deposition compares favorably to observed wet deposition in both source areas, while in January, wet deposition is overpredicted in North America by about a factor of 1.5 and under predicted in Europe by about the same factor. Remote area deposition predictions (shown in Table 7) appear reasonable relative to observations.

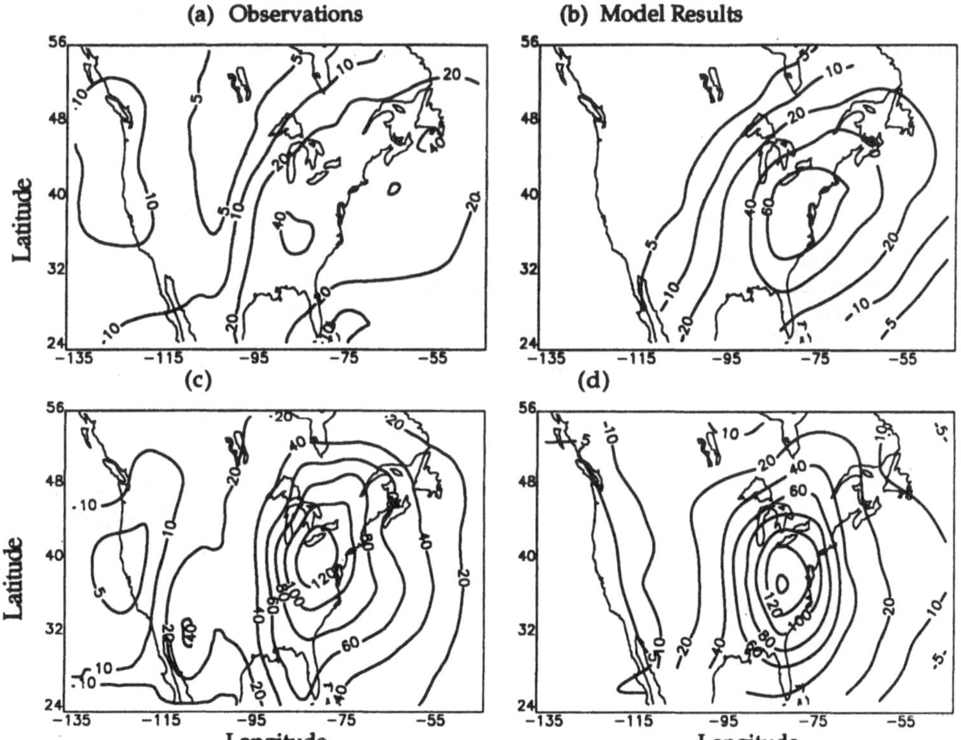

Figure 6. Observed and calculated wet deposition of sulfate over North America. The observations shown are an average from several years of the NADP monthly observations for January and for July for the U.S. and Canada (Watson and Olsen, 1984). These were averaged and interpolated to the model grid. Panel **(a)** gives observations and **(b)** model calculations for January. Panel **(c)** gives observations and **(d)** model calculations for July. Contour intervals are 5, 10, 20, 40, 60, 80, 100, 120, 140, and 160 kg S km^{-2} month^{-1}.

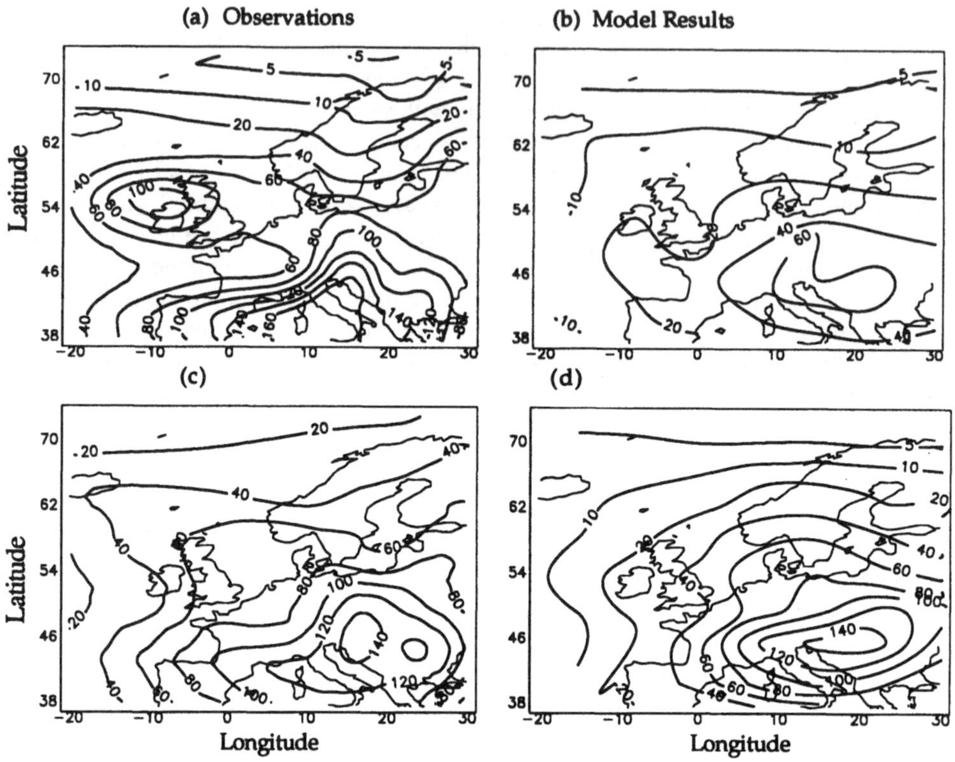

Figure 7. Observed and calculated wet deposition of sulfate over Europe. The observations shown are an average from several years of EMEP monthly observations for January and for July (J. Schaug, Norwegan Institute of Meteorological Studies, private communication, 1991). These were averaged and interpolated to the model grid. Panel (a) gives observations and (b) model calculations for January. Panel (c) gives observations and (d) model calculations for July. Contour intervals are 5, 10, 20, 40, 60, 80, 100, 120, 140, and 160 kg S km^{-2} $month^{-1}$.

Table 7. Observed and calculated non-seasalt sulfur wet deposition at regional and remote sites (kg S km^{-2} yr^{-1}). Observations by several investigators as referenced in Langner and Rodhe (1991).

Location	Observation	Model
Northern Hemisphere		
Poker Flat, Alaska	32	45
Weathership Lima	152	131
Bermuda	450	257
San Carlos, Venezuela	170	141
East Africa	150	82

Location	Observation	Model
Southern Hemisphere		
Lake Calado, Brazil	120	124
Am. Samoa	66	63
Katherine, N. Australia	64	47
Amsterdam Island	100	49
New Zealand	67	92
Macquarie Island	37	33
South Pole	1	2

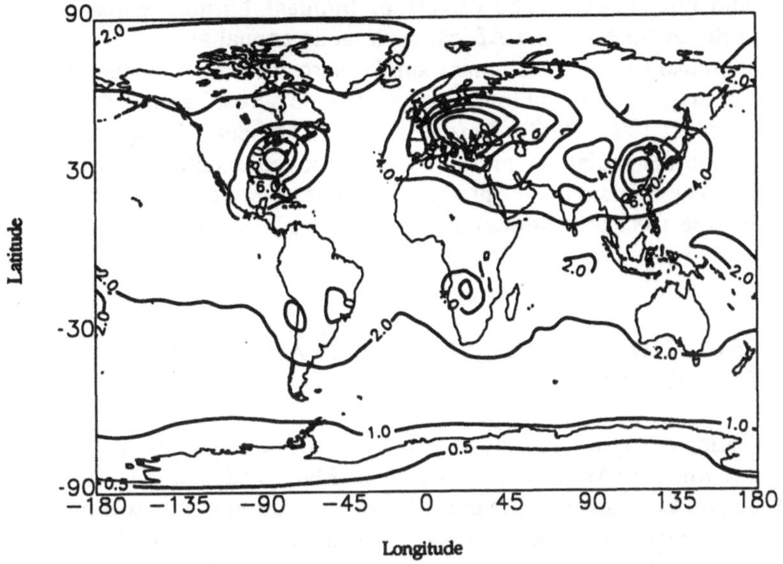

Figure 8. Calculated annual-average column abundance of aerosol sulfate (mg m^{-2}). Contour intervals are 0.5, 1.0, 2.0, 4.0, 6.0, 8.0, 10.0, 12.0, and 14.0 mg m^{-2}.

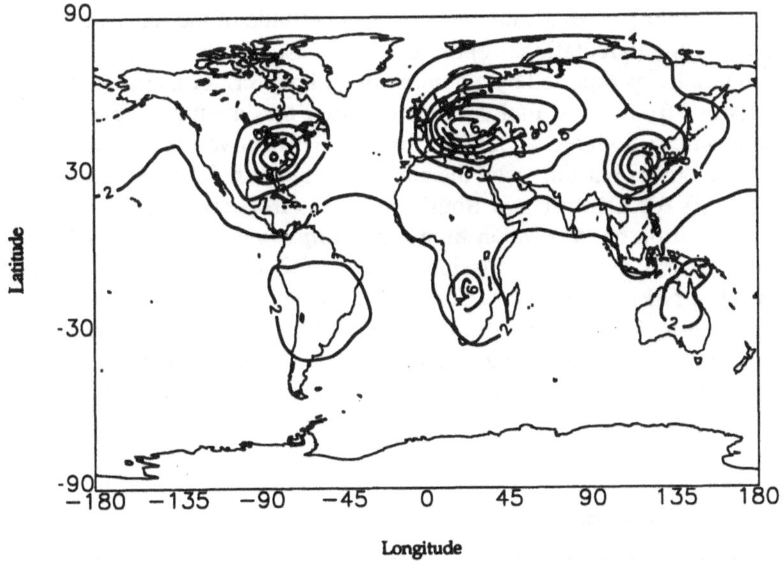

Figure 9. Ratio of the annual-average column abundance of aerosol sulfate for all sources of reactive sulfur (fossil fuel, biomass burning, terrestrial soils, forest, crops, degassing volcanoes and oceans) and for natural sources only (terrestrial soils, forest, crops, degassing volcanoes and oceans). Contour intervals are 2, 4, 6, 8, 10, 12, 14, and 16.

The model may be used to assess the relative importance of anthropogenic emissions in determining the concentrations of sulfate aerosol. Figure 8 shows the calculated annual average total sulfate aerosol column abundance, while Figure 9 shows the ratio of the predicted sulfate aerosol column abundance for all sources and for natural sources only. Areas of strong anthropogenic (mainly fossil fuel) emissions in North America, Europe, and Asia stand out clearly. The effects of biomass burning emissions of SO_2 are discernible in South America and Africa, even on an annual average basis. We note that the source of (mostly organic) biomass smoke aerosols have not been included in this simulation, but may almost double the anthropogenic aerosol abundance expected in the model (Penner *et al.*, 1991; 1992). This version of the model is being coupled to a climate model to evaluate the radiative forcing by anthropogenic sulfur emissions (Penner and Taylor, 1993). The climate model predicts a climate forcing from anthropogenic sulfate aerosols (using the 1980 inventories) of approximately -0.9 Wm^{-2} on an annual average basis.

MODEL PREDICTIONS OF THE EFFECTS OF ANTHROPOGENIC NO_x EMISSIONS ON O_3 AND OH

The formation of O_3 in the troposphere results from oxidation of CO, CH_4, and non-methane hydrocarbons (NMHCs) in the presence of NO_x and sunlight. As noted above, we have fixed the concentrations of CO and CH_4 in the present model while NMHCs are not yet included. Model calculations of the tropospheric O_3 and OH abundance require reactive nitrogen emissions from five main sources: fossil fuel combustion, biomass burning (including wood fuel burning), soil microbial activity, lightning discharges, and production of reactive nitrogen in the stratosphere by reaction of $O(^1D)$ with N_2O. A separate category of fossil fuel emissions from aircraft is considered because of its injection at high altitude. Table 8 gives a summary of the magnitude of the reactive nitrogen sources that were used in the calculations reported here. Also shown are the range in possible emissions magnitudes as reported by a number of authors (cf., Penner, 1993).

The emissions from fossil fuel combustion used the inventory developed by Dignon (1992) except where updated according to the GEIA inventories noted above. The emissions from biomass burning used the amount of biomass burned monthly compiled by Hao *et al.* (1989) for Central and South America, tropical Africa and tropical Asia. Additional information on Australian biomass burning was compiled by Dignon (private communication). We assumed emission factors of 2.5 g N kg^{-1} dry matter for forest burning and 1.9 g N kg^{-1} dry matter for savanna burning based on the analysis of fuel nitrogen published by Dignon and Penner (1991). The aircraft source was injected on an annual average basis and used a 1989 inventory compiled by the Warren Spring Laboratory (1989). This source was distributed vertically according to prevailing flight corridors and altitudes.

Table 8. Sources of reactive nitrogen (Tg N yr^{-1}).

Source	Total	Range
Anthropogenic		
Fossil fuel burning	22.4	15 – 25
Biomass burning (land clearing)	10	2 – 20
Aircraft emissions	0.23	0.1 – 0.7
Wood burning	0.2	0.1 – 0.8
Natural		
Soil microbial emissions	5	1 – 20
Lightning discharge	10	2 – 100
Production in the stratosphere	1.0	0.5 – 1.5
Total source strength	48.8	20 – 168

The natural sources of reactive nitrogen were each varied on a monthly basis. The total annual NO_x emission for cloud to cloud and cloud to ground lightning discharges was adjusted to 10 Tg yr^{-1} based on previous analysis and fitting to observed nitrate deposition and nitric acid concentrations (Atherton *et al.*, 1991). The distribution of source intensity was based on observed thunder days (Handbook of Geophysics, 1960) and used the methodology of Hameed *et al.* (1981). NO_x emissions from soil activity used an inventory compiled by Dignon *et al.* (1991) and were based on relationships which accounted for vegetation type, soil moisture and soil temperature.

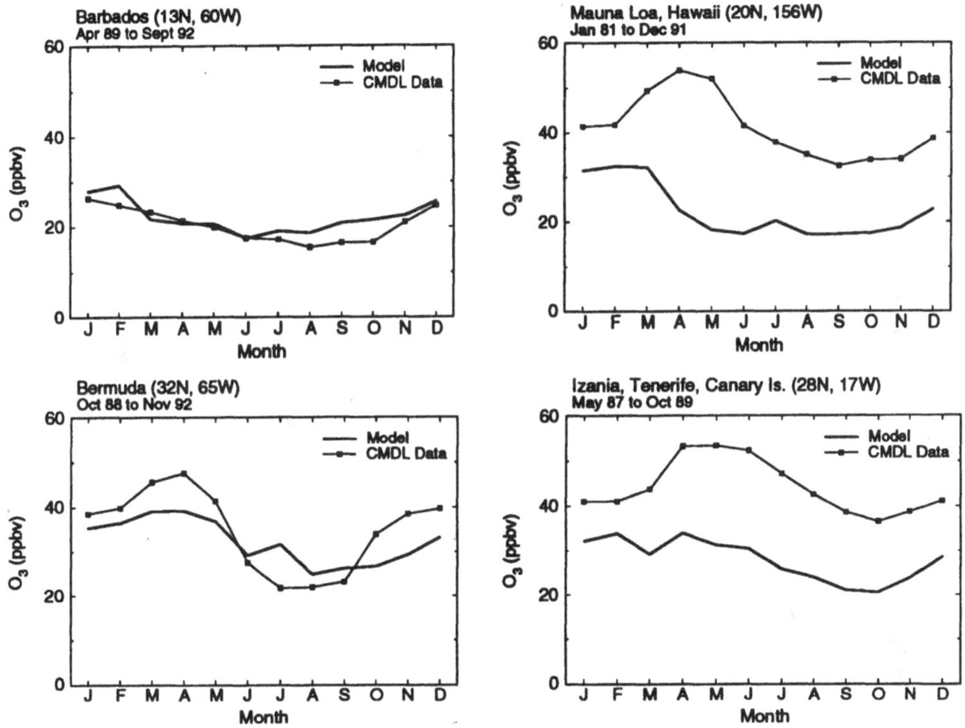

Figure 10. Observed (squares) and calculated (smooth line) monthly mean ozone concentrations at the surface at Barbados, Mauna Loa, Bermuda, and Izania. Observations are from the NOAA Climate Monitoring and Diagnostics Laboratory (S. Oltmans, private communication, 1993).

The adequacy of the current model may be judged by comparison of model predictions with measurements of air concentration and wet deposition. Comparisons for wet deposition of nitrate are only slightly high relative to observations in North America in summer and are 25% lower than observations in winter. Model predictions of nitrate at remote Pacific Ocean sites are improved relative to those presented earlier as well (cf., Penner *et al.*, 1991). However, free tropospheric concentrations are still too high. Recent observations of nitrogen compounds provide support for a substantial fractional abundance of peroxyacetyl nitrate (PAN) which is not calculated in the current simulation. Future simulations which include nonmethane hydrocarbon sources will allow a more robust calculation of the nitrogen cycle.

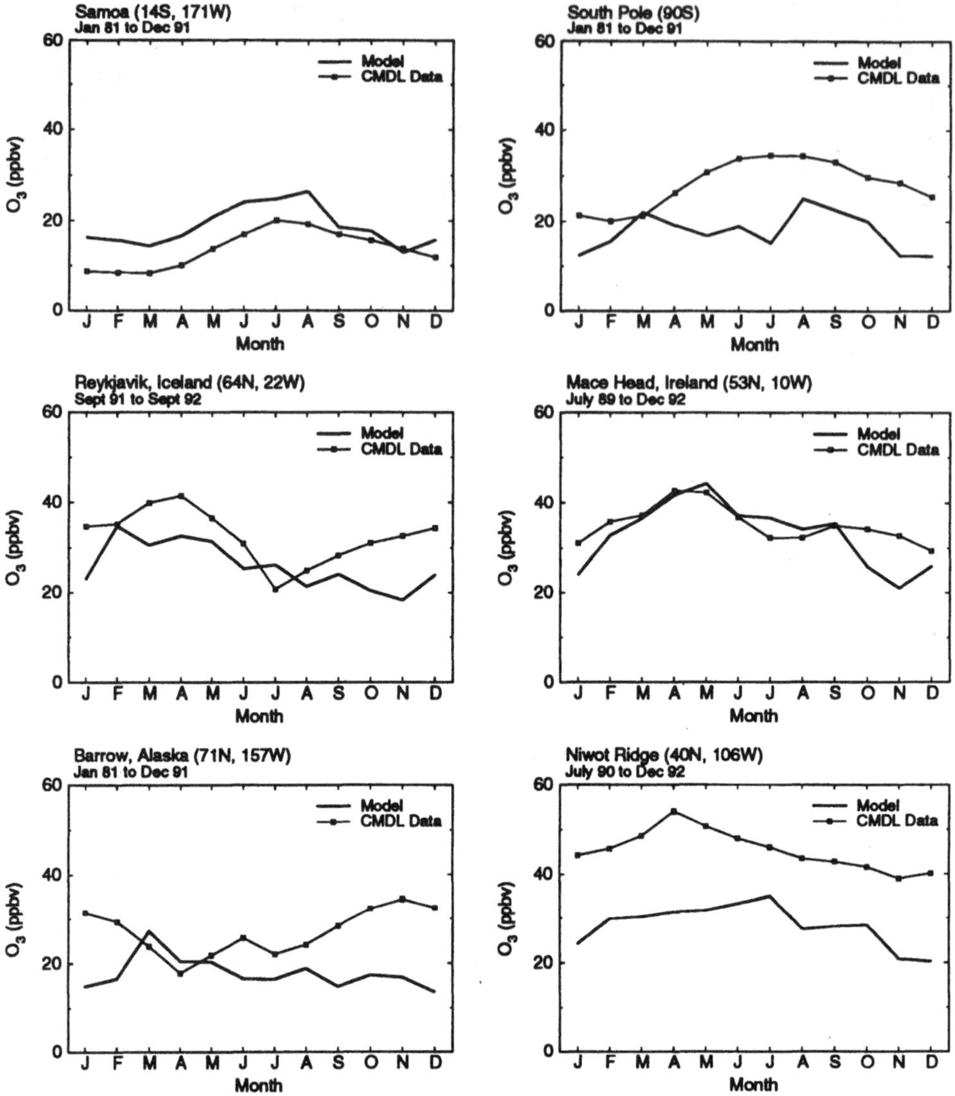

Figure 10 (continued). Observed (squares) and calculated (smooth line) monthly mean ozone concentrations at the surface at Samoa, the South PoleReykjavik, Mace Head, Barrow and Niwot Ridge. Observations are from the NOAA Climate Monitoring and Diagnostics Laboratory (S. Oltmans, private communication, 1993).

Calculations for O_3 may be compared to monthly mean surface concentrations and to O_3 profiles compiled by the NOAA Climate Monitoring and Diagnostics Laboratory (Oltmans, private communication, 1993; Oltmans and Levy, 1993; Komhyr *et al.*, 1989). Figure 10 compares the predicted monthly mean surface O_3 concentration to data from Reykjavik (64°N, 22°W), Mace Head (53°N, 10°W), Barrow (71°N, 157°W), Niwot Ridge (40°N, 106°W), Barbados (13°N, 60°W), Mauna Loa (20°N, 156°W), Bermuda (32°N, 65°W), Izania (28°N, 17°W), Samoa (14°S, 171°W) and near the South Pole (90°S, 25°W). At four of the sites (Barbados, Bermuda, Mace Head and Samoa) the agreement for both total abundance and seasonal cycle is quite good. However, at six sites (Barrow, Reykjavik, Mauna Loa, Izania, Niwot Ridge, and the South Pole), model values average about 15 ppbv less than observations. The lack of agreement may be partly due to dynamics (model predictions at Barrow, for example, are poor in spring, even for species that do not undergo photochemical processing [cf., Penner *et al.*, 1993a]). Part of the under prediction, however, may be caused by the fact that the current model does not treat NMHCs and the resulting production of PAN.

Model predicted profiles of O_3 partial pressures are compared to ozonesonde observations in Figure 11. Figure 11 shows the annual average observed and model predicted profiles between the surface and 200 mbars at five northern hemisphere sites (Resolute [74°N, 95°W], Barrow [71°N, 157°W], Edmonton [53°N, 114°W], Boulder

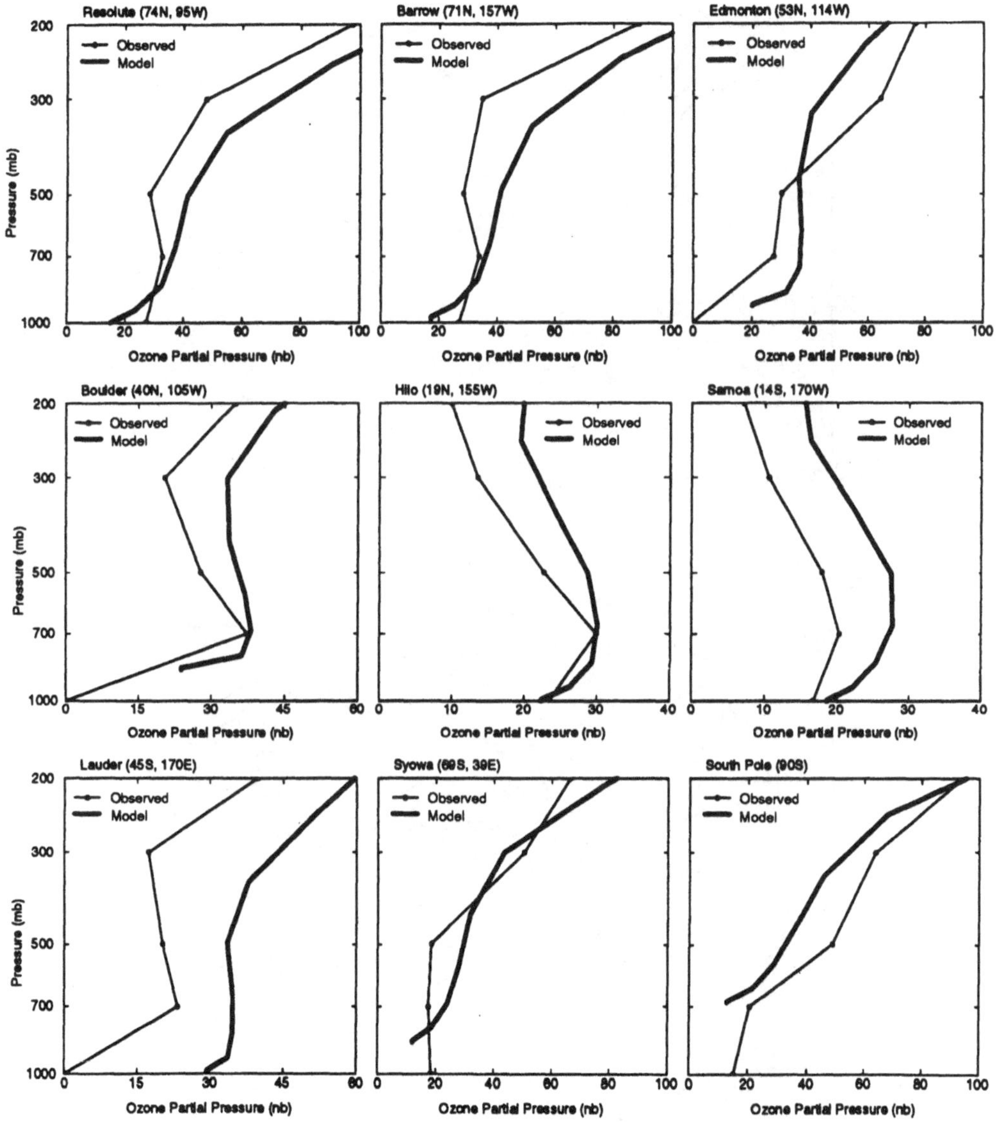

Figure 11. Observed and calculated tropospheric ozone partial pressures as a function of altitude at Resolute (74°N, 95°W), Barrow (71°N, 157°W), Edmonton (53°N, 114°W), Boulder (40°N, 105°W), Hilo (19°N, 155°W), Samoa (14°S, 170°W), Lauder (45°S, 170°E), Syowa (69°S, 39°E) and South Pole (90°S, 25°W) (Komhyr *et al.*, 1989).

[40°N, 105°W], and Hilo [19°N, 155°W]) and at four southern hemisphere sites (Samoa [14°S, 170°W], Lauder [45°S, 170°E], Syowa [69°S, 39°E] and near the South Pole [90°S, 25°W]). Values above 200 mbars were adjusted from the LLNL 2-D model to match the measured concentrations approximately. As shown in Figure 11, the increase in O_3 with latitude and altitude is captured by the model, but, overall, predicted concentrations are higher than observations in the free troposphere.

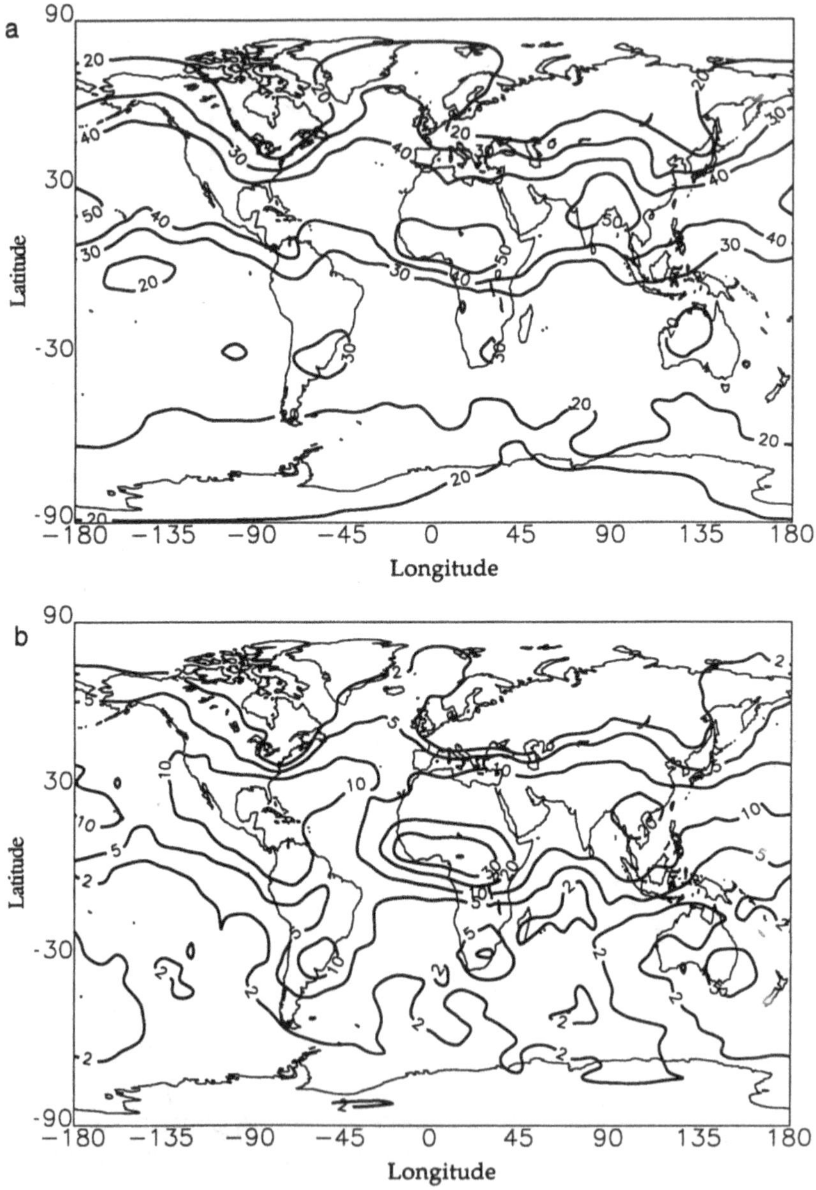

Figure 12. (a, top panel) Calculated ozone concentrations at the surface in January with all sources of reactive nitrogen emissions included. **(b, bottom panel)** The difference between a calculation with all sources and with natural sources only. Contour intervals are 20, 30, 40, and 50 ppbv for top panel, and 2, 5, 10, 20, and 30 ppbv for bottom panel.

The relative importance of anthropogenic NO_x emissions to surface O_3 concentrations is shown in Figures 12 and 13. Figure 12a and 13a, respectively, show O_3 concentrations for January and July with all sources of reactive nitrogen included in the simulation while Figures 12b and 13b show the difference between the predicted concentration for all sources and natural sources only. As indicated there, concentrations of O_3 increase by between 10 and 30 ppb over regions of important biomass burning in January and by

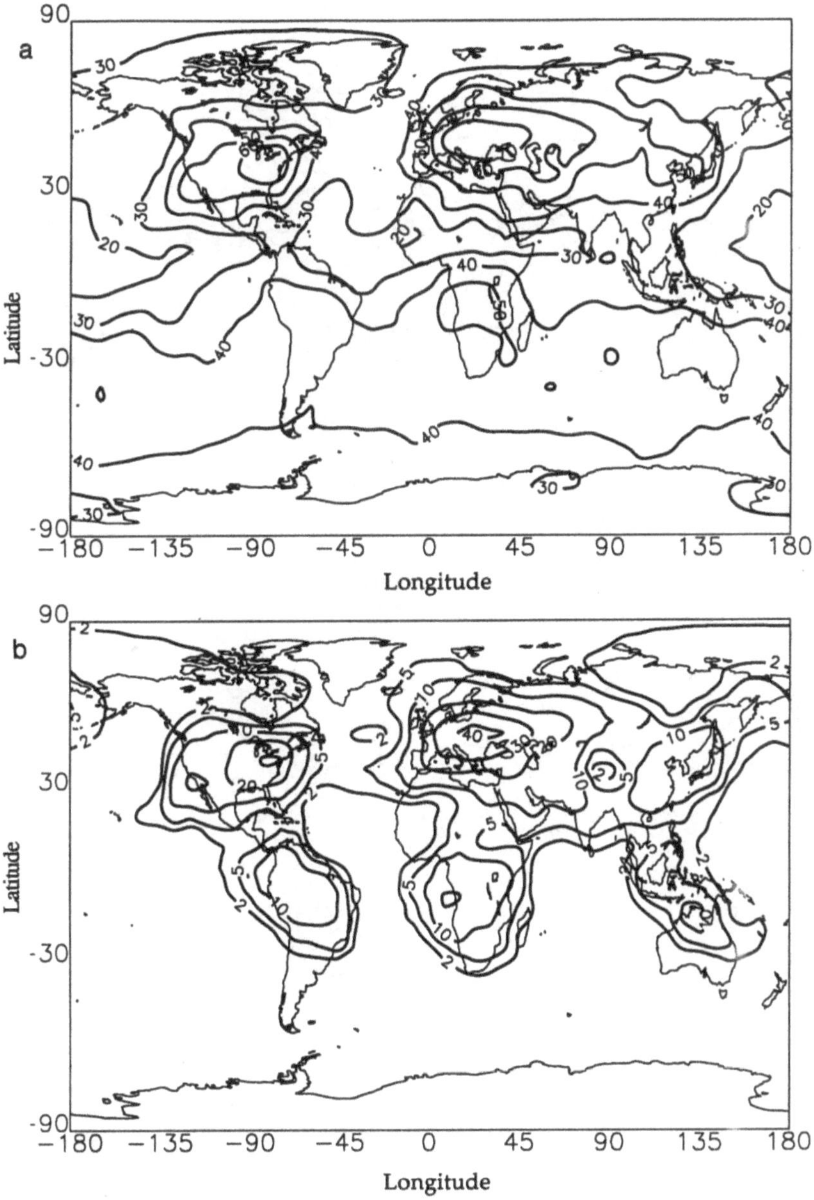

Figure 13. (a, top panel) Calculated ozone (O_3) concentrations at the surface in July with all sources of reactive nitrogen emissions included. **(b, bottom panel)** The difference between a calculation with all sources and with natural sources only. Contour intervals are 20, 30, 40, 50, 60, and 70 ppbv for top panel, and 2, 5, 10, 20, 30, and 40 ppbv for bottom panel.

between 20 and 40 ppb over regions of important fossil fuel emissions in Europe and the U.S. in July. Concentrations of tropospheric OH for July are shown in Figure 14. Concentrations increase from average concentrations near 1.7×10^6 cm^{-3} and 8.0×10^5 cm^{-3} for the northern hemisphere and southern hemisphere, respectively, in July for natural sources only, to concentrations of 1.9×10^6 cm^{-3} and 8.7×10^5 cm^{-3}, respectively, for July when all sources of reactive nitrogen are included. Figure 14 indicates that larger changes occur near the source regions.

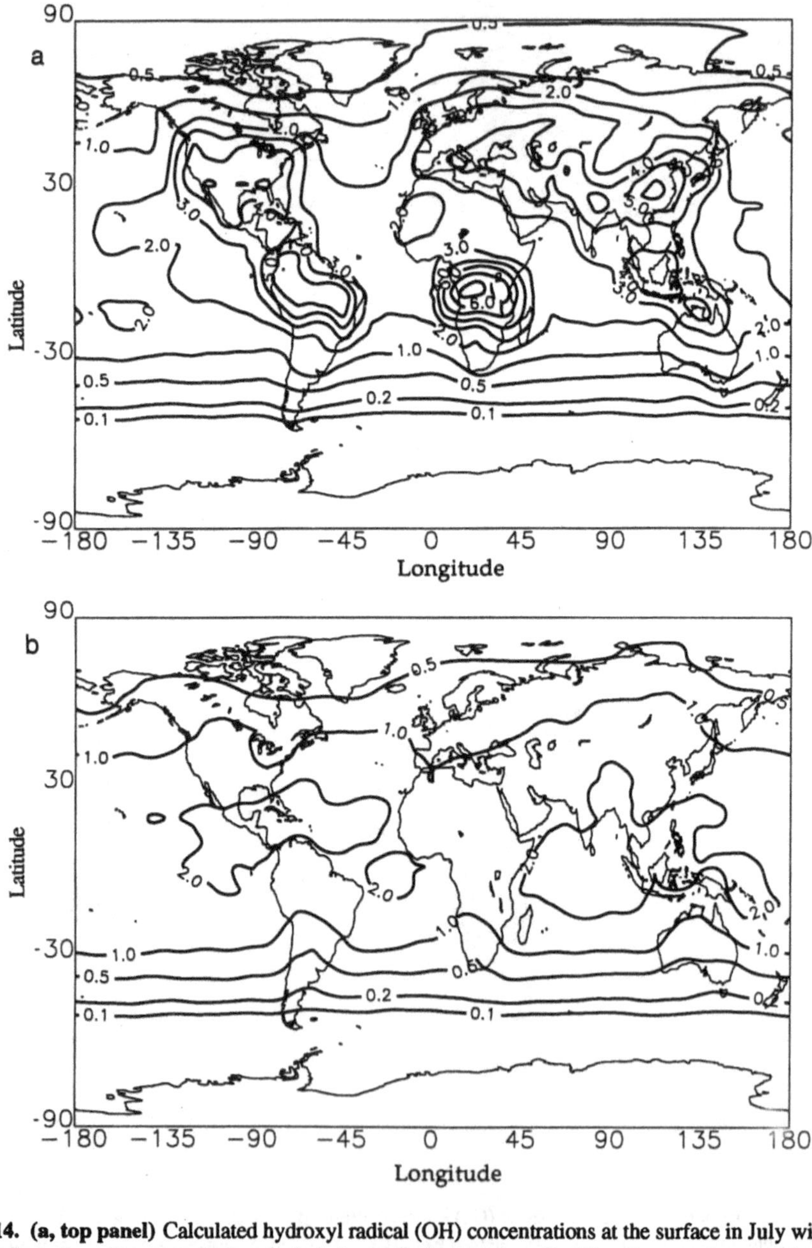

Figure 14. (a, top panel) Calculated hydroxyl radical (OH) concentrations at the surface in July with all sources of reactive nitrogen emissions included, and **(b, bottom panel)** with only natural sources included. Contour intervals are in units of 10^6 cm^{-3}.

CONCLUSIONS

Accurate emissions inventories are a crucial element in the further advancement of three-dimensional photochemical models. These emissions, when combined with models may be used to check the accuracy of our understanding of photochemical processes, transport, transformation, and removal. Comparison of the models with data show a number of similar features, but O_3, in particular, is over predicted in the free troposphere, pointing to the need for improved models. It is important to develop our understanding of emissions inventories (both natural and anthropogenic) further because the models have demonstrated that anthropogenic sulfate emissions may have increased the burden of sulfate aerosol by a factor of 10 or more near source regions and that anthropogenic NO_x emissions may have increased ozone concentrations by between 10 and 40 ppb in surface air near source regions.

ACKNOWLEDGEMENTS

Joyce Penner is grateful for the efforts of Dan Bergmann, Hal Eddleman, Chuck Molenkamp, and John Walton who each helped with model simulations or the figures for this paper. Carmen Benkovitz kindly provided the NO_x and SO_2 inventories prior to publication and Sam Oltmans provided the data used in Figure 10. This work was performed under the auspices of the U.S. Department of Energy under Contract W-7405-Eng-48. Partial support was provided by the U.S. EPA under Interagency Agreement DW89932676-01-0, by NASA under grant NAGW-1827, and by the DOE Quantitative Links Program. The LLNL LDRD program provided support for the coding of the O_3 model.

REFERENCES

Atherton, C.S., 1993, Thesis, University of California, Davis, CA.

Atkinson, R., D.L. Baulch, R.A. Cox, R.F. Hampson Jr., J.A. Kerr, and J. Troe, 1992, Evaluated kinetic and photochemical data for atmospheric chemistry: Supplement IV, *Atmos. Environ.*, 26A:1187-1230.

Benkovitz, C., 1982, Compilation of an inventory of anthropogenic emissions in the United States and Canada, *Atmos. Environ.*, 16:1551-1563.

Covey, C., S.J. Ghan, J.J. Walton, and P.R. Weissman, 1990, Global environmental effects of impact-generated aerosols: Results from a general circulation model, *in:* "Global Catastrophes in Earth History: An Interdisciplinary Conference on Impacts, Volcanism and Mass Mortality," V.L. Sharpton and P.D. Ward (eds).,Geological Society of America, Boulder, CO, pgs. 263-270.

Crutzen, P.J., and P. Zimmerman, 1991, The changing photochemistry of the troposphere, *Tellus*, 43AB:136-151.

DeMore, W.B., S.P. Sander, D.M. Golden, R.F. Hampson, M.J. Kurylo, C.J. Howard, A.R. Ravishankara, C.E. Kolb, and M.J. Molina, 1992, "Chemical Kinetics and Photochemical Data for Use in Stratospheric Modeling," Jet Propulsion Laboratory, California Institute of Technology Publication 92-20.

Dignon, J., and S. Hameed, 1985, Global emissions of nitrogen and sulfur oxides from 1860 to 1980, *J. Air Poll. Control Assoc.*, 39:180-186.

Dignon, J., and J.E. Penner, 1991, Biomass burning: A source of nitrogen oxides in the atmosphere, *in:* "Global Biomass Burning," J. Levine (ed.), MIT Press, Cambridge, MA, pgs. 370-375.

Dignon, J., 1992, NO_x and SO_x emissions from fossil fuels: A global distribution, *Atmos. Environ.*, 26A:1157-1163.

Dignon, J., J.E. Penner, and J.J. Walton, 1993, Simulation of ^{222}Rn and ^{210}Pb in a three-dimensional chemical transport model, in preparation.

Erickson III, D.J., J.J. Walton, S.J. Ghan, and J.E. Penner, 1991, Three-dimensional modeling of the global atmospheric sulfur cycle: A first step, *Atmos. Environ.*, 25A:2513-2520.

Fishman, J., V. Ramanathan, P.J. Crutzen, and S.C. Liu, 1979, Tropospheric ozone and climate, *Nature*, 282:818-820.

Galloway, J.N., (and 35 others), 1992, Sulfur and nitrogen levels in the North Atlantic Ocean's atmosphere: A synthesis of field and modeling results, *Global Biogeochemical Cycles*, 6:77-100.

Ghan, S.J., M.C. MacCracken, and J.J. Walton, 1988, Climatic response to large atmospheric smoke injections: Sensitivity studies with a tropospheric general circulation model, *J. Geophys. Res.*, 93:8315-8337.

Graedel, T.E., T.S. Bates, A.F. Bouwman, D. Cunnold, J. Dignon, I. Fung, D.J. Jacob, B.K. Lamb, J.A. Logan, G. Marland, P. Middleton, J.M. Pacyna, M. Placet, and C. Veldt, 1993, A compilation of inventories of emissions to the atmosphere, *Global Biogeochemical Cycles*, 7:1-26.

Johnston, H.S., D.E. Kinnison, and D.J. Wuebbles, 1989, Nitrogen oxides from high-altitude aircraft: An update of potential effects on ozone, *J. Geophys. Res.*, 94:16,351-16,363.

Kato, N., and H. Akimoto, 1992: Anthropogenic emissions of SO_2 and NO_x in Asia: Emissions inventories, *Atmos. Environ.*, 26A:2997-3017.

Kato, N., and H. Akimoto, 1993, Errata, *Atmos. Environ.*, in press.

Komhyr, W.D., S.J. Oltmans, P.R. Francois, W.F.J. Evans, and W.A. Matthews, 1989, The latitudinal distribution of ozone to 35 km altitude from ECC ozone sonde observations, *in:* "Ozone in the Atmosphere," R.D. Bojkov and P. Fabian (eds.), A. Deepak Publishing, Hampton, VA, pgs. 147-150.

Langner, J., and H. Rodhe, 1991, A global three-dimensional model of the tropospheric sulfur cycle, *J. Atmos. Chem.*, 13:225-264.

Lelieveld, J., and P.J. Crutzen, 1990, Influences of cloud photochemical processes on tropospheric ozone, *Nature*, 343:227-233.

Lelieveld, J., and P.J. Crutzen, 1991, The role of clouds in tropospheric photochemistry, *J. Atmos. Chem.*, 12:229-267.

Oltmans, S., and H. Levy II, 1993, Seasonal cycle of surface ozone over the western North Atlantic, *Nature*, 358:392-394.

Oort, A.H., 1983, "Global Atmospheric Circulation Statistics: 1958-1973," National Oceanic and Atmospheric Administration Professional Paper 14, U.S. Government Printing Office, Washington, D.C.

Pacyna, J., 1992, "Point Source Emissions of SO_2 and NO_x in the Asian Part of the Former USSR," Norwegian Institute for Air Research.

Penner, J.E., 1990, Cloud albedo, greenhouse effects, atmospheric chemistry and climate change, *J. Air and Waste Management Assoc.*, 40:456-461.

Penner, J.E., 1993, The role of human activity and land use change in atmospheric chemistry and air quality, *in:* "Proceedings of the 1991 Global Change Institute on Global Land Use/Cover Change," B. Turner (ed.), UCAR/OIES, Boulder, CO.

Penner, J.E., C.S. Atherton, J. Dignon, S.J. Ghan, J.J. Walton, and S. Hameed, 1991, Tropospheric nitrogen: A three-dimensional study of sources, distribution, and deposition, *J. Geophys. Res.*, 96:959-990.

Penner, J.E., S.J. Ghan, and J.J. Walton, 1991, The role of biomass burning in the budget and cycle of carbonaceous soot aerosols and their climate impact, *in:* "Global Biomass Burning," J. Levine (ed.), MIT Press, Cambridge, MA, pgs. 387-393.

Penner, J.E., R. Dickinson, and C. O'Neill, 1992, Effects of aerosol from biomass burning on the global radiation budget, *Science*, 256:1432-1434.

Penner, J.E., H. Eddleman, and T. Novakov, 1993a, Towards the development of a global inventory of black carbon emissions, *Atmos. Environ.*, in press.

Penner, J.E., R.J. Charlson, J.M. Hales, N. Laulainen, R. Leifer, T. Novakov, J. Ogren, L.F. Radke, S.E. Schwartz, and L. Travis, 1993b, Quantifying and minimizing uncertainty of climate forcing by anthropogenic aerosols, *Bull. Amer. Met. Society*, submitted.

Penner, J.E., and K. Taylor, 1993, "Aerosols, Atmospheric Chemistry and Climate: The Need for Coupled Simulations," presented at the CHAMMP Science Team Meeting, March 15-17, 1993, Monterey, CA.

Sandnes, H., and H. Styve, 1992, "Calculated budgets for airborne acidifying components in Europe, 1985, 1986, 1987, 1988, 1989, 1990 and 1991," Meteorological Synthesizing Centre-West, The Norwegian Meteorological Institute. EMEP/MSC-W 1/92, Oslo, Norway.

Saeger, M., J. Langstaff, R. Walters, L. Modica, D. Zimmerman, D. Fratt, D. Dulleba, R. Ryan, J. Demmy, W. Tax, D. Sprague, D. Mudgett, and A.S. Wernerk, 1989, "The 1985 NAPAP Emissions Inventory (version 2): Development of the Annual Data and Modelers' Tapes," U.S. Environmental Protection Agency, EPA-600/7-89-012a, Washington, D.C.

Savoie, D., and J. Prospero, 1989, Comparison of oceanic and continental sources of non-seasalt sulfate over the Pacific Ocean, *Nature*, 339:685-687.

Spiro, P.A., D.J. Jacob, and J.A. Logan, 1992, Global inventory of sulfur emissions with 1° × 1° resolution, *J. Geophys. Res*, 97:6023-6036.

Tilton, B.E., 1989, Health effects of tropospheric ozone, *Environ. Sci. Technol.*, 23:257-263.

Wagner, J., R.A. Walters, L.J. Maiocco, and D.R. Neal, 1986, "Development of the 1980 NAPAP Emissions Inventory," Environmental Protection Agency, EPA-600/7-86-057a, Washington, D.C.

Walton, J.J., M.C. MacCracken, and S.J. Ghan, 1988, A global-scale Lagrangian trace species model of transport, transformation, and removal processes, *J. Geophys. Res.*, 93:8339-8354.

Watson, C.R., and A.R. Olsen, 1984, "Acid Deposition System (ADS) for Statistical Reporting: System Design and User's Code Manual," Rep. EPA-600/8-84-023, U.S. Environ. Prot. Agency, Research Triangle Park, NC.

Wuebbles, D.J., and D.E. Kinnison, 1989, A two-dimensional model study of past trends in global ozone, *in:* "Ozone in the Atmosphere," R.D. Bojkov and P. Fabian (ed.), A. Deepak Publishing.

Wuebbles, D.J., D.E. Kinnison, K.E. Grant, and P.S. Connell, 1993, Effects of the Mt. Pinatubo eruption on the chemistry, radiative and transport processes in the stratosphere, *in:* "Proceedings of the Conference on Atmospheric Chemistry," American Meteorological Society, January 17-22, 1993, Anaheim, CA, pgs. 1-8.

ACCOMPLISHMENTS AND FUTURE CHALLENGES: AN IGAC PANEL DISCUSSION

The First IGAC Scientific Conference ended with a final panel session intended to assess current progress and future prospects. Ronald Prinn moderated the discussion, introducing the six panelists:

- Paul Crutzen, Max-Plank Institut fur Chemie, Germany
- Ian Galbally, C.S.I.R.O., Australia
- Pamela Matson, University of California, Berkeley, USA
- Henning Rodhe, University of Stockholm, Sweden
- Eugenio Sanhueza, I.V.I.C., Venezuela
- Ronald Prinn, Massachusetts Institute of Technology, USA

The Panel Discussion began with each of the above participants being given about five minutes each to answer the following two questions:

1. As reported at this conference and elsewhere, what are some special IGAC results or accomplishments worth highlighting?
2. What are the challenges for the future and the prospects for significant progress?

The following provides an edited transcript of the panelists' replies.

Paul Crutzen

I think we have had a very good week here. The most important thing that I notice is that our science is progressing extremely well. We have had, on the whole, very high class presentations, and it is appropriate that the most important thing coming out of IGAC is very good science. Several activities are moving along extremely well. For example, good work is being done in NARE (North Atlantic Regional Experiment), APARE (East Asian/North Pacific Regional Experiment), MAC (Multiphase Atmospheric Chemistry), and BIBEX (Biomass Burning Experiment: Impact on the Atmosphere and Biosphere). I think we have seen very impressive results from these activities. Also, the paper that was given near the end of the conference on NOMHICE (Non-Methane Hydrocarbon Intercomparison Experiment), is of extreme value and critical for progress. The work on atmospheric modeling is also coming along well. I've seen some fine work presented here and more people are getting into the act.

When it comes to the future progress of IGAC, of course, we all know that we are struggling to get recognition in our countries for the needed support. I think this should be a main item of the discussion here; how can we improve our interactions with the funding agencies? We have been discussing this in the IGAC Steering Committee, but I think it is very important to hear the opinions of all the participants here. The only concern I have about IGAC, and here I'm really thinking specifically of the situation in Germany, is that it can lead to very large programs. This may result in the grassroots science, the simple science, which you do with simple means, with a small proposal, not getting the recognition in the future that it should have. I see signs of this in Germany. Those Germans here, who disagree with me, please speak up. But if that is a development that occurred in IGAC, I would be very worried about it. There is an element of danger here. That's where I will finish my comments for the moment.

Ian Galbally

When I reflected on this I wondered how far we had come since Dookie. At that time, we wanted to do two things. We wanted to create a method for international coordination in atmospheric chemistry and to provide new impetus to the field. I think IGAC has been highly successful in that regard. It has brought together and accelerated work in atmospheric chemistry, and there is a great degree of inter-laboratory and international cooperation. This first IGAC conference on global atmospheric and biospheric chemistry was very much focused, I think, on sources and sinks. I expect that in future conferences there will be changes in the balance, and that on some occasions atmospheric chemistry modeling, carbon dioxide studies, and other subjects, will have a greater prominence. I think it's natural that these changes will occur with time.

As a personal opinion, I think that some of the papers presented here have been excellent contributions to science, while some of the papers didn't appear to have such excellence based on their presentations. I think that excellence is something we have to pay a lot of attention to. I'd also say that with IGAC we have an undertaking of a size and nature that none of us have probably ever been involved with before. I think the International Geophysical Year of 1957 and 1958 was, perhaps, the only previous undertaking of such a large and international nature.

And so, looking forward, I think the things we have to do are to maintain the excellence of the work that's being carried out, maintain the high level of collaboration and communication, and maintain our enthusiasm. Basically, our job in IGAC should be to ensure that these activities and this level of excellence continue over the next five years, or more. In this way atmospheric chemistry will evolve from a young discipline into a mature discipline. Thank you very much.

Pamela Matson

The mark for me of a good meeting is one that makes me think, and that I learn from. Given those criteria, I definitely consider this to have been a good conference. And you have no idea what a strong statement that is since I've been drugged-out on cold medicine for the whole week, and yet I've still learned something! So I thank you for teaching me. I could go through and list all the different talks that I appreciated especially, but rather than doing that, I would like to reflect on what this meeting tells me we've accomplished in the IGAC framework.

When IGAC first came into being, it was largely devoted to atmospheric chemistry. Then, thanks to Paul Crutzen and the IGBP committee (many of those people are in this room), important biospheric aspects were merged into the IGAC program. And that meant bringing in biologists, ecologists, biogeochemists, and a number of other disciplines to work with atmospheric chemists and atmospheric modelers. We've actually succeeded at doing that. When we developed the programs four or five years ago, we thought it would be essential to bring these disciplines together, but we didn't have any guarantee that it would work. In fact, I think this meeting absolutely shows that it has worked. It *is* multi-disciplinary. We *do* talk each other's language. We have learned from each other in our various disciplines. The various disciplines are addressing many of the same questions now, for many of the same reasons. So that's a very positive step.

I think there are some marked successes. The BIBEX (Biomass Burning Experiment: Impact on the Atmosphere and Biosphere) activity is a multi-disciplinary activity with atmospheric chemists, biogeochemists, ecologists, and remote sensing people all working together at regional and global scale, and it's been very fruitful. And there are others: the HESS (High Latitude Ecosystems as Sources and Sinks of Trace Gases) activity is another example. I definitely agree with Bill Reeburgh that five years ago we really didn't have a very good idea about the spatial distribution of methane sources and sinks, or the about the processes that control them. I think this is an example of the fact that if you take a lot of people and resources and channel them into a focused area, you can make incredible progress in just a few years. So we've gone a long way with that. The RICE (Rice Cultivation and Trace Gas Exchange) activity is another example. Five years ago Hans-Ulrich Neue was the only person, or one of a very few people, actually looking at trace gas emissions from flooded agricultural systems. Now there are many people in the developing world, and in the developed world, working on that issue, and we've come a long way in our understanding. So there are, I think, also great successes here. And those are just a few examples. I think there are notable accomplishments in many other activities.

But, I also think it's not time to be complacent. We've actually passed up some really good opportunities. We haven't done things as well as we could. We haven't used each other's knowledge as well as we could. I'll pick on myself and use, as an example, the BATGE (Biosphere-Atmosphere Trace Gas Exchange in the Tropics) activity that I work in. We are a very small group of people and we have very little funding actually, and we are working in an area where very few people have worked before. But that's really no excuse. In designing our studies within this field we have not used the knowledge that's come from previous work. There's no need to make the same mistakes that the temperate trace gas people have made, now that we are working in the tropics. We can use their experience. We can use their understanding of the processes that control fluxes in order to carefully design our studies. I think we can learn from that more than we have already. And, there are other examples of shortcomings. I get the sense that in some of the activities, there are an awful lot of measurements going on. But we can't just do more and more measurements, without integration, without synthesis of the data, without putting the measurements into the context of biosphere-atmosphere interactions or biospheric processes. So, I think we all need to very carefully look at what measurements we're making, and ask if we really need to make more, or whether it's time now for synthesis.

I think the modeling and synthesis activity is coming along in some areas and we've had some great presentations on that today. I would really like to see us integrate across all these different activities in the process of developing those models. For example, I would like to see us develop spatially explicit biogeochemical models for the ocean or land surface, that estimate gas sources and sinks, and couple these with models of atmospheric chemistry and transport. I think we had one or two examples today of how to do that. But I think we need to do more. As another example, the temperate, tropical and boreal activities should be working together to develop global monitoring networks—we haven't done that very much yet.

So these are things to look forward to in the future. I think when we meet in another year or another two years, there's going to be an even greater need for emphasis on synthesis of the information and knowledge we've already gained. Thanks.

Henning Rodhe

My general impression is that the quality of presentations, both oral and poster have been very high. And, to me, IGAC is very much alive. Especially, I'm most pleased with the high quality of interdisciplinary interaction that has become so obvious here. A goal that has been achieved here is to have atmospheric chemists and biologists work together in ways they didn't use to do some years ago. I think IGAC really is accomplishing something as an organization; it is specifically helping to foster this cooperation.

I have some specific comments on the activities that I am involved with. I have seen a lot of interesting results from aerosol studies. Also, the MAC (Multiphase Atmospheric Chemistry) activity has continued to organize and prepare for experiments. BATGE (Biosphere-Atmosphere Trace Gas Exchange in the Tropics) and DEBITS (Deposition of Biogeochemically Important Trace Species) are presently two rather small activities, which I think are extremely important, and very interesting. But maybe we should perhaps combine these two activities. One is looking at species going up and the other at species going down, but they are related species.

One activity that I would like to emphasize is GEIA (Global Emissions Inventory Activity). I think that it is an extremely important accomplishment of IGAC. I have been so pleased to see the presentations today reporting data sets, which are very important information for modelers. And finally, I want to address the Global Integration and Modeling activity, GIM. So far it has been in the conceptual phase, but now we are ready to develop this activity, and I think that's going to be very important for IGAC. IGAC needs to try to bring things together now, and through GIM have its research focus more on the overall understanding of the whole system.

I have a critical comment on the presentations. Next time I'd like to see more time for discussion. I am not referring to the usual questions on the papers, but to the general discussion of the progress of each activity. Each activity presentation could be followed by an open discussion and evaluation. To allow this we would need to have fewer oral presentations. The poster sessions here have been very good, and very useful. Maybe we should have more poster presentations, fewer oral presentations, and more discussions.

Where are we going in the future? I think it's clear that we are moving ahead. But I think it may also be timely, as we have discussed in the Steering Committee, to sit back and evaluate again the goals that we formulated at the beginning of IGAC. We should define better how we should work for the next five or ten years.

We still have a lot of very interesting problems to solve in the upcoming years. Well, I hope to see you in Fuji-Yoshida in Japan in early September of 1994. That's where and when the next IGAC scientific symposium will be held. We can continue these discussions at that time.

Eugenio Sanhueza

I am also very impressed with the progress in most of the IGAC activities. Some of them are going very rapidly. I didn't expect that progress was going to be so fast. Important contributions to understanding in this field are being made by the study of long- and short-lived gases in many changing ecosystems: tropical forest clearing, savanna cultivation, effect of fertilization in temperate agriculture soils, effects of increasing temperature in a temperate forest, the role of tundra and other high-latitude ecosystems, and so on.

Most of the subject is really well-covered, so I feel, in general, IGAC is going very well at the present time. One problem or gap that I want to stress, in determining our IGAC directions for the future: I note that, at present, participation of research groups or even individual researchers from developing countries, for the most part, is at a very low level. For example, the TRAGEX (Trace Gas Exchange: Mid-Latitude Terrestrial Ecosystems and Atmosphere) project is a very active program. They are showing very good progress and already have a network in the United States. Later this year they will set up another one in Europe. But, for example, it seems that nothing is occurring in the southern part of South America, where large areas are used in agriculture. We know that the soil fluxes have very large variability, and results that are being obtained in the northern hemisphere are not necessarily valid to extrapolate to the southern hemisphere. Another example: there are several projects looking at the chemistry of the troposphere in the northern hemisphere. But with the exception of some campaign types of measurement like TROPOS and SAFARI, and some permanent activity in Australia, very little is happening in the rest of the tropics and southern hemisphere. There are no permanent regional projects characterizing the atmosphere of the south of South America or Africa that look, for example, for changes in the oxidation capacity of the atmosphere there. Since many or most of the studies dealing with global or climate change are long-term studies, research bases in developing regions should, for several reasons, include participation of scientists from the developing countries. Ideally, in the future, independent research teams from developing regions should actually run the studies in the developing world. I think that this is a very important undertaking for the future progress of IGAC. Thank you.

Ronald Prinn

It's getting more and more difficult to say something original, of course, now that I'm the sixth speaker! So I'll take the liberty of repeating some things that have been said, but put them in a somewhat different way. I certainly think that IGAC is off to a good start, and that's been demonstrated here with the quality of the oral and poster papers. I was most impressed with a number of the poster papers. Looking to the future, I think more of us should volunteer to give our papers in poster form, and not always expect to come to a conference and give an oral presentation. That would free up more time for discussion, which I think Henning Rodhe mentioned the need for earlier.

I certainly agree with Pamela Matson that we need to go into the business of integration of these results. That includes developing large three dimensional models, and using these models for this purpose. However, when you look at the progress in developing these models, you realize that many of the critical data bases needed to test them don't really exist. As one example, we know very little about vertical profiles of ozone throughout most of the world, and particularly this is true in the tropics. How are we going to believe these models, when for one of the most critical atmospheric components that they will predict, namely ozone, we don't have a database in the tropics? How can we tell if we are doing it right or doing it wrong? Now, that leads into my second point concerning something that Eugenio Sanhueza just brought up.

There are significant gaps in the research program as demonstrated at this conference. And I'm sure, unfortunately, that the same statement is going to be true a year or two down the road, but we do have to do something about it. The research in the tropics is not proceeding at the needed pace. But most of the countries in the tropics are developing countries, and a significant reason why the research is not proceeding there is that they don't have the resources to proceed. We must, as a community, think of ways in which we can bring researchers in developing countries into the IGAC program. We must provide them with critical resources like instrumentation, for example, and work with them as collaborators. Because that is the only way, I think, looking to the future that we are really going to fill in this massive data gap involving developing countries in general, and the tropical countries in particular. We will all have to work together to try to get around that major difficulty.

So much for the current state of IGAC. Another point I wanted to make was one regarding IGAC's relationship to society. We are not addressing a topic that is merely of academic interest. As inadequate as the funding is at the present time, it would be even less if the work weren't important to society. So we must remember that there are some things we do in IGAC that should serve the people outside of our community in important ways. Through understanding the cycles of all the critical chemical components in the atmosphere, both the biological and purely chemical parts of them, we, at the end, want to be able to predict the composition of the atmosphere going into the future. We shouldn't forget that that's our task for the broader community, and we must carry out a research program that, at the end, produces models that have that capability, and ones that we, as a community, are proud of.

A further point involves the relation between understanding fundamental processes and mitigation of detrimental effects. There was an interesting point made by Heinz-Ulrich Neue in his conference presentation. If you understand the processes by which rice cultivation leads to methane emissions, you could also address the issue of whether rice can be grown using highly productive strains, but in a way that decreases the emissions of methane? Similarly, after understanding all of the relevant processes, can we develop energy sources that will lead to minimal local and regional air pollution, and minimal emissions of globally important gases? We should remember that understanding processes, like those that produce emissions, potentially helps achieve mitigation. If we are going to do something about greenhouse gas emissions we need to understand the processes by which they are produced. We are not going to stop people from producing energy. We are not going to stop growing rice. So we must ask instead how can those activities proceed with minimal impact on the environment.

So, with that, I'll close the remarks of the panel, and now open it to any member of the audience who wishes to say something.

Pat Zimmerman, NCAR, U.S.

I think that IGAC is at its best when it builds bridges, and at its worst when it builds walls by overly defining projects. When you do, it tends to give the community the impression of specific territory, yet many of the questions that need to be solved can only be solved by bridging the walls. For example, if we want to understand global carbon cycling we need to have coordination among the relevant groups to design an experimental approach to solve that problem. That's my first comment.

The second comment is that I think one of the important activities of IGAC is to recruit scientific talent in the quest of understanding the biogeochemical cycles. If the meetings were held at places where students could come, and if forums were provided so that the students didn't feel self-conscious about asking probing questions and contributing to the discussion, then I think that we might have more vibrant energy. And, in the long run, we might have a research force that could be more productive.

Guy Brasseur, NCAR, U.S.

I would like to make three quick remarks. First of all, I also think that IGAC has improved the visibility of atmospheric chemistry and biogeochemistry. I am somewhat concerned, however, by a particular tendency. Certain countries, including the USA, have tended to push the agenda from a broad perspective of global change toward a more narrow perspective on climate change. This creates a spirit in which the research is really driven by things like the IPCC (Intergovernmental Panel on Climate Change) Report. However, there are a number of problems that are chemically or biologically related that are not leading to the greenhouse effect or to ozone depletion, but still need to be recognized as being important to society and to the political world.

I am a little disappointed, this is my second remark, by the fact that representation from developing countries is pretty small in this conference. I think, for example, that there are very few representatives from Africa. If we are to understand what global change is about, we have to find a way in which we can bring some of these people to meetings in the future. I think that if the Steering Committee has the opportunity to invite people in the future, the invitees should be representing the continents in a more equal way. I also believe laboratories in the developed countries could potentially help very much. What we are trying to do at NCAR is to identify a number of sister laboratories in the developing world and to try to work with them and establish long-term collaborations. I think that if other laboratories did the same thing, we could probably have much better communication and representation.

Finally, the problem of education in IGAC was not really mentioned, except by Ron Prinn at the beginning of the conference, and I don't know exactly what the status is. Perhaps in relation to this problem of better participation by developing countries, the developed countries should be helping in the education of a younger generation of scientists who could participate in these meetings in the future.

Wei-Han Su, People's Republic of China

Thank you. I just have a few short remarks. I think the IGAC activity is very important for the study of global changes. Also, I agree with the suggestion that we have to pay more attention to the developing countries. I think both the tropical regions and the middle-latitude regions are important for our study. For example, in the west coast of the Pacific Ocean, there is a high population density and, with rapid industrialization ongoing, the nitrogen emission increases in this region will be very significant in the future. So I would like to suggest that IGAC pay more attention to the rapidly changing mid-latitude regions, and particularly East Asia and Southeast Asia.

I also would like to give you some information about the Chinese work in this area. We have formed an IGAC committee, and now the Chinese Academy of Sciences is setting up more than 50 sites to do the needed long-term monitoring tasks. Finally, I would like to use this chance to say that we endorse long-term monitoring in developing countries, and we endorse the extension of the modeling work into these regions. That's all.

Jos Lelieveld, Germany

I would like to ask a question about quality. Many projects are proceeding quite rapidly and producing a number of sub-projects. I wonder how we can preserve the quality of the work in these projects. It is very necessary.

Ronald Prinn

Do you mean quality assurance for the data obtained, the modeling, or across the board?

Jos Lelieveld

I refer to the quality of the overall project or sub-project itself. It could be that some projects are not keeping their focus, or some sub-projects are not high quality.

Ronald Prinn

Let me attempt a reply. First, concerning the review process, it is the responsibility of the IGAC Steering Committee to review and to try to evaluate the state of progress in each of the activities as they proceed. We, unfortunately, only meet once a year, so we try to do things in the interim, by communicating using faxes, electronic mail, and phones. But that doesn't work very well. At each Steering Committee meeting we do go over each of the projects and try to, at least in a broad-brush way, assess the state of progress, and recognize any important gaps. And, of course, I think you can tell something on this subject from this conference. The fact that there are certain of the IGAC activities that really had no papers here indicates that some of those activities are not proceeding very fast.

The other problem you bring up is quality assurance. Is the work being done of high quality? Assuring quality is also a responsibility of the IGAC Steering Committee, but more importantly it is a primary responsibility for the Convenor and Committees of the individual IGAC activities. One thing that can be said at the moment, of course, is that we are right at the beginning. There's not that much that we can look at and evaluate, although we certainly can try to do that. But, as I have noted, we rely to a great extent on the Convenor of each activity, and the committee that is under them, to do this quality assurance. The Convenors have been carefully chosen, and the committees under them, in order to achieve that. So it is a community responsibility to ensure quality. It's something that the Steering Committee and Convenors can try to assess, but in the end it really must be the entire community that is responsible for quality. Each individual scientist should be making very sure that what they do is done carefully, and is properly inter-compared and calibrated relative to other investigations.

Hajime Akimoto, University of Tokyo, Japan

I relate very much to the comments of Professor Su. The problem is not always the involvement of the developing countries. Whether developed or developing, many non-North American and non-European countries were neither involved in, nor informed about, the initial structure of IGAC. I was very lucky. I could attend the Dookie meeting, and had the chance there to propose APARE (East Asian/North Pacific Regional Experiment), so that program has been incorporated in the IGAC structure. But there are many scientists in Japan now studying things like CO_2,biosphere interactions, etc., who are not aware of the relevant IGAC activity at all. So they are not coordinated, and that's always a problem. Many of the current IGAC activities should be trying to involve these people in the international program. And so at some stage, you need to open the Steering Committee and activity committees to some new non-North American and non-European members. We need to broaden the international program. That is my comment.

Ronald Prinn

Yes, this is good advice to us, and I will try to see that we heed it.

There being no further questions from the conference participants, let me just close this discussion by repeating briefly some remarks I made last night. I would like to thank the Israeli Institute for Biological Research and the OHOLO Conference organization for hosting this meeting. Special thanks go to Dr. Samuel Cohen, Director of that Institute, and Amram Golombek, who worked very hard on all the local arrangements. Let me thank the panel and the audience for the discussions here, the hotel for its hospitality, and each of you who presented papers, or, equally important, presented posters for your contributions. I hope you leave this conference with better knowledge and greater enthusiasm for attacking the problems that we face in the future. I remind you that the second IGAC conference will be held in Fuji-Yoshida, Japan in September of 1994. We expect probably twice the number of participants that we had at this meeting at the next one. I don't know what we'll do about keeping down the number of oral presentations. I think we'll all have to be prepared to have many more posters, and I think I will set an example by having my paper as a poster.

With that, I'd like to close this first IGAC scientific conference. I hope in looking back that we will all see this as an historic first step. The participants here were reporting science results from IGAC for the first time. As we all know, we had a lot of planning meetings in IGAC and it is a pleasure now to see the plans being implemented. This is just the first in a series of scientific conferences, which, I hope, as time goes on will blossom and become more and more impressive events. So, with that, thank you, and I hope your travels home are safe.

INDEX

*Page numbers in **bold** denote appearance of subject in figure or table on cited page.*

Page numbers in bold denote appearance of subject in figure or table on cited page.

*Page numbers in **bold** denote appearance of subject in figure or table on cited page.*

FSC
www.fsc.org
MIX
Papier | Fördert
gute Waldnutzung
FSC® C083411

Kolibri 360 GmbH
Ferdinand-Jühlke-Straße 7
99095 Erfurt, Deutschland
produktsicherheit@kolibri360.de